AN INTRODUCTION TO
CLASSICAL DYNAMICS

*Mechanics are the Paradise of mathematical science,
because here we come to the fruits of mathematics.*

LEONARDO DA VINCI

AN INTRODUCTION TO
CLASSICAL DYNAMICS

GARRISON SPOSITO

University of California, Riverside

JOHN WILEY & SONS, Inc.

New York / London / Sydney / Toronto

Library of Congress Cataloging in Publication Data

Sposito, Garrison, 1939–
    An introduction to classical dynamics.

    Includes bibliographical references and index.
    1. Dynamics.   I. Title.
QA845.S73   1976   531'.11   75–23472
ISBN 0–471–81756–2

Printed in the United States of America

10 9 8 7 6 5 4 3 2 1

To the memory of
Professor Roy Overstreet

and

To Mary

Preface

This book is intended for use in courses of one semester or two quarters on the dynamics of particles and assemblies of particles, including rigid bodies. Although the book begins with the most basic concepts, a useful assimilation of its contents requires a previous exposure to introductory mechanics based in calculus. No fine details of that exposure must be remembered; it is assumed only that the student has enough familiarity with the fundamental concepts to appreciate the need for careful discussions of the physical and mathematical aspects of classical dynamics. Differential equations and partial derivatives are used throughout but are applied in "cookbook" fashion. Vectors also are essential, beginning with Chapter 4, and a detailed discussion of them is given in the Mathematical Appendix along with material on rotations and curvilinear coordinate systems.

The central theme is that classical dynamics is a self-consistent discipline of physics, with many applications in a number of its modern subfields, which today are of increasing importance. These include geophysics, atmospheric physics, space physics, physical oceanography, and environmental physics, all of which are represented significantly in the textbook and in the problems. Some examples are the pendulum seismograph, atmospheric and oceanic circulation, artificial Earth satellites, and the settling of particulates. I hope that their inclusion will broaden the perspective of the student and stimulate the lectures of the teacher.

The plan of the book is to discuss the physical bases of dynamics—Newton's Laws and the conservation theorems—and to illustrate them extensively through a discussion of one-dimensional motions in the first three chapters. In Chapters 4 and 5 the essential features of motion in space are surveyed, including the use of curvilinear coordinates, the concept of angular momentum, the two-particle problem, the Kepler problem, and the theory of elastic scattering. Chapters 6 through 9 discuss important topics from which the teacher may choose material to complete his lectures: assemblies of particles (including molecular mechanics), rotating frames of reference, special relativity, and an introduction to Lagrangian and Hamiltonian mechanics. To provide the broadest possible choice of classroom subject matter, four "Special Topics" also have been added following Chapters 1 to 3, and 8. Problems relating to these topics have been included in the sets at the ends of those chapters. Answers and hints for *all* of the problems are given at the end of the book.

The manuscript was diligently typed by Ms. Linda G. Smith, to whom I am greatly indebted. I also express my appreciation to Professors Eugen Merzbacher, John McCullen, and Robert Eisberg for their helpful reviews. The responsibility for errors or unclear passages is my own, and I would appreciate students calling them to my attention.

Garrison Sposito

Contents

1

The Foundations of
Classical Dynamics

1.1 THE NEWTONIAN PROGRAM FOR DYNAMICS

Classical dynamics began with the publication, in 1687, of Isaac Newton's monumental treatise, *Philosophiae Naturalis Principia Mathematica.*[1] In this book Newton set down three postulates which he believed would make possible the mathematical description of the motion of any single particle or collection of particles. These famous postulates are familiar from introductory physics as Newton's Laws of Motion. We know today that they are only very good approximations to the true axioms of particle dynamics, which must account for the quantization of energy and the Einsteinian principles of relativity. Nonetheless, because the degree to which it differs in its predictions from what is known to be correct theory is in practice often insignificant, and because its conceptual framework is easily related to common experience, we shall be on excellent ground to

---

[1] *The Mathematical Principles of Natural Philosophy*, revised translation by F. Cajori, University of California Press, Berkeley, 1960.

1

begin our discussion by considering in detail what we shall call *Newton's program for dynamics*.

In order that Newton's Laws of Motion be understood with the least ambiguity, we shall give a few prefatory definitions of terms that should be recalled from introductory mechanics. These terms are as follows.

Particle   A particle is any physical object whose motion can be described fully by its position in space and by its velocity as functions of the time. Particles are the direct concern of Newton's Laws. An assembly of particles, which usually requires more to describe its motion than just a single position and velocity, is an indirect concern of Newton's Laws of Motion in that its behavior may be considered by extending the Laws.[2]

Position   We shall consider this quantity to be a continuous function of the time that possesses at least two derivatives. The position of a particle is measurable, in the simplest case, by means of a ruler and a clock and may be represented geometrically by a succession of points in a suitably chosen reference frame. Ordinarily each of these points is specified by a set of three real numbers called *coordinates*. However, we shall suppose initially, for the sake of mathematical simplicity, that just one spatial coordinate is enough to determine the position of a particle. In this way the position becomes what is called a "scalar function" and the rules of conventional algebra apply. Whenever the more general situation obtains, the position becomes a "vector function" and the rules of vector algebra must be used.[3] We shall consider this mathematical complication in Chapter 4.

Velocity   This quantity is the first derivative of the position with respect to the time. In the special case of a scalar position function we have, therefore,

$$v(t) = \lim_{\Delta t \to 0} \left[\frac{x(t + \Delta t) - x(t)}{\Delta t} \right] \equiv \frac{dx}{dt} \tag{1.1}$$

in a conventional calculus notation, where $x(t)$ is the position. The velocity $v(t)$ is a continuous function of the time t and possesses at least one derivative. A sense of direction is always associated with it and, in the case of $v(t)$ given by Equation 1.1, that direction is designated by an implicit plus or an explicit minus sign appearing before the numerical value. The plus sign will be taken generally to mean either "up" or "to the right," depending on the orientation of the frame of reference, and the minus sign will mean either of the opposite directions relative to the origin of the coordinate $x(t)$. The velocity can be measured by assigning its direction to be

---

[2] For example, Newton himself showed (*Principia*, pp. 193–195, Cajori translation) that it is possible to regard a planet as a particle when we consider its orbit about the sun, if the planet is a perfect sphere.

[3] These rules are discussed in section A.1 of the Appendix.

the direction of motion of the particle and by calculating the slopes of the lines tangent to the graph of $x(t)$, in keeping with the usual geometric interpretation of a derivative, in order to get its numerical values (see Figure 1.1).

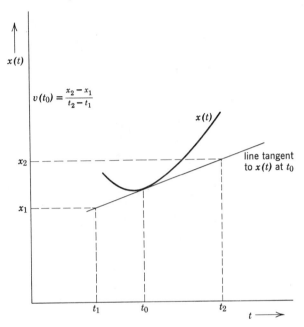

FIGURE 1.1. The velocity $v(t_0)$ is the slope of the line tangent to $x(t)$ at $t = t_0$.

Acceleration   Acceleration is the first derivative of the velocity with respect to time and is also a continuous function of the latter variable. Conventionally we have

$$a(t) = \lim_{\Delta t \to 0} \left[\frac{v(t + \Delta t) - v(t)}{\Delta t} \right] \equiv \frac{dv}{dt} \qquad (1.2)$$

for the acceleration $a(t)$. As with the velocity, acceleration has both a numerical magnitude and an associated direction. The magnitude is measured in a way analogous to that for determining the velocity. The direction assigned is that direction in which the velocity is *changing*. The direction of the acceleration, therefore, need not have anything to do with the observed direction of motion of the particle.

The quantities position, velocity, and acceleration are referred to as *kinematical quantities* because their definitions involve only the fundamental ingredients of motion: length and time. In Table 1.1 the kinematical quantities are listed along with a recounting of their dimensions and their units of measure in the three most commonly used systems. The definitions of the systems of units and the relations among them are assumed to be known from introductory physics.

TABLE 1.1 The Kinematical Quantities Important in Classical Mechanics

Quantity	Dimensions	Unit		
		SI	CGS	American Engineering
Length	L	meter (m)	centimeter (cm)	foot (ft)
Time	T	second	second (s)	second
Position	L	m	cm	ft
Velocity	LT^{-1}	m/s	cm/s	ft/s
Acceleration	LT^{-2}	m/s²	cm/s²	ft/s²

1 foot = 0.3048 meter (exactly). 1 centimeter = 0.01 meter.

Now we are in a position to state and explain Newton's Laws of Motion. We shall not be presenting these laws exactly as they were given in the *Principia*, but instead we shall express them in a form that reflects the contemporary status of classical dynamics. A discussion of Newton's original conceptions regarding the Laws of Motion will be found in Special Topic 1 at the end of this chapter.

THE FIRST LAW   *It is always possible to find a frame of reference in which a particle, free of influence from matter and radiation, is moving with a constant velocity.*

A glance at the First Law in this form shows that it deals with an ideal situation, since it is not possible to isolate completely any physical object from the rest of the universe. However, no difficulty in interpretation should develop as long as it is remembered that every physical law or postulate has validity only insofar as it agrees with experience within some predetermined level of tolerance for error. In practice the influence of surrounding matter and radiation can be minimized to whatever extent is desired and the First Law can be understood as an extrapolation to vanishing influence. Then it states that we can always find a "preferred" frame of reference in which the particle under consideration will be moving with a constant velocity (i.e., with a constant speed in a fixed direction). Evidently, if one such frame of reference exists, an infinite number of them exists. This is because all frames moving at constant velocity with respect to *one another* would yield an observation of constant velocity (including, possibly, zero velocity) for a particle moving that way in any one of them. Newton (and before him Galileo) believed that this uniform motion was the "natural" or "equilibrium" motion of a free particle and that it would be observed at least in a limiting sense within a certain class of frames of reference. These special frames of reference are called *inertial frames. Inertia* is the name given to the "inherent tendency" of a free particle to move with constant velocity.

It should be clear, even from these few remarks, that the First Law has much the same character as a definition. When a free particle is observed to move uniformly, that motion is defined as the natural one and the frame of reference in which it is observed is deemed a member of the preferred inertial class. What remains quite vague about the First Law is the reason the inertial frames are preferred and the precise meaning of the phrase "free of influence." These questions are, in fact, only two faces of the same problem and are answered admirably by the Second Law.

THE SECOND LAW   *The influence of matter and radiation on a particle is manifest in the form of a force. A force causes the velocity of the particle to change in such a way that the time rate of change of the velocity is in the direction of and is proportional to the force.*

The Second Law tells us that forces are the purveyors of all the interactions in the universe. These quantities possess both magnitude and direction. They bring about *changes in the velocity* of a particle, that is, they cause a particle to *accelerate*. It is now clear why inertial frames of reference are the preferred ones. Relative to those frames, a particle unaffected by forces moves with constant velocity. If a force should act on the particle, its velocity changes and we may attribute all of that change to the force, according to the Second Law. In some other, noninertial frame of reference, the free particle would not move with constant velocity and we could not unambiguously determine the influence of matter and radiation on it by invoking the concept of force.

The force acting on a particle generally may be represented as a function of the position, velocity, and time. In the special case we now are considering, this function is a scalar one, but it is associated with a direction according to the sign convention that we previously discussed for acceleration. With this stipulation we can write the Second Law in the mathematical form

$$F(x, v, t) = ma(t) \qquad (1.3)$$

or, by Equation 1.2,

$$F(x, v, t) = m \frac{dv}{dt} \qquad (1.4)$$

where m is simply a constant of proportionality. As is well known, the constant m is supposed to be a number characteristic of the particle itself and is called the *mass*. Since the force $F(x, v, t)$ and the acceleration $a(t)$ always have the same direction the mass is always a positive number. Moreover, the algebraic form of Equation 1.3 suggests that, for a given, fixed magnitude of the force, small accelerations are associated with large masses, and vice versa. It follows that the mass must represent the relative degree to which a particle can resist force, a large mass being characteristic of a particle difficult to accelerate. But unaccelerated motion reflects the inertia of a particle, according to what was stated previously. Therefore, *the mass of a particle should be a numerical measure of its inertia.* Newton was aware of the common experience that bulky objects were difficult to accelerate and was led to equate mass with the quantity of matter in an object. But the better definition is the one related to inertia, since it develops in a straightforward way from the first two Laws of Motion and is independent of whatever range of experience with accelerated particles one may have had. In addition to this argument, it is pointed out that the modification of the Laws of Motion brought on by the Special Theory of Relativity makes the association of mass with the quantity of matter completely untenable.

It is apparent that there is no absolute value for the mass of a particle, for a change in the numerical magnitude of m by a multiplicative constant will only change the unit of mass and not the ratios of mass between different particles. But it is these ratios solely that are of physical significance since, for a given force, they determine the ratios of acceleration between different interacting particles. (We shall return to this point later in our discussion of the Third Law of Motion.) It is on this basis that one of the most common ways to measure the mass has evolved. The procedure simply amounts to defining a unit of mass, then deducing the mass of a given object by comparison with another object that is said to possess one unit of mass. The usual method for objects directly accessible in the laboratory is called *weighing* and involves the experimental fact that the force exerted by the gravitational field of the Earth on a particle close to the ground (i.e., the weight of the particle) is expressed by

$$F = -mg \tag{1.5}$$

The quantity g is an empirical parameter, "the acceleration due to gravity," and the minus sign indicates that the force is directed downward. By experiment[4] it is known, to a very high degree of precision, that the acceleration g does not depend on the mass m in Equation 1.5. This means that we may compare the weights of two objects, for example, on an equal-arm balance, as a means of determining the ratio of their masses. For particles that are not directly accessible in the laboratory, such as electrons or planets, the mass is usually measured by observing the position of the particle as a function of the time under a known applied force. The theoretical predictions of $x(t)$ are then compared with experiment and "turned around" to yield a calculated value of the mass.

Once the unit of mass has been set, the unit of force follows directly from Equation 1.3. The unit of mass in the Système International is the *kilogram*, which is very nearly the mass of 10^{-3} m^3 of pure water. Then, by definition, a force of one *newton* causes a particle of mass one kilogram to accelerate at the rate of one meter per second per second. This information, along with alternate units for mass and force, is summarized in Table 1.2.

The Third Law of Motion differs from its predecessors in being heavily dependent on experiment. In presenting it we shall pay respect to its limitations more so than is customary in order to bring out this dependence.

---

[4] This fact was discovered by Galileo and is equivalent to the statement that m in Equation 1.3 ("inertial mass") is proportional to the mass that appears in Newton's law of gravitation ("gravitational mass"). This proportionality has been demonstrated experimentally to within an error of one part in 10^{11}. For an excellent discussion, see R. H. Dicke, *Gravitation and the Universe*, American Philosophical Society, Philadelphia, PA., 1970, Lecture I.

TABLE 1.2 The Dynamical Quantities Important in Classical Mechanics

	Dimensions		Unit		
Quantity	SI-CGS	American Engineering	SI	CGS	American Engineering
Mass	M	FT^2L^{-1}	kilogram (kg)	gram (g)	slug
Force	MLT^{-2}	F	newton (N)	dyne	pound (lb)

1 slug ≃ 14.59 kilogram. 1 gram = 0.001 kilogram.
1 pound ≃ 4.45 newton. 1 dyne = 10⁻⁵ newton.

THE THIRD LAW   *For an isolated pair of particles that exert forces upon one another along their line of centers, the mutual forces are equal in magnitude and opposite in direction.*

A careful examination of the Third Law in the context of the theory of fields indicates that this postulate can be fulfilled only if the forces involved are propagated *instantaneously*. (When the force exerted by particle 1 propagates to particle 2, does the force go to where particle 2 was when propagation began or to where particle 2 will be when the force finally affects that particle?) But this requirement stands in contradiction with the Theory of Relativity, which sets a finite upper limit on the speed of propagation of the effects of force fields. Therefore, it cannot be met exactly in any real physical interaction. On the other hand, the extent to which the velocity of propagation of forces enters into the observed motions of particles is quite frequently small enough to be ignored.

The mathematical form of the Third Law may be written

$$F_{12} = -F_{21} \tag{1.6}$$

where F_{ij} is the force particle j exerts on particle i ($i, j = 1, 2$). This equation holds the possibility of giving further information about the mass of a particle. If we substitute Equation 1.3 into Equation 1.6, we get

$$m_1 a_{12} = -m_2 a_{21} \tag{1.7}$$

or

$$\frac{m_1}{m_2} = \left|\frac{a_{21}}{a_{12}}\right|$$

Equation 1.7 clearly shows the impossibility of an absolute unit of mass, since a change in the numerical values of m_1 and m_2 by a constant factor would not alter the Third Law. Stated another way, all three of the Laws of Motion are invariant under a change in the unit of mass and thus cannot be used to define an absolute unit. In addition to this fact, we point out that Equation 1.7 could be used in a measurement of mass wherein two particles isolated from their surroundings interact through some force and the resulting accelerations are determined. This method is free of the appeal to extraneous experiments that is necessary in weighing, but suffers from the fact that it is quite difficult to carry out for most objects.

Weighing seems by far to be the simpler alternative, despite the possible objection that may be made against it on grounds of theoretical purity.

If three particles are brought together in isolation from the rest of the universe, we can describe their motions only after appending Newton's Laws with an important assumption. Newton himself considered this assumption just a corollary to the Laws of Motion (*Principia*, pp. 14–17, Cajori translation), but there is good reason to elevate its status at least to that of the Third Law. As with the Third Law, the truth of the assumption we are about to make must in every case of its application be verified empirically. In practice, we find that its use is justified completely to about the same degree that the use of classical dynamics itself is justified. Therefore, we shall state it here and adopt it as a part of our basic theory. It is given as follows.

Assumption *The total force on a particle interacting with two or more other particles is the sum of the forces exerted on it individually by those other particles.*

For example, in the case of three particles, the Second and Third Laws may be expressed

$$m_1 a_1 = F_{12} + F_{13}$$
$$m_2 a_2 = F_{21} + F_{23} \tag{1.8}$$
$$m_3 a_3 = F_{31} + F_{32}$$

and

$$F_{12} = -F_{21}$$
$$F_{13} = -F_{31} \tag{1.9}$$
$$F_{23} = -F_{32}$$

respectively, where a_i ($i = 1, 2, 3$) stands for the acceleration of the ith particle caused by all of the forces acting on it. If we combine Equations 1.8 and 1.9 we get the general result

$$m_1 a_1 + m_2 a_2 + m_3 a_3 = 0 \tag{1.10}$$

Now suppose that F_{23} is a very strong force, strong enough to bind particles 2 and 3 together as a rigid unit. It follows that, in this case, $a_2 = a_3$ and Equation 1.10 becomes

$$m_1 a_1 = -(m_2 + m_3) a_2$$

But this equation is just a statement of the Third Law applied to a body of mass m_1 and a body of mass $(m_2 + m_3)$. We conclude from this result that the behavior of a pair of particles rigidly bound together is the same as that of a single particle whose mass is the sum of the masses of the pair. Therefore, we are in a position to suggest that mass is an *additive* quantity and, by a straightforward extension of the argument given above, that the mass of a collection of particles is just the sum of the masses of all the particles in the collection.

Having stated the Laws of Motion, we can now examine the program for dynamics that emerges from them. For now we shall consider only the motion of a single particle and, therefore, we need bring in only the Second Law. The program in outline is as follows (see Figure 1.2).

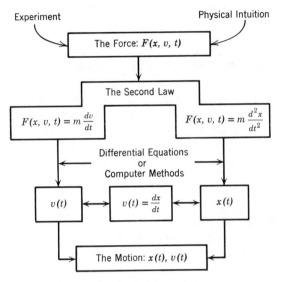

FIGURE 1.2. The Newtonian Program for classical dynamics.

THE NEWTONIAN PROGRAM

1. From a careful observation of the motion and, with the further help of experiments, physical intuition, or a guess, we write down an expression for the force, generally as a function of the position, velocity, and the time.

2. Next we introduce the force expression into the Second Law and consider the resulting equation as a *differential equation* of the first order for $v(t)$, as given by Equation 1.4. If the form of the force expression makes it more appropriate to do so, we use Equation 1.1 to transform the Second Law into

$$F(x, v, t) = m \frac{d^2 x}{dt^2} \tag{1.11}$$

which is a differential equation of the *second* order for $x(t)$.

3. We solve either Equation 1.4 for the velocity or Equation 1.11 for the position, by employing analytical methods in the theory of ordinary differential equations or numerical methods involving the use of a digital computer. If Equation 1.4 is solved, we apply Equation 1.1 to the solution to get a first-order differential equation for $x(t)$ and then repeat the solution process. If Equation 1.11 is solved, we differentiate the result to calculate $v(t)$.

The retrieving of the position and velocity of a particle from the Second Law yields all the information needed to describe the motion of a particle. The Newtonian Program makes it evident that this information derives necessarily both from physical law[5] and from the available mathematics pertinent to differential equations. We shall see briefly in the next section, and in great detail throughout subsequent chapters, how these two theoretical structures work together to solve the problems in classical dynamics.

1.2 AN EXAMPLE: THE FREELY FALLING PARTICLE

We can get a deeper understanding of the Newtonian program by considering the well-known problem of a particle moving vertically in the gravitational field of the Earth. Under the conditions that the position of the particle is never far from the ground and that no forces other than the gravitational force are acting, the Second Law may be written

$$-mg = m \frac{dv}{dt} \qquad (1.12)$$

where $g = 9.7804 \text{ m/s}^2$ on the equator at sea level and remains within 0.5 percent of this figure elsewhere on the surface of the Earth. Thus we may take g as an effectively constant parameter and rewrite Equation 1.12 as a first-order differential equation for $v(t)$:

$$\frac{dv}{dt} = -g \qquad (1.13)$$

The solution of (1.13) follows easily from the general method for handling linear differential equations of the first order. This method we shall outline briefly for future reference.[6]

A differential equation of the general form

$$\frac{dy}{dt} + p(t)y(t) = q(t) \qquad (1.14)$$

where $p(t)$ and $q(t)$ are arbitrary integrable functions of t, may be solved by calculating the *integrating factor* $\exp(\int p(t)\, dt)$, then multiplying it on both sides of Equation 1.14 to produce the expression

$$\frac{d}{dt}[y(t) \exp(\int p(t)\, dt)] = q(t) \exp(\int p(t)\, dt) \qquad (1.15)$$

---

[5] R. P. Feynman has aptly characterized the Newtonian Program by the phrase: *Pay attention to the forces!*

[6] For details, see E. D. Rainville and P. E. Bedient, *Elementary Differential Equations*, The Macmillan Co., New York, 1969, Chapter 2.

Equation 1.15 is then integrated with respect to t to obtain

$$y(t) = \frac{\int q(t) \exp\left(\int p(t)\, dt\right) dt + C}{\exp\left(\int p(t)\, dt\right)} \tag{1.16}$$

the solution for $y(t)$, where C is an arbitrary constant of integration. (All of the integrals indicated here are indefinite integrals.)

In Equation 1.13 we have $p(t) = 0$ and $q(t) = -g$. Therefore, the integrating factor is simply the number 1 and

$$v(t) = -gt + C \tag{1.17}$$

is the solution for the velocity of the particle. The arbitrary constant C that always appears may be evaluated in whatever manner is most appropriate to the physical problem under study. One of the more useful ways to evaluate the constant is to put $t = 0$ on both sides of Equation 1.17. Doing so, we find $v(0) = C$ and, consequently,

$$v(t) = v(0) - gt \tag{1.18}$$

Equation 1.18 comprises a theoretical prediction for the velocity that is subject only to a prior knowledge of the *initial* velocity of the particle.

To calculate the position of the particle, we use Equation 1.1 to rewrite Equation 1.18 in the form

$$\frac{dx}{dt} = v(0) - gt \tag{1.19}$$

This equation is a first-order differential equation for $x(t)$ with $p(t) = 0$ and $q(t) = v(0) - gt$. It follows directly from Equation 1.16 that

$$x(t) = v(0)t - \tfrac{1}{2}gt^2 + C \tag{1.20}$$

or

$$x(t) = x(0) + v(0)t - \tfrac{1}{2}gt^2 \tag{1.21}$$

if we choose to evaluate the constant of integration by setting $t = 0$ on both sides of Equation 1.20. Equation 1.21 is a complete theoretical prediction of the position, dependent only on a full knowledge of the *initial* position and velocity of the particle. It should be evident from our procedure that two arbitrary constants such as $x(0)$ and $v(0)$ will always appear in the equation for $x(t)$, since two integrations are required in order to compute the latter function from the Second Law. The predictive ability of the Newtonian Program, then, resides not solely in the Law of Motion (1.4) but in that Law and accurate information about *the initial conditions* on the particle under examination. This special blend of physical and mathematical requirements is what distinguishes Newtonian determinism: the behavior of a particle is fully determined whenever the initial position and velocity of the particle are known along with the appropriate from of the Second Law. To whatever extent we do not possess totally accurate knowledge of the initial conditions we have less than a determinate theory of the motion we wish to describe and predict.

The theory we have developed through Equations 1.8 and 1.21 predicts that the velocity of a particle moving vertically in a uniform, constant gravitational field will decrease (become more negative) linearly with the time, and that the position will decrease quadratically with the time. A precise estimate of the growth of $v(t)$ or $x(t)$ depends, of course, on what are the initial conditions. In Figure 1.3 are graphs of $x(t)$, for example, plotted according to the three general possibilities for the initial velocity, given a positive initial position:

$$v(0) > 0 \quad \text{The particle is thrown upward}$$
$$v(0) = 0 \quad \text{The particle falls from rest}$$
$$v(0) < 0 \quad \text{The particle is thrown downward}$$

Notice that, in the first case, the particle actually increases its position above the ground before falling. This motion is the expected result of its early upward thrust and is illustrative of the fact, pointed out in Section 1.1, that the direction of motion and the direction of the acceleration need not always coincide.

If Equations 1.18 and 1.21 are correct, it should be possible, given the initial conditions and the value of the parameter, g, to predict exactly the motion of a

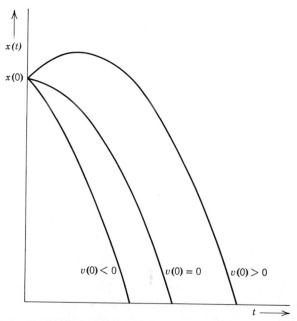

FIGURE 1.3. The position of a particle moving vertically in a uniform, constant gravitational field under different initial conditions for the velocity.

particle subjected to terrestrial gravitation. However, this kind of comparison of the theory with experiment is seldom carried out in practice. What is usually done is to employ numerical data taken on $x(t)$ to compute the values of the constant parameters that appear in the theory. In the present case we could use such data to calculate a value for the gravitational acceleration g. In Table 1.3, for example, are listed some laboratory data on the free fall in air of a solid sphere. These data are believed to be accurate to within one percent, but they have not been corrected for the effect of air resistance on the motion of the sphere. However, this effect should be quite small for the velocities involved here and can be neglected at least to the accuracy of the position measurements. The velocities given in the third column of the table were not obtained graphically from the position data, as suggested in the preceding section, but instead were determined more accurately with the expression

$$v(t + \tfrac{1}{2}\varepsilon) = \frac{[x(t + \varepsilon) - x(t)]}{\varepsilon}$$

where $\varepsilon = 1/30$ second and the time is advanced in intervals of duration ε. This interpolation formula for $v(t)$ is a finite-difference form of Equation 1.1, which associates the velocity in the middle of a time interval with the average velocity computed for that interval and, therefore, should be a nearly exact relation in the limit of very small ε.

A graph of the position data in Table 1.3 against the time will not be a simple quadratic curve because "time zero" for the data is actually about 0.2 second after the sphere was dropped from rest. This complication can be dealt with by rewriting Equation 1.21 to read

$$\frac{-[x(t) - x(0)]}{t} = -v(0) + \tfrac{1}{2}gt \tag{1.22}$$

TABLE 1.3 Laboratory Data on the Free Fall
in Air of a Solid Sphere

$-[x(t) - x(0)]$ (cm)	t (s)	$-v(t)$ (cm/s)	t (s)
		231.0	.0167
7.70	.0333	262.5	.0500
16.45	.0667	294.0	.0833
26.25	.1000	325.5	.1167
37.10	.1333	359.7	.1500
49.09	.1667	392.7	.1833
62.18	.2000	425.4	.2167
76.36	.2333	456.6	.2500
91.58	.2667	489.3	.2833
107.89	.3000	523.5	.3167
125.34	.3333	555.6	.3500
143.86	.3667		

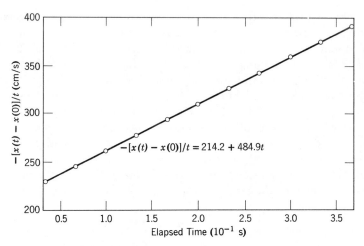

FIGURE 1.4. A graph of the position data in Table 1.3. The equation is the result of a least-squares fit of a line to the points shown.

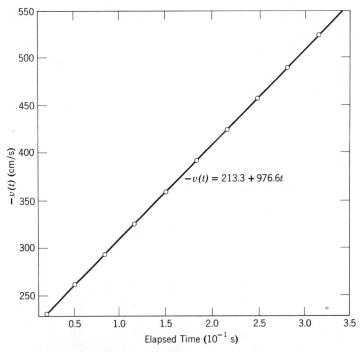

FIGURE 1.5. A graph of the velocity data in Table 1.3. The equation is the result of a least-squares fit of a line to the points shown.

[Remember that $v(0) < 0$ here because $t = 0$ is actually a point in time *after* the sphere was dropped.] A plot of Equation 1.22 may be compared directly with a plot of $-v(t)$ against the time, which also will be a straight line but with a slope equal to g itself. Graphs of Equations 1.22 and 1.18 are shown in Figures 1.4 and 1.5, respectively. The data in both instances are seen to lie very nicely along straight lines that follow the mathematical expressions

$$\frac{-[x(t) - x(0)]}{t} = 214.2 + 484.9t$$

$$-v(t) = 213.3 + 976.6t$$

in CGS units. The values for $v(0)$ derived from these graphs (their y intercepts) agree within 0.4 percent. The values for g are 970 cm/s^2 and 977 cm/s^2, respectively. These differ from the accepted value for g quoted above at most by about 0.8 percent, which is consistent with the experimental accuracy of the position data.

1.3 CONSERVATION THEOREMS AND THE NEWTONIAN PROGRAM

As it stands, Equation 1.4 is quite sufficient to describe the motion of a single particle within the scope of the Newtonian Program. But perhaps it has become apparent, from our discussion of the Second Law and from its application in the preceding section, that a broad physical insight into the nature of the many possible motions of a particle subjected to arbitrary forces will not be forthcoming without experience from a large number of examples. In order to understand comprehensively what is the character of force and the motion it causes it seems as if we are destined to investigate in detail a superfluity of cases and to proceed by induction. How else can we come by a unified view of the importance of initial conditions and the relations between velocity and position permitted by the Second Law? Fortunately, this prophecy of tedium turns out in practice to be largely untrue, for purely mathematical reasons. It is, in fact, possible to state with certainty some very general and far-reaching conclusions about the motions of particles on the basis of a set of analytical deductions known as the *conservation theorems*. These theorems, as their name suggests, involve in each instance the definition of a quantity of physical significance whose numerical value can be demonstrated to be independent of the time during the motion of a particle (or collection of particles). For this reason the conserved quantity can tell us much which is general about dynamics according to the Newtonian Program: it represents a theoretical entity that remains unperturbed while the particle or assembly of particles it describes may be changing in its motion quite drastically. In this sense a conserved quantity characterizes the motion and can present us with an additional relation among dynamical properties that is of great value in applications. Stated another way, the conserved quantity makes it possible for us to glean physical information from the Second Law without having to solve it directly for $v(t)$ and $x(t)$. The price we pay for this advantage, as will be clear soon, is the requirement of interpreting operationally a quantity whose original meaning and derivation are wholly mathematical in character. Thus we must accept the fact that conserved quantities are necessarily abstract quantities without the direct relation to experience attendant with the concept of force.

The first conservation theorem we shall prove is actually most useful when an assembly of particles is being investigated. Our purpose in stating it here only for a single particle moving in one spatial dimension is to illustrate in the simplest way the manner in which conservation theorems are proved. A more general form of the theorem will be developed in Chapter 4.

The Linear Momentum Theorem *The linear momentum $p(t)$ of a particle is defined by*

$$p(t) \equiv mv(t) \tag{1.23}$$

where m is the mass of the particle and $v(t)$ is the velocity. If the force acting on a particle vanishes, then $p(t)$ is a constant of the motion.

Proof According to Equations 1.4 and 1.23, the Second Law may be expressed in the form

$$F(x, v, t) = \frac{dp}{dt} \tag{1.24}$$

If the force is zero this equation reduces to

$$\frac{dp}{dt} \equiv 0$$

for *all* values of the time. But this vanishing of the derivative is just the necessary and sufficient mathematical condition for the function $p(t)$ to have a fixed numerical value, that is, for the linear momentum to be a "constant of the motion."

We see from this exercise the general structure of all conservation theorems. The first step is the definition of a new and, possibly, somewhat abstract physical quantity. The definition indicates the measurability of the new quantity. The conservation theorem or "law" is then proved by relating the new quantity to the force and by prescribing the specific conditions under which the first time-derivative of the quantity will vanish *identically*. In this example we see that the linear momentum is a direct indicator of the presence of force and that its measurability is assured by that of the mass and the velocity.

The second conservation theorem to be considered is quite useful even in the description of the motion of a single particle. The connection between it and the Second Law, however, is less direct than that found in the Momentum Theorem.

The simplest logical bridge is built by giving the two preliminary definitions that follow.

Kinetic Energy *The change in kinetic energy undergone by a particle moving between the positions x_1 and x_2 is formally defined by*

$$T_2 - T_1 \equiv \int_{x_1}^{x_2} F(x)\, dx \tag{1.25}$$

where T_2 is the kinetic energy evaluated at the point x_2, etc., and $F(x)$ is the force acting on the particle and is assumed explicitly independent of the velocity and the time.

Upon introducing the Second Law into Equation 1.25, we can further write the definition of kinetic energy as

$$T_2 - T_1 = m \int_{x_1}^{x_2} \frac{dv}{dt}\, dx = m \int_{v_1}^{v_2} v\, dv = \tfrac{1}{2}mv_2{}^2 - \tfrac{1}{2}mv_1{}^2 \tag{1.26}$$

where in the second step we have used Equation 1.1 to set $dx = v\, dt$. Equation 1.26 suggests a general expression for the kinetic energy as a function of the velocity:

$$T(v) = \tfrac{1}{2}mv^2 \tag{1.27}$$

This expression requires the permissible but arbitrary convention $T(0) \equiv 0$.

The kinetic energy is a function that depends explicitly on the velocity and implicitly on the time. Because of its explicit form (1.27), it is often referred to as *the energy a particle possesses by virtue of its motion*. Notice once again that the absolute value of this energy is not defined. Equations 1.25 and 1.26 make it clear that any constant could be added to $T(v)$ without affecting its meaning and, therefore, that only differences in kinetic energy have a physical significance. The units of kinetic energy are those of force times distance and are given the name *joule* in the Système International. A summary of dimensions and units for the conserved quantities is given in Table 1.4.

Potential Energy *The change in potential energy undergone by a particle moving between positions x_1 and x_2 is defined by*

$$V(x_2) - V(x_1) \equiv -\int_{x_1}^{x_2} F(x)\, dx \tag{1.28}$$

TABLE 1.4 Some Conserved Quantities Important in Classical Mechanics

Quantity	Dimensions		Unit		
	SI-CGS	American Engineering	SI	CGS	American Engineering
Linear momentum	MLT^{-1}	FT	kg-cm/s	g-cm/s	lb-s
Energy	ML^2T^{-2}	FL	joule	erg	ft-lb

where $V(x)$ is the potential energy at the position x and the force $F(x)$ is assumed explicitly independent of the velocity and the time.

The complete definition of the potential energy is provided by Equation 1.28. The precise form of this function, unlike that of the kinetic energy, can be known only after the force has been specified in detail. The force must depend on position alone[7] in order that the potential energy be an explicit function of $x(t)$. Forces that show a velocity or time dependence cannot be employed to define $V(x)$. Because of this stipulation we may interpret the quantity $V(x)$ as *the energy a particle possesses by virtue of its position in a field of force.* Once again the absolute value of this energy is not meaningful, although we may always choose a zero of potential energy at some value of x in order to given an operational sense to $V(x)$ as "the potential energy at position x."

Finally, let us point out that the concept of the integral as an antiderivative leads immediately, through Equation 1.28, to the relation

$$F(x) = -\frac{dV}{dx} \tag{1.29}$$

between a conservative force and the potential energy. This expression, with the help of the concept of *work* (mechanical energy transfer), often is used as part of the basis of a more physical derivation of the conservation theorem for energy than what is implied directly by Equations 1.25 and 1.28. In terms of work, Equation 1.28 would be an equality between the work done by the force $F(x)$, when the particle moves from x_2 to x_1, and the change in potential energy. The kinetic energy of the particle simply would be *defined* by Equation 1.27, and Equations 1.25 and 1.26 would comprise an exercise relating kinetic energy to the work integral. In our discussion, we have chosen instead to define kinetic and potential energy changes so as to make the Energy Theorem a simple mathematical exercise and to yield Equation 1.27 as a *result* of our basic definitions. The important physical problem is then only the empirical interpretation of the conservation law.

Now we may state

The Total Energy Theorem *The total mechanical energy E of a particle is defined by*

$$E(x, v) \equiv T(v) + V(x) \tag{1.30}$$

If the force acting on a particle is a function of its position alone, then the total energy is a constant of the motion.

---

[7] Forces of this kind are called *conservative*. A conservative force acting in one spacial dimension need only be velocity and time independent. In three dimensions this condition is necessary but not sufficient. This fact will be discussed further in Chapter 4.

Proof   The theorem is established by the chain of equalities

$$\frac{dE}{dt} = \frac{d}{dt}[T(v) + V(x)] = \frac{dT}{dv}\frac{dv}{dt} + \frac{dV}{dx}\frac{dx}{dt}$$

$$= mv(t)\frac{dv}{dt} - F(x)v(t)$$

$$= v(t)F(x) - F(x)v(t) \equiv 0$$

where we have differentiated implicitly with respect to the time and have invoked Equations 1.1, 1.27, and 1.29 along with the Second Law.

The Energy Theorem applies whenever the force is conservative. It provides a valuable relation between the position and the velocity that is seemingly extraneous to but yet consistent with the Second Law. The theorem states that, no matter how these functions vary with the time, throughout the motion the position and the velocity always will be such as to make the combination on the right-hand side of Equation 1.30 equal to a fixed number. Equation 1.30 is, therefore, an effective condition on the motion that may be used to obtain information about $x(t)$ and $v(t)$ without recourse to a differential equation.

To see how this works out in detail, consider again the simple problem of a particle moving vertically in a uniform and constant gravitational field. In this case Equations 1.5 and 1.28 tell us that the potential energy difference between two points in the field is

$$V(x_2) - V(x_1) = mg \int_{x_1}^{x_2} dx = mgx_2 - mgx_1$$

or, in general,

$$V(x) = mgx \tag{1.31}$$

relative to $V(0)$. Therefore, we have

$$E = \tfrac{1}{2}mv^2 + mgx \tag{1.32}$$

for the total energy of the particle at any time t. For the sake of concreteness we shall set $x(0) > 0$ and $v(0) > 0$. Then we know that somewhere during its motion the particle will change its direction in response to the gravitational force and begin to fall. At the point in space where this change occurs, the velocity must be zero because it is a continuous function of the time and continuous functions cannot change sign without going through zero. It follows that at this point we may write Equation 1.32 in the form

$$E = mgx_m \tag{1.33}$$

where x_m is the position of the particle when it begins to fall. Now, because E does not depend on the sign of the velocity, the point where the velocity goes to zero must correspond to the *maximum* value of $x(t)$ (given that E is a constant). Moreover, we can always write

$$E = \tfrac{1}{2}mv^2(0) + mgx(0) \qquad (1.34)$$

in terms of the initial conditions. The Energy Theorem says that Equations 1.33 and 1.34 are numerically identical. Therefore, we may equate those expressions and go on to write

$$h \equiv x_m - x(0) = \frac{v^2(0)}{2g} \qquad (v(0) > 0) \qquad (1.35)$$

which is the well-known relation between the maximum height h attained by the particle and its initial velocity. Without even having looked at the Second Law we have derived a nontrivial expression relating two important dynamical parameters of the motion, h and $v(0)$, which could, in fact, be used to estimate the value of g. This is a very typical (if simple) example of the Energy Theorem employed to its best advantage.

What may yet remain a little vague about our application is its physical meaning. How do we explain what the particle is *doing* in terms of the concept of energy? To answer this, let us point out first of all that the definitions (1.25) and (1.28) make it evident that changes in energy have to do with forces applied over the distance between two positions of a particle. If a force acts to decrease the speed of a particle, the simultaneous change in its position will produce a decrease in kinetic energy according to Equation 1.26. But then the potential energy must *increase* in obedience to Equation 1.30 and the Energy Theorem. In regard to the example considered above, these mathematical requirements translate simply into conditions on the velocity and the height of the particle. The absolute value of the velocity initially decreases to zero as the kinetic energy is given up in favor of potential energy and, according to Equation 1.31, a greater height above the ground. After the particle changes direction and rushes downward, its potential energy diminishes and its kinetic energy rises sharply, but always in agreement with Equation 1.32. What occurs during the motion, then, is a continual transfer of *energy form* from kinetic to potential and *vice versa*. The constancy of the total energy is maintained by this transfer and the balance of each kind of energy reflects to a significant extent the observed behavior of the particle at each instant of its motion.

FOR FURTHER READING

A. P. French — *Newtonian Mechanics*, W. W. Norton, New York, 1971. Chapter 6 presents Newton's Laws from a point of view quite similar to that taken in this book. Pages 367–381 provide a good introduction to the concept of energy.

D. Kleppner and	*An Introduction to Mechanics*, McGraw-Hill, New York, 1973.
R. J. Kolenkow	Chapter 2 contains a fine discussion of force, mass, and inertial frames of reference.
R. P. Feynman,	*The Feynman Lectures on Physics*, Addison-Wesley, Reading, Mass.,
R. B. Leighton, and	1962, Volume I. Chapter 4 presents a classic introduction to the idea
M. Sands	of energy conservation.
C. Kittel,	*Mechanics*, McGraw-Hill, New York, 1973, 2nd ed. revised by
W. D. Knight, and	A. C. Helmholz and B. J. Moyer. Chapter 5 is a very careful intro-
M. A. Ruderman	duction to the concept of energy.
I. Newton	*The Mathematical Principles of Natural Philosophy*, revised trans-
	lation by F. Cajori, University of California Press, Berkeley, 1960.
	It is well worth the time to read Newton's original statement of the
	Laws of Motion. The UC Press edition is available in paperback.

PROBLEMS

1. Discuss the assumptions underlying the following alternative statement of the First Law: *Uniform motion requires no cause.*
2. Is the First Law merely a special case of the Second Law?
3. Consider a noninertial frame of reference that has a constant acceleration a_0 along the positive x direction. A particle at rest in this noninertial frame is maintained in its quiescent state by the restraining force

$$F(x) = -kx$$

where k is a positive constant and x is the distance the particle is from the origin. Show that

$$a_0 = \frac{-k}{m} x$$

Why is the particle at rest *behind* the origin ($x < 0$)?
4. The moon is in free fall toward the Earth and, therefore, provides a noninertial frame of reference for an astronaut. Should the astronaut then be "weightless" when on the moon?
5. Draw the three graphs of $v(t)$ corresponding to those of $x(t)$ in Figure 1.3. What point on the graph for $v(0) > 0$ is associated with the position of the particle when it turns around?
6. The buoyant force on an object of fixed total volume V completely immersed in a fluid (gas or liquid) of constant density ρ_f is found from experiment to be

$$F_B = \rho_f V g$$

where g is the acceleration due to gravity. Consider a parcel of material of mass m that is subject to this force and to the force of gravity. (Two practical examples which may approximate this situation are a parcel of hot smoke released into the atmosphere by a coal-fired electric powerplant and sewage effluent released underwater by a pipe running into the sea.) Apply the Newtonian Program to find the motion of the parcel. What is the physical condition for a uniform motion (including rest) of the parcel?
7. A particle of mass m is acted on by the force ("gravity rheostat")

$$F(t) = -mg(1 - e^{-t/\tau})$$

where g is the acceleration due to gravity and τ is a positive "time constant." Apply the Newtonian Program to find the motion of the particle. Show that the velocity of the particle is *quadratic* in the time for t very small. (*Hint*: A Taylor expansion will be useful.)

8. An electron in the ionosphere is subject to the force

$$F(t) = F_0 \sin\left(\frac{2\pi t}{T}\right)$$

when a monochromatic radio wave of period T passes through, where F_0 is a positive constant. Calculate the motion of the electron given the initial conditions $x(0) = 0$, $v(0) = v_0 > 0$. Why does the electron gain the *constant* drift velocity $(F_0 T/2\pi m)$ from the strictly oscillatory applied force?

9. A particle is accelerated by the discontinuous force

$$F(t) = \begin{cases} F_0 & 0 < t < \tfrac{1}{2}T \\ -F_0 & \tfrac{1}{2}T < t < T \end{cases}$$

during the time interval T. Calculate and draw a graph of the resulting velocity as a (continuous) function of the time. The initial velocity is zero.

10. Prove that the sum of linear momenta for an isolated pair of particles is a constant of the motion whenever the Third Law is valid.

11. Show that the sum of kinetic and potential energies is constant anywhere along the path of a moving particle, given *only* the defining relations (1.25) and (1.28). What then can be said about the relation between energy conservation and the Second Law?

12. Verify the Energy Theorem for a freely falling particle by computing the total energy explicitly as a function of time.

13. Use the empirical equations given in Figures 1.4 and 1.5 to check the laboratory data in Table 1.3 for conformity with the Energy Theorem. Take into account the error in the data when evaluating your computation.

14. Calculate the potential energy function $V(x)$ corresponding to the following conservative forces. In each case you must choose a convenient value of x at which to set the potential equal to zero. All of the constant parameters that appear are positive.

(a) $F(x) = -e^2/x^2 \qquad (x \geqslant 0)$

(b) $F(x) = \dfrac{24\varepsilon}{\sigma}\left[2\left(\dfrac{\sigma}{x}\right)^{13} - \left(\dfrac{\sigma}{x}\right)^{7}\right] \qquad (x \geqslant 0)$

(c) $F(x) = e^{-\alpha x}\left(\dfrac{1 + \alpha x}{x^2}\right) \qquad (x \geqslant 0)$

(d) $F(x) = \dfrac{V_0\pi}{a}\sin\left(\dfrac{2\pi x}{a}\right)$

15. Calculate the "energy function"

$$E(t) = \tfrac{1}{2}mv^2(t) + mgx(t)$$

for the particle subject to the force given in Problem 7. The initial conditions may be taken to be $x(0) = 0$, $v(0) = 0$. Show that $E(\infty)$ is less than the total energy of a particle subject to the force $F = -mg$. Why is this so? Finally, consider $E(t)$ in the limit $\tau \to \infty$, which corresponds to imposing the force $F = -mg$ with infinite slowness. Compare your result with the total energy of a free particle under the initial conditions given above. (The two energies should be the same. This is an example of the Adiabatic Theorem, which states that a perturbing force imposed infinitely slowly will not change the total energy of a particle.)

SPECIAL TOPIC 1
Newton's Conception of Force

In the early part of 1685, Isaac Newton composed the final revisions of the third version of the tract *De Motu* ("On Motion"), which he had written in response to the urging of a friend, the astronomer Edmond Halley.[1] Prior to Halley's catalytic visit, in August of 1684, Newton apparently had not considered problems in mechanics seriously for approximately 15 years. He was, in fact, known academically for his published work concerning light and color, and not for his privately circulated contributions to mathematics. However, Halley's enthusiastic reception of the first version of *De Motu* was sufficient encouragement to prepare a larger manuscript for general publication. This effort culminated with the appearance of the *Principia* in 1687.

The revised *De Motu* contains a listing of what Newton ultimately called his Laws of Motion. (Initially he had used the term "hypothesis" instead of "law," but a change was made in the third version.) These Laws were as follows:[2]

LAW 1   *By reason of its innate force every body preserves in its state of rest or of moving uniformly in a straight line unless in so far as it is obliged to change its state by forces impressed on it.*

LAW 2   *The change of motion is proportional to the force impressed and takes place along the straight line in which the force is impressed.*

LAW 3   *As much as any body acts on another so much does it experience in reaction.*

LAW 4   *The relative motion of bodies enclosed in a given space is the same whether that space rests absolutely or moves perpetually and uniformly in a straight line without circular motion.*

---

[1] A complete account of Halley's role in the publication of Newton's *Principia* is given by I. B. Cohen, *Introduction to Newton's "Principia,"* Harvard University Press, Cambridge, Mass., 1971, Chapter III.

[2] An English Translation of *De Motu* is in J. Herivel, *The Background to Newton's "Principia,"* Oxford University Press, London, 1965. We have used this translation here.

LAW 5   *The common center of gravity of a number of bodies does not change its state of rest or motion by reason of the mutual actions of the bodies.*

LAW 6   *The resistance of a medium is jointly proportional to the density of that medium, the area of the moved spherical body and the velocity.*

What is immediately striking about this list is that there are six Laws instead of the three that finally appeared in the *Principia*.[3] In the latter, Newton has granted Laws 4 and 5 only the status of corollary (Corollaries V and IV, respectively, of the Laws of Motion) and has removed Law 6, which clearly is empirical and approximate, from the axiomatic level altogether. These changes do not seem so drastic if *De Motu* is taken only to be a work marking a point of transition in the development of the mature dynamical concepts that appeared in the *Principia*.

The first three Laws given in *De Motu* are easily recognized as the ones that are today universally identified with Newton. Law 3, for example, reads in the *Principia*:[4]

LAW III   *To every action there is always opposed an equal reaction: or, the mutual actions of two bodies upon each other are always equal, and directed to contrary parts.*

The wording there is a little changed, but the physical meaning is certainly the same. This cannot be so easily said, however, in the case of Law 1. For a close scrutiny of that axiom suggests that there may be a significant, even if subtle, difference in its formulation from that of Law I in the *Principia*:[4]

LAW I   *Every body continues in its state of rest, or of uniform motion in a right line, unless it is compelled to change that state by forces impressed upon it.*

The statement of Law 1 that appears in *De Motu* indicates that in 1685 Newton regarded what is called inertia—the inherent tendency in matter to do what is described above in Law I—as the direct manifestation of some kind of *force*. This "innate force" was defined by Newton[2] as "the power by which a body preserves in its state of rest or of moving uniformly in a straight line." Thus it clearly was distinct from "impressed force," which he defined[2] as "that by which a body is urged to change its state of moving or resting." Newton evidently believed at the time that

---

[3] Most of the discussion that follows is derived from Chapters VII and VIII in R. S. Westfall, *Force in Newton's Physics*, American Elsevier, New York, 1971. A concise essay covering the same ideas is R. S. Westfall, "Stages in the Development of Newton's Dynamics" in *Perspectives in the History of Science and Technology*, edited by D. H. D. Roller, University of Oklahoma Press, Norman, 1971. This essay is a fine starting point for the student wishing to know more about the history of the Newtonian Program.

[4] The English translation is from I. Newton, *The Mathematical Principles of Natural Philosophy*, revised translation by F. Cajori, University of California Press, Berkeley, 1960, p. 13.

two kinds of force operated in nature. The first was a universal property of matter along with extension in space and durability. It was proportional to the quantity of matter in a body and was the means through which that body resisted changes in its state of uniform motion. The second type of force arose with actions *outside* a body that attempt to alter its motion.

In modern terms we might say that Newton's concept of innate force is equivalent to the mathematical statement $F = mv$. This point of view seems strange, not just because we are used to calling the product mv linear momentum, but also because attributing a "power" or innate force to matter appears out of character with the more passive role given it in the concept of inertia, and especially in the Second Law. On the other hand, in order to achieve a quantitative particle dynamics, there had to be a way to express the constant ratio between impressed force and the change in the state of motion that it causes. Newton's final compromise, between the possibility of a confusing proliferation of concepts of force and the impossibility of a view of matter as totally inert substance, was the statement of Law I that we find in the *Principia*. This idea of inertia—surely a stroke of genius—is a beautiful optimization: a viewpoint, adequate for the foundation of a general, mathematical dynamics, that does not pretend to explain but only sets aside gently an inherent paradox in the mechanical concept of matter.[5]

The statement of Law 2 in *De Motu* is essentially identical, both in language and in content, with what appears in the *Principia* as Law II. It is evident that these two axioms would translate into mathematics in the form $F = \Delta mv$, instead of $F = ma$ that appears in every modern textbook on mechanics. That is, Newton says in his Second Law that impressed force brings about a change in the motion (Δmv), not a rate of change in the motion (ma). This fact is not difficult to understand, since Newton wrote most often about impulsive forces, the result of which the quantity Δmv is a direct measure. It is understood simply that one is considering the application of force over a certain period of time. The ambiguity in choosing the length of that period, in the case of a continuously applied force, is overcome by passing to the limit of a series of infinitesimal impulses. (Recall Newton's famous derivation of Kepler's Law of Equal Areas.) On the other hand, when discussing other specific motions (e.g., circular motion), Newton seems to be employing the concept of a continually applied force causing a continual acceleration; that is, $F = ma$. A survey of Newton's work[3] indicates that the term "force" was used quite generally without discretion as to when it meant Δmv and when it meant ma.

This difficulty, which is at the least one of dimensional confusion, was resolved finally some 20 years after Newton's death by the mathematician Leonhard Euler.[6]

[5] That Newton did not, in fact, formulate his Laws of Motion until the age of 42 is another counter-example, standing beside those offered by Kepler, Galileo, Huygens, and others, to the myth that physicists do their most notable work when they are under 40.

[6] A discussion of Euler's role in completing the Second Law is given by C. Truesdell, "Reactions of Late Baroque Mechanics to Success, Conjecture, Error, and Failure in Newton's *Principia*," in *The "Annus Mirabilis" of Sir Issac Newton*, edited by R. Palter, MIT Press, Cambridge, Mass., 1970.

In a paper on celestial mechanics he showed for the first time that the equation of motion for any paticle had the general form

$$m \frac{d^2 x_i}{dt^2} = F_i \qquad (i = 1, 2, 3)$$

where x_i is the ith spatial coordinate and the impressed force coordinate F_i is given. We would not hesitate today to call this expression "Newton's Second Law" and, indeed, we can readily see it in the context of the *Principia*. But the relation does not appear anywhere in Newton's work before or after 1687. It is Euler's equation.

2

The Motion of a Particle in One Dimension

I. CONSERVATIVE FORCES

2.1 MOTION IN A UNIFORM FIELD

Auniform field is a region of space wherein at every point the force acting on a particle may.be expressed, for a single spatial coordinate,

$$F = ma_0 \tag{2.1}$$

where m is the mass of the particle and a_0 is an empirical parameter usually called "the acceleration due to the field." In general a_0 will depend on the mass and other properties of the particle as well as on the character of the field. It will never depend on the position (or the time, since the force F is conservative). We have looked already at one example of a uniform field in Section 1.2, where the vertical motion of a particle near the surface of the Earth was investigated. In that case a_0 became $-g$, the acceleration due to the gravitational field. Another common example occurs when a charged particle moves along a line parallel with the direction of a uniform, constant electric field in which it is imbedded. Then we have

$$a_0 = \left(\frac{q}{m}\right) E_0 \tag{2.2}$$

where the first factor is the charge-to-mass ratio of the particle and the second is the field intensity. Other instances where Equation 2.1 applies (although they do not involve fields) include the uniform deceleration of an automobile as it approaches a stop sign and the testing of a rocket motor along a horizontal track.

The dynamics that results from Equation 2.1 and the Second Law is in every way analogous to what was described in Section 1.2. By employing the procedure developed there we find, with no difficulty,

$$v(t) = v(0) + a_0 t \tag{2.3}$$

$$x(t) = x(0) + v(0)t + \tfrac{1}{2}a_0 t^2 \tag{2.4}$$

for the motion of a particle in a uniform field. In doing applications it is often convenient to eliminate the time variable between these two equations in order to produce the following direct relation between velocity and position:

$$v^2(t) = v^2(0) + 2a_0[x(t) - x(0)] \tag{2.5}$$

The most interesting special case of motion in a uniform field is when the direction of the initial velocity is opposite to that of the acceleration. Then one of the constants in Equation 2.3 is negative and we know immediately that for some value of the time the velocity will go to zero. This value is

$$t = \frac{-v(0)}{a_0}$$

Moreover, Equation 2.4 then holds the possibility of an extremum (maximum or minimum) value for $x(t)$. To compute it we need only invoke the usual condition for an extremum,

$$\left(\frac{dx}{dt}\right)_{t=t_m} = 0 \tag{2.6}$$

to find from (2.4)

$$t_m = \frac{-v(0)}{a_0}$$

$$x_m(t_m) = x(0) - \tfrac{1}{2}\frac{v^2(0)}{a_0} \tag{2.7}$$

The result for t_m should not surprise us, since Equation 2.6 is also just a statement that the velocity is zero at t_m! The equation for x_m produces a maximum when $a_0 < 0$ and a minimum when $a_0 > 0$. This expression is, of course, the analog of Equation 1.35, which we deduced by another means.

The Energy Theorem applied to the uniform field states

$$E = \tfrac{1}{2}mv^2(t) - ma_0x(t) \tag{2.8}$$

by analogy with Equation 1.32. The further development of this equation to show that it is an alternative to the Second Law is quite instructive. We begin by substituting Equation 1.1 into Equation 2.8 to get

$$E = \tfrac{1}{2}m\left(\frac{dx}{dt}\right)^2 - ma_0x(t)$$

or

$$\left(\frac{dx}{dt}\right)^2 - 2a_0x(t) = \frac{2E}{m} \tag{2.9}$$

Equation 2.9 is, technically, a nonlinear differential equation of the first order for $x(t)$. (It is nonlinear because the derivative is raised to a power.) We can solve it formally with the rearrangement

$$\frac{dx}{\pm\left(\dfrac{2E}{m} + 2a_0x\right)^{1/2}} = dt$$

along with the integration of both sides to yield the result

$$\int_{x(0)}^{x(t)} \frac{dx}{\pm\left(\dfrac{2E}{m} + 2a_0x\right)^{1/2}} = t \tag{2.10}$$

Equation 2.10 is a complete, if implicit, solution for the position, as long as the velocity does not vanish. The positive sign is used when the velocity remains positive as the particle travels from $x(0)$ to $x(t)$; the negative sign is used in the opposite case. The integral is done in parts if the particle changes direction during its motion. To continue further we assume the velocity to be positive, then look to a table of integrals, where we find

$$\int \frac{dx}{(a + bx)^{1/2}} = \frac{2}{b}(a + bx)^{1/2}$$

With this information we can reduce Equation 2.10 to

$$\left[\frac{2E}{m} + 2a_0x(t)\right]^{1/2} - \left[\frac{2E}{m} + 2a_0x(0)\right]^{1/2} = a_0t$$

Now we turn to Equation 2.8 again and note that, since E is a constant of the motion, we can always write

$$E = \tfrac{1}{2}mv^2(0) - ma_0x(0) \tag{2.11}$$

This equation can be used to simplify our result in the form

$$\left[\frac{2E}{m} + 2a_0 x(t)\right]^{1/2} - v(0) = a_0 t$$

On squaring both sides of this expression and referring to (2.11) once more, we find

$$x(t) = \tfrac{1}{2}\frac{v^2(0)}{a_0} - \frac{E}{m a_0} + v(0)t + \tfrac{1}{2}a_0 t^2$$
$$= x(0) + v(0)t + \tfrac{1}{2}a_0 t^2$$

We see that the Energy Theorem can reduce a dynamical problem to a problem of quadrature (i.e., the doing of an integral). The entire solution for $x(t)$ has been obtained including the initial conditions. Equation 2.3 for the velocity may be derived easily from this solution by differentiation. Even Equation 2.5 can be recovered, since it is the immediate result of putting together Equations 2.8 and 2.11. These examples, despite their simplicity, further underscore the great importance of the concept of energy in particle dynamics.

2.2 TURNING POINTS

When an electron flies into a uniform electric field, moving along the field direction, or when a solid sphere is thrown upward in the gravitational field of the Earth, the motion is distinguished by a return toward the initial position. The location of the particle when this change in direction occurs has been considered previously and, for the two examples just mentioned, is given by Equation 2.7. This place of velocity change is known formally as a *turning point* of the motion. As should be evident from the derivation of (2.7), turning points exist whenever the velocity of a particle is zero while the acceleration is not. This requirement is quite general and in no way depends on the particular form of the force. Indeed, translated into purely mathematical terms, the requirement is

$$\left(\frac{dx}{dt}\right)_{t=t_m} = 0 \qquad \left(\frac{d^2x}{dt^2}\right)_{t=t_m} \neq 0 \qquad (2.12)$$

which is just the necessary and sufficient condition that the function $x(t)$ take on an extremum value at time t_m. But extremum values of the position are precisely what we mean by turning points: they represent the greatest penetration of the particle into a region while traveling along a given direction from its initial position.

The Energy Theorem is brought to bear on this problem when we note that the two conditions (2.12) imply

$$E = V(x_m) \qquad \left(\frac{dV}{dx}\right)_{x=x_m} \neq 0 \qquad (2.13)$$

according to Equations 1.29 and 1.30, at a turning point x_m. Equations 2.13 state that one condition for a turning point is that the total energy of a particle be *potential* energy. The physical interpretation of this condition is quite simple. When a moving particle has completely transferred its kinetic energy, motion in the current direction of travel is no longer possible because there is no longer any energy to perpetrate it. The particle does not remain stopped for more than an instant, however, because a force still acts on it. It begins to move again in response to this force, but in the opposite direction. (If the force caused it to move in the *same* direction as it was going, it would not have been *losing* kinetic energy.)

It should be understood that Equations 2.13 are of less generality than Equations 2.12, since the latter are valid regardless of whether the force is conservative. On the other hand, Equations 2.13 require much less information in order to apply them and are invaluable to the qualitative analysis of motion when the potential energy is a complicated function of the position. It is the description of this situation that we shall now consider.

2.3 MOTION IN A NONUNIFORM FIELD

When we must describe the motion of a particle subjected to a force that is an arbitrary function of the position, the appropriate form of the Second Law is

$$\frac{d^2x}{dt^2} = \frac{1}{m} F(x) \tag{2.14}$$

This differential equation for $x(t)$ may not be at all easy to solve if the force $F(x)$ is a complicated function. In cases like that, however, we may still appeal to the Energy Theorem to reduce the problem formally to a quadrature. Even if the resulting integral is difficult to evaluate, the conservation of energy can provide us with a thumbnail sketch of the motion under given initial conditions and we may hope to extract enough information from the physical picture we construct to suggest the form of $x(t)$. If we cannot perceive sharply from our analysis the time dependence of the position, often we can get sufficient insight to determine whether the theoretical motion and, therefore, the form of the force chosen, correspond reasonably well with what is known from experiment. In this way the analysis at least can serve to narrow the possibilities whenever the mathematical expression for the actual force on a particle has not been determined completely.

To see how the Energy Theorem is used to analyze motion in a nonuniform field, consider the force given by

$$F(x) = 8F_0 \left(\frac{x}{x_0}\right) \left[1 - \left(\frac{x}{x_0}\right)^2\right] \exp\left[-2\left(\frac{x}{x_0}\right)^2\right]$$

This expression corresponds to the potential energy function

$$V(x) = V_0 \left[2\left(\frac{x}{x_0}\right)^2 - 1\right] \exp\left[-2\left(\frac{x}{x_0}\right)^2\right] \tag{2.15}$$

where $V_0 = F_0 x_0$ and x_0 are parameters to be fixed by experiment (see Figure 2.1). The potential function $V(x)$ is a very rough approximation to the potential energy of a positively charged particle, such as an alpha particle, inside and near the nucleus of a large atom. If we were to insert $F(x)$ into Equation 2.14 we would learn nothing of the motion without some kind of approximation scheme. Rather than take that alternative just yet we shall discuss the motion qualitatively using the Energy Theorem. Five different cases corresponding to the five values of the total energy shown in Figure 2.1 may be distinguished.

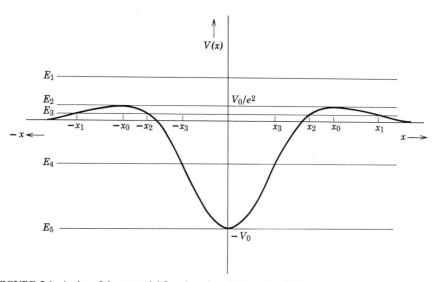

FIGURE 2.1. A plot of the potential function given in Equation 2.15.

(1) $E_1 > V(x)$. We shall suppose that the particle with this total energy begins its motion at a point far out in the negative x direction with a velocity directed to the right. At this point its energy will be virtually all kinetic energy owing to the smallness of $V(x)$ for large $|x|$. As the particle moves toward the origin, its velocity decreases while $V(x)$ grows to its maximum value, V_0/e^2. At $-x_0$ the velocity is least and is given by

$$v^2 = v_1{}^2(0) - \frac{2V_0}{me^2} \qquad (x = -x_0) \qquad (2.16)$$

where we have noted that

$$E_1 = \tfrac{1}{2}mv_1{}^2(0)$$

under the initial conditions we have chosen. After passing $-x_0$ the particle rapidly increases its speed to the maximum value

$$v^2 = v_1^2(0) + \frac{2V_0}{m} \qquad (x = 0) \tag{2.17}$$

which occurs at the potential energy minimum. Thereafter the velocity decreases again to what it was at $-x_0$, then increases slightly as the particle moves past the potential energy maximum at x_0 and off to infinity in the positive x direction. We see that in this case the only effect of the potential function is to attract the particle momentarily into the neighborhood of the origin. Even if we were to place the particle initially right at the origin, the constancy of the total energy guarantees that the particle will have a large initial speed there and will quickly scramble up one side or the other of $V(x)$, depending which way it is moving. The language of our conclusion here may seem picturesque, but it is conventional and has led to the names *potential barrier*, for $V(x)$ near a maximum, and *potential well*, for the function near a minimum. The analogy with hills and valleys in a uniform gravitational field is quite apparent and may be used to advantage in a first acquaintance with the kind of analysis we are carrying out here.

(2) $E_2 \geqslant V(x)$. Here the motion is something like what has been just described with the exception of the behavior near the tops of the potential barriers. At first glance it might appear that a particle coming in from the left will turn around at $-x_0$; for, we have $E_2 = V_0/e^2$ there and this supposedly is a condition for a turning point. On the other hand, it seems equally likely that the particle could fall into the potential well after reaching the top of the barrier or perhaps remain there forever at rest. This peculiar situation has arisen because only *one* of the conditions (2.13) is satisfied at $-x_0$ for the special value of the total energy we have chosen. Equation 1.29 and the necessary condition for a minimum value of $V(x)$ show that the force vanishes at $-x_0$, in contradiction to the second of Equations 2.13. Thus the force changes sign at the top of the potential barrier and, in fact, is such on either side as to push a particle *away* from the barrier top after even an infinitesimal displacement. We conclude that a particle coming in from the left to the point $-x_0$ will stop at the top of the barrier, since its energy there is entirely potential energy, but that the slightest displacement from its resting position will cause it to leave and never return. This behavior has suggested the term *unstable equilibrium* for a particle at rest on a point of potential energy maximum. We have equilibrium because the force is zero, but it is an unstable one because the force to either side will not act as a buffer against disturbances of the equilibrium situation.

(3) $V_0/e^2 > E_3 > 0$. A particle coming in from the left with total energy E_3 will decrease its velocity until it reaches the point $-x_1$ marked in Figure 2.1.

There it will stop and turn around to return to infinity in the negative x direction. In this case we have a *scattering process* and we may pinpoint the site of the scattering by solving the turning point equation

$$E_3 = \tfrac{1}{2}mv_3{}^2(0) = V(-x_1)$$

for x_1. It should be evident from this result that scattering processes carried out at different total energies (initial velocities) could provide information about the shape of a potential barrier. For example, the height of the barrier could be determined by measuring the amount of scattering as a function of the velocity of the particles incident on the barrier.

If the particle begins its motion inside the potential well it is destined to oscillate forever between the points $-x_2$ and $+x_2$, since the line representing E_3 crosses the graph of $V(x)$ at those points. This repetitive motion must occur because the particle has not enough energy to surmount the potential barrier. Therefore, we may state the very general conclusion that *the motion of a particle of sufficiently low total energy in a potential well will always be periodic motion.*

(4) $0 > E_4 > -V_0$. When the total energy of the particle is negative it cannot exist anywhere but inside the potential well. This statement derives immediately from the Energy Theorem and the fact that the kinetic energy is a positive function of the velocity: if the particle is subject to

$$\tfrac{1}{2}mv^2 + V(x) < 0$$

and were to be in a region where $V(x)$ is positive, its kinetic energy would have to be negative, which is not possible. Based on what was concluded above, we must have it that, in one spatial dimension, a particle with negative total energy is capable only of periodic motion, in this instance between the turning points $-x_3$ and $+x_3$. Incidentally, it should not be puzzling at all that the total energy of a particle can take on negative values. This property is a direct result of where we choose the potential energy to be zero—a choice that is arbitrary. Should we wish to renormalize $V(x)$ by adding to it V_0 energy units, for example, every permissible value of the total energy would be positive or zero, but all of our basic conclusions would remain unchanged. The same would be true, of course, were we to renormalize the potential function to make it everywhere negative.

(5) $E_5 = -V_0$. A particle with a total energy equal to $-V_0$ cannot reside anywhere but at the origin with zero initial velocity. The constancy of the total energy and the fact that $V(x)$ is minimum at the origin then combine to insure that a particle in this circumstance will remain at rest forever. If it should be displaced slightly from its resting position, the force, which on either side of it acts opposite to the direction of displacement, will push it back to its starting point. Here we have an example of a point of *stable equilibrium*: a condition of zero force along with a restoring tendency for slight perturbations from the position of rest.

The foregoing analysis can be carried out for any potential energy that is a continuous and differentiable function of the position. Although the procedure is largely qualitative, it does provide for numerical evaluations of the turning points as well as a relation between the velocity and the position that is the general form of Equations 2.5, 2.16, and 2.17:

$$v^2(t) = v^2(0) - \frac{2}{m}\left[V(x(t)) - V(x(0))\right] \tag{2.18}$$

where it is assumed that $V(x)$ is an explicitly known function of the position.

To understand the motion of a particle on a more quantitative basis we would require a solution of Equation 2.14. On the face of it, this would be difficult indeed, as the Second Law is usually a complicated differential equation. A systematic, direct method of attack does exist, however, through which the Second Law with an arbitrary conservative force may be solved by successive approximations to whatever degree of accuracy is desired. This procedure is especially useful when the potential function $V(x)$ from which the force is derived displays a minimum and the motion near that point is what interests us most. In that case the approximation technique begins as follows.

We assume that $V(x)$ has a minimum at $x = a$ and that the function may be expanded about the minimum in a *Taylor series*:

$$V(x) = V(a) + \frac{1}{2}\left(\frac{d^2V}{dx^2}\right)_{x=a}(x-a)^2 + \cdots + \frac{1}{n!}\left(\frac{d^nV}{dx^n}\right)_{x=a}(x-a)^n + \cdots. \tag{2.19}$$

where $V(a)$ is the minimum value of $V(x)$. Equation 2.19 does not contain the first power of x because

$$\left(\frac{dV}{dx}\right)_{x=a} = 0$$

is a necessary condition for a minimum of $V(x)$. Now, the series (2.19) is equivalent to the original potential function and consequently does not in itself provide a simplification of our problem. What does simplify matters is the prospect of using (2.19) in a well-defined manner to extract $x(t)$ from Equation 2.14. This we can accomplish by a term-by-term examination of the series. Our task is made easier if we first renormalize $V(x)$ relative to its minimum and if we introduce into (2.19) the new variable

$$\xi(t) \equiv x(t) - a$$

which is the displacement from the equilibrium point. Thus we have

$$V(\xi) = \frac{1}{2}\left(\frac{d^2V}{d\xi^2}\right)_{\xi=0}\xi^2 + \cdots + \frac{1}{n!}\left(\frac{d^nV}{d\xi^n}\right)_{\xi=0}\xi^n + \cdots. \tag{2.20}$$

which is a Maclaurin series for $V(\xi)$. Now we shall *approximate* $V(\xi)$ by

$$V(\xi) \simeq V_H(\xi) \equiv \tfrac{1}{2}k\xi^2 \tag{2.21}$$

where

$$k \equiv \left(\frac{d^2 V}{d\xi^2} \right)_{\xi = 0}$$

is a parameter called the *force constant* and $V_H(\xi)$ is known as the *harmonic oscillator potential function*. As we shall see in the next section, the motion of a particle whose potential energy is $V_H(\xi)$ can be calculated exactly. Therefore, to the extent that $V_H(\xi)$ accurately represents $V(\xi)$ in the region of interest, we have at hand a complete solution of our problem. As a typical example, consider $V(x)$ given by Equation 2.15. For this potential function $a = 0$ and we find

$$V_H(x) = 4V_0 \left(\frac{x}{x_0} \right)^2 \tag{2.22}$$

on evaluating the second derivative at $x = 0$. The function $V_H(x)$ is compared with an appropriately renormalized $V(x)$ in Figure 2.2. The behavior of the true potential function is reproduced closely by the approximate potential only for positions close to the origin, as expected. For the student wishing full details, the remaining parts of the general approximation procedure for solving Equation 2.14 are discussed in Special Topic 2 at the end of this chapter. For now we shall turn to the important and classic problem of calculating the motion of a particle under $V_H(\xi)$.

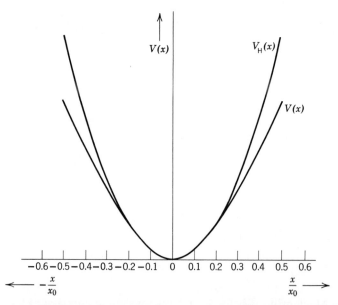

FIGURE 2.2. A comparison of $V(x)$, given in Equation 2.15, with its harmonic-oscillator approximation, given in Equation 2.22.

2.4 THE LINEAR HARMONIC OSCILLATOR

A linear harmonic oscillator is a particle whose position is specified by a single coordinate and whose motion is governed by the force

$$F_H(\xi) = -k\xi \tag{2.23}$$

where $\xi(t)$ is the displacement of the oscillator from a position of stable equilibrium. Equation 2.23 comes directly from Equation 2.21. It states that the force $F_H(\xi)$ *opposes* the displacement of the oscillator and that the magnitude of this opposition is directly proportional to the displacement. This relation between force and displacement has been observed in a large variety of experiments on both macroscopic and microscopic particles. Among the former are the "mass on a spring" and the "mass on a wire" familiar from introductory mechanics (see Figures 2.3

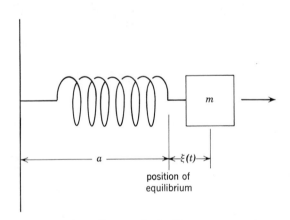

FIGURE 2.3. The linear harmonic oscillator as the familiar mass on a spring.

and 2.4). The microscopic particles for which Equation 2.23 is valid include the important category of atom pairs bound together to form diatomic molecules. Some measured values of force constants for these oscillators are listed in Table 2.1.

The Second Law (2.14) applied to the harmonic oscillator is

$$\frac{d^2\xi}{dt^2} + \omega_0{}^2\xi(t) = 0 \tag{2.24}$$

where

$$\omega_0 = \left(\frac{k}{m}\right)^{1/2} \tag{2.25}$$

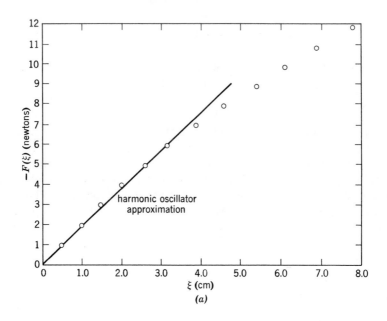

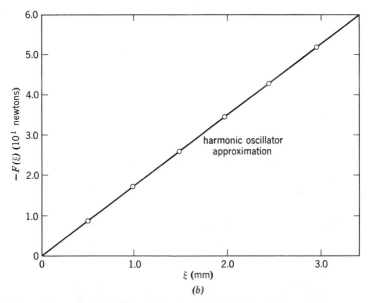

FIGURE 2.4. (*a*) Some data on the force applied to a spring and its resultant elongation. (*b*) The same kind of data for a long copper wire.

TABLE 2.1 Force Constants Observed for the Interaction of Two Atoms in a Diatomic Molecule

Molecule	k 10^5 dyne/cm	ω_0 10^{14} rad/s
CH	4.484	5.3903
CD	4.484	3.9576
Cl_2	3.286	1.0641
HBr	4.116	4.9911
HCl	5.158	5.6317
DCl	4.903	3.9383
HF	9.655	7.7956
DF	9.643	5.6477
HI	3.142	4.3504
OH	7.791	7.0359
OD	7.801	5.1253

is a constant parameter that conveniently gathers together the mass and the force constant. Equation 2.24 is a linear differential equation of the second order for $\xi(t)$. Its solution can be obtained readily through the method of the auxiliary equation, which we describe in its essentials now for convenience of reference.[1]

A differential equation of the form

$$a \frac{d^2 y}{dt^2} + b \frac{dy}{dt} + cy(t) = 0 \tag{2.26}$$

is called a linear, homogeneous differential equation of the second order with real-valued, constant coefficients a, b, and c. It is solved for $y(t)$ by writing down immediately the *auxiliary equation*

$$am^2 + bm + c = 0 \tag{2.27}$$

where m is any root of Equation 2.27:

$$m = \frac{-b \pm (b^2 - 4ac)^{1/2}}{2a} \tag{2.28}$$

If the two roots m_1 and m_2 of Equation 2.27 are distinct real numbers, then the solution of Equation 2.26 is

$$y(t) = c_1 \exp(m_1 t) + c_2 \exp(m_2 t) \tag{2.29}$$

where c_1 and c_2 are two arbitrary constants of integration. If the two roots m_1

---

[1] For a discussion with all the proofs see E. D. Rainville and P. E. Bedient, *Elementary Differential Equations*, The Macmillan Co., New York, 1969, Chapter 6.

and m_2 are real numbers both equal to the same number m_0, then the solution is

$$y(t) = (c_1 + c_2 t) \exp(m_0 t) \tag{2.30}$$

Lastly, if m_1 and m_2 are *complex* numbers respectively equal to[2]

$$m_1 = \alpha + i\beta \qquad m_2 = \alpha - i\beta$$

where $i = \sqrt{-1}$, then the solution is

$$y(t) = c_1 \exp(\alpha t) \cos \beta t + c_2 \exp(\alpha t) \sin \beta t \tag{2.31}$$

Returning to Equation 2.24, we see that it has the form of (2.26) with $a = 1$, $b = 0$, and $c = \omega_0^2$. Equation 2.28 then tells us that $m_1 = i\omega_0$ and $m_2 = -i\omega_0$. Thus it is Equation 2.31 to which we must turn for the solution of our problem. The result is

$$\xi(t) = c_1 \cos \omega_0 t + c_2 \sin \omega_0 t \tag{2.32}$$

since $\alpha = 0$ and $\beta = \omega_0$ here. The arbitrary constants in Equation 2.32 may be evaluated as usual in terms of the initial conditions. The general relations are

$$\xi(0) = c_1$$
$$v(0) = c_2 \omega_0$$

the second of which follows after computing the first derivative of (2.32). If the oscillator is set into motion by pulling it out to a position on the right of the equilibrium point and then releasing it from rest, the general initial conditions reduce to

$$\xi(0) = A > 0$$
$$v(0) = 0$$

and we have $c_1 = A, c_2 = 0$. Accordingly, the motion of the oscillator is given by

$$\xi(t) = A \cos \omega_0 t$$
$$v(t) = -A\omega_0 \sin \omega_0 t \tag{2.33}$$

Equations 2.33 are plotted in Figure 2.5. We see quite readily from the graphs that both $\xi(t)$ and $v(t)$ are *periodic* functions of the time. This fact we could have predicted on the basis of the Energy Theorem, since $F_H(\xi)$ never vanishes except at the origin and

$$E = \tfrac{1}{2}k\xi^2 \qquad (E > 0) \tag{2.34}$$

leads to a *pair* of turning points

$$\xi_1 = \left(\frac{2E}{k}\right)^{1/2} \qquad \xi_2 = -\left(\frac{2E}{k}\right)^{1/2} \tag{2.35}$$

---

[2] It is possible to prove that the complex-valued roots of Equation 2.27 must be the conjugate pairs given here. See E. D. Rainville and P. E., Bedient, op. cit., p. 105.

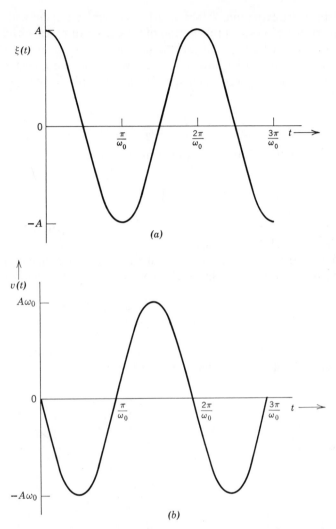

FIGURE 2.5. (a) The displacement from equilibrium of the harmonic oscillator as a function of time. (b) The velocity of the same oscillator as a function of time.

regardless of the (positive) value of the total energy E. Thus the particle oscillates forever between the two sides of the potential well $V_H(\xi)$. Because the velocity is zero at a turning point, the second of Equations 2.33 helps transform Equations 2.35 into the more explicit statements

$$\xi_1 = \xi(0) = A \qquad \xi_2 = \xi\left(\frac{\pi}{\omega_0}\right) = -A \qquad (2.36)$$

These, of course, are extremum values of $\xi(t)$, in perfect agreement with Figure 2.5.

The graph of $\xi(t)$ shows that the motion of the oscillator repeats itself in successive time intervals of length $2\pi/\omega_0$. (This fact derives mathematically from the periodic character of the cosine function.) It follows that we should see the oscillator take up its initial position once again after $2\pi/\omega_0$ time units have elapsed and that we may call

$$T \equiv \frac{2\pi}{\omega_0} \tag{2.37}$$

the *period of oscillation*. Equation 2.25 permits us to rewrite this expression as

$$T = 2\pi \left(\frac{m}{k}\right)^{1/2} \tag{2.38}$$

which is a theoretical relation between three measurable parameters. Equation 2.38 provides a straightforward means of checking our theory, since m and k can be determined independently to compare with a direct measurement of the period (see Figure 2.6). Conversely, the mass, along with the period, or its inverse,

$$v \equiv \frac{1}{T} = \frac{1}{2\pi}\left(\frac{k}{m}\right)^{1/2} \tag{2.39}$$

the *frequency of oscillation*, can be employed to calculate the force constant. Data such as those in Table 2.1 are often obtained in this way.

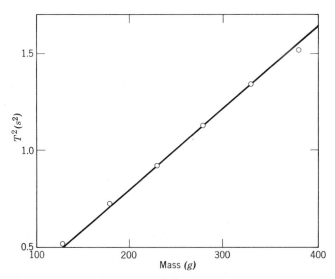

FIGURE 2.6. An experimental verification of Equation 2.38.

The *amplitude of oscillation* during a given period of oscillation is defined as the maximum displacement of the oscillator from its equilibrium position during that period. In the present case the maximum displacement is fixed in value as time passes and we have simply: amplitude $= A$. The physical significance of the amplitude may be seen from combining Equations 2.35 and 2.36 to yield

$$E = \tfrac{1}{2}kA^2 \tag{2.40}$$

The total energy of the oscillator is proportional to the square of the amplitude.

The quantity $\omega_0 t$ is referred to as the *phase angle* of the oscillation. The phase angle is always expressed in radian measure and suggests the name *angular frequency* for ω_0 by virtue of the "units" of the latter parameter, radian/second, and its proportionality to the frequency of oscillation v through Equations 2.37 and 2.39. It should be understood that this name is largely a figure of speech and is not to imply in any way that ω_0 is a directly observable frequency. A look at Figure 2.5 again shows that the phase angles of the position and velocity at comparable points on their graphs (e.g., at their first extremum values) differ by the constant amount of $\pi/2$ radians. This difference in phase angle is the natural consequence of the facts that the motion is periodic and that the velocity must vanish when the position takes on an extremum value. Indeed, at a turning point the force $F_H(\xi)$ has dragged the oscillator to a standstill; but, one quarter of a cycle later, it is the position that has come down to the origin while the velocity has gone on to increase to its maximum absolute value. This causes the particle to overshoot the equilibrium position, thereby bringing on the retarding force once again and inducing oscillation. It follows that the velocity must always run ahead of the position by one quarter of a cycle, that is, by $\pi/2$ radians.

The concept of phase angle plays an especially important part in the theory of driven oscillatory motion and has prompted for this reason a common alternate form for Equation 2.32. If we define

$$c_1 \equiv C \cos \delta \qquad c_2 \equiv C \sin \delta \tag{2.41}$$

and introduce these equations into (2.32), we get

$$\xi(t) = C \cos (\omega_0 t - \delta) \tag{2.42}$$

upon invoking the familiar trigonometric identity

$$\cos (\theta - \varphi) \equiv \cos \theta \cos \varphi + \sin \theta \sin \varphi$$

The constant parameter δ is called the *phase constant*. For the initial conditions we have dealt with as an example, the phase constant is zero and the parameter C is just the amplitude. In general we would have the equations

$$C^2 = \xi^2(0) + \left(\frac{v(0)}{\omega_0}\right)^2 \qquad \tan \delta = \frac{v(0)}{\omega_0 \xi(0)} \tag{2.43}$$

2.5 THE LINEAR OSCILLATOR IN A UNIFORM FIELD

One of the most cited examples of a linear harmonic oscillator is the hooked weight hanging on a flexible spring. When this apparatus is set up in the laboratory and the weight is pulled downward and released, we do indeed observe it in oscillatory motion and, provided the oscillations are of small amplitude, we are able to verify quantitatively[3] Equations 2.33 and 2.38. This experience is familiar to many from introductory physics and would suggest nothing surprising were it not for the fact that a weight hanging on a spring *cannot* be an harmonic oscillator as we have described it in the previous section. The reason for this conclusion is simple. If we ignore air resistance and the internal friction of the spring, the total force acting on the oscillator must be

$$F(\xi) = -k\xi - mg \tag{2.44}$$

where $-g$ is the acceleration of the oscillator due to the gravitational field of the Earth. Equation 2.44 is obviously different from Equation 2.33 and would not necessarily lead to the theoretical results of the previous section. Thus the well-known "mass on a spring" is not an harmonic oscillator, but instead is an harmonic oscillator in a uniform field. The motion of the latter particle, then, is what ought to be considered. This we shall do now with the expectation of discovering how it is that the "mass on a spring" seems to move as if it were unaware of the presence of gravity.

The Second Law for the force $F(\xi)$ is

$$\frac{d^2\xi}{dt^2} + \omega_0{}^2\xi(t) = -g \tag{2.45}$$

Equation 2.45 is an example of a linear, *inhomogeneous* differential equation of the second order for $\xi(t)$. Usually, special methods are required to solve this type of equation. But, in the present case, the simple form of the inhomogeneous part, $-g$, provides us with the opportunity to use a modified version of the method of the auxiliary equation. We begin by defining

$$z(t) \equiv \xi(t) + z_0 \tag{2.46}$$

where z_0 is a constant parameter whose value is yet undetermined. With the insertion of Equation 2.46 into Equation 2.45 we get

$$\frac{d^2z}{dt^2} + \omega_0{}^2z(t) - \omega_0{}^2z_0 = -g$$

---

[3] We must, of course, make an allowance for the mass of the spring first.

Now we choose the value of z_0 in order to remove the inhomogeneous term:

$$z_0 = \frac{g}{\omega_0{}^2} \tag{2.47}$$

This choice leaves us with

$$\frac{d^2 z}{dt^2} + \omega_0{}^2 z(t) = 0 \tag{2.48}$$

which can be solved in the usual way by the method of the auxiliary equation!
Using the results of the Section 2.4, we easily find

$$z(t) = C \cos(\omega_0 t - \delta) \tag{2.49}$$

and

$$\xi(t) = -\frac{g}{\omega_0{}^2} + C \cos(\omega_0 t - \delta) \tag{2.50}$$

Equation 2.50 is graphed in Figure 2.7. It is evident from the plot that, although
its periodicity remains the same, $\xi(t)$ is no longer symmetric about the origin as was
$\xi(t)$ in Figure 2.5a. But, if we take note of the fact that all measurements of the
position of an oscillator are made relative to the point of stable equilibrium, we
realize that it is $z(t)$, not $\xi(t)$, which is observed in the laboratory. In other words,
insofar as the periodic character of the motion is concerned, we may conclude that

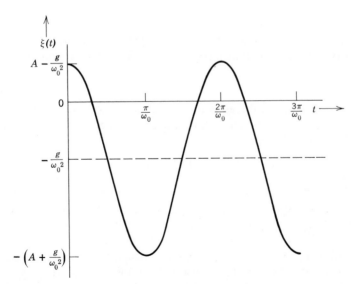

FIGURE 2.7. The displacement of an harmonic oscillator in a uniform gravitational field, as a function
of time.

the harmonic oscillator in a uniform gravitational field will carry out a motion that is *physically indistinguishable* from that of an oscillator in the absence of such a field. In particular, it follows that Equations 2.33 and 2.38 can be verified in principle by observing the motion of a weight hung on a spring.

It is worthwhile to inquire a little more deeply into the physical basis of the difference between Equations 2.42 and 2.50 by examining the potential energy functions for both kinds of oscillator. For the "unperturbed" oscillator we have

$$V_H(\xi) = \tfrac{1}{2}k\xi^2$$

which is a function invariant in value under the replacement of $-\xi$ for ξ. This means that the potential energy is completely symmetric about the origin of $\xi(t)$ and, consequently, that a particle moving under V_H may be expected to trace out beneath the origin the mirror image of its path above the origin (Figure 2.5a). Now we "switch on" the uniform gravitational field. The potential energy is suddenly no longer $V_H(\xi)$ but instead is

$$V(\xi) = \tfrac{1}{2}k\xi^2 + mg\xi + mga \tag{2.51}$$

where the parameter a, it will be recalled, is the equilibrium position of the unperturbed oscillator relative to the origin of $x(t)$, its absolute position. The potential function in Equation 2.51 is manifestly *not* invariant under the replacement of $-\xi$ for ξ. The cause of this loss of invariance is the constant, unidirectional gravitational force that has imposed on the problem a preferred direction—downward. Therefore, we cannot suppose that the motion of a particle according to $V(\xi)$ will display the reflection symmetry it had according to $V_H(\xi)$; and, indeed, it does not (Figure 2.7).

On the other hand, the mathematical device we used to solve Equation 2.45 can be put to work here as well. If we introduce Equation 2.46 into Equation 2.51, we get

$$V(z) = \tfrac{1}{2}kz^2 - kz_0 z + mgz + (\tfrac{1}{2}kz_0{}^2 - mgz_0 + mga)$$

where the term in parentheses is an unimportant constant. This expression is transformed into one symmetric about the origin of $z(t)$ when Equation 2.47 has been used. Then we have

$$V(z) = \tfrac{1}{2}kz^2 + \text{constant}$$

and the reflection invariance of the problem is restored. Because of this symmetry we can return to the theory of the unperturbed oscillator with the only memory of the gravitational field appearing in the new position of stable equilibrium, which is at $-g/\omega_0{}^2$ relative to the origin of $\xi(t)$. It should be understood clearly that our shifting of the origin is not an empty mathematical artifice for solving a differential equation. The new position of stable equilibrium is located precisely at the minimum of the new potential function $V(\xi)$, as can be readily demonstrated by differentiating Equation 2.51 with respect to ξ and setting the result equal to zero.

FOR FURTHER READING

T. C. Bradbury *Theoretical Mechanics*, Wiley, New York, 1968. Chapter 4 offers an extensive discussion of motion in one spatial dimension.

T. W. B. Kibble *Classical Mechanics*, Halsted Press, New York, 1973. The ideas discussed in this chapter may be found in Chapter 2 of Kibble's book.

K. R. Symon *Mechanics*, Addison-Wesley, Reading, Mass., 1971. As with the text by Kibble, an alternative discussion of the basics of conservative, linear motion may be found in Chapter 2.

J. R. Barker *Mechanical and Electrical Vibrations*, Barnes and Noble, New York, 1964. This little book is a very readable introduction to all kinds of vibrating systems: masses on springs, pendula, electric circuits, and the like.

A. A. Andronov, *Theory of Oscillators*, Addison-Wesley, Reading, Mass., 1966. Part
A. A. Vitt, and I presents a very complete, advanced discussion of harmonic
S. E. Khaikin oscillators.

PROBLEMS

1. Use the Energy Theorem to derive the fact that the turning point for a particle decelerated in a uniform field depends only on the ratio of the total energy to the strength of the force.

2. Suppose that the position and velocity of a particle each may be written as a Maclaurin series in powers of the time. Give physical arguments to show that these expansions reduce to Equations 2.4 and 2.3, respectively, in the case of motion in a uniform field.

3. One of the standard methods for determining the depth of water in a well involves a small detonation at the wellhead and the detection there of the sound reflected from the water surface below. If creating a detonation is inconvenient, a small projectile can be dropped instead and the sound of its splash into the water may be detected. Suppose a chart recorder indicates a total elapsed time of 1.5 s between the instant of fall of the projectile and the reception of the sound of its splash. How deep does the water surface lie? (The speed of sound may be taken as 335 m/s for the air temperature in the well.)

4. A driver determines that, in order to pass safely a truck moving ahead of him, he must come alongside after the truck has gone no more than 100 m past the place it is when passing begins. Given that the constant speed of the truck is 22 m/s and that the driver is doing 27 m/s at a point 30 m behind the truck when he begins to pass, compute the minimum acceleration he must have for safe passing.

5. An electron enters a uniform, decelerating electric field of 20 N/coulomb. It moves along a field line and is observed to stop its forward motion in 10^{-6} s, after traveling a distance of 1.76 m. Using these data, calculate the charge-to-mass ratio of the electron.

6. Derive Equations 2.13 from Equations 2.12.

7. The Morse potential function,

$$V(x) = V_0 \left\{ \exp\left[2c\left(1 - \frac{x}{x_e} \right) \right] - 2\exp\left[c\left(1 - \frac{x}{x_e} \right) \right] \right\} \qquad (x \geqslant 0)$$

where V_0, c, and x_e are positive constants, is used very often to describe the potential energy of one atom near another in a diatomic molecule.

(a) Sketch a graph of $V(x)$.

(b) Describe the motion of a particle according to $V(x)$, given the conditions: $0 < E < V_0 e^{2c}$, $v(0) < 0$, $x(0) \simeq +\infty$.

(c) Repeat the process for the conditions: $-V_0 < E < 0$, $v(0) = 0$, $x(0) > x_e$ but finite.

(d) Calculate the frequency of small oscillations about the point x_e.

8. The Woods-Saxon potential function

$$V(x) = \frac{-V_0}{1 + a \exp(x/x_0)} \qquad (x \geq 0)$$

where V_0, a, and x_0 are positive constants, has been employed successfully to represent the potential energy of a neutron in the vicinity of a heavy nucleus.

(a) Sketch a graph of $V(x)$.

(b) Describe the motion of a particle according to $V(x)$ given the conditions:

$$-\frac{V_0}{1 + a} < E < 0, v(0) > 0, x(0) = 0$$

(c) Expand $V(x)$ in a Maclaurin series to first order in (x/x_0). Introduce the result into the Second Law and calculate the motion of a particle subject to the conditions in (b). Estimate the range of E over which the calculation should be valid.

9. Expand the Woods-Saxon potential in a Maclaurin series to second order in (x/x_0). Incorporate the result into the Second Law and calculate the motion under the initial conditions given Problem 8 (b).

10. Calculate the minimum speed an alpha particle must have in order to escape to $+\infty$ from the point $x = 0$ in the potential function given by Equation 2.15.

11. A potential function, better than that given in Equation 2.15, for an alpha particle near and inside a large nucleus has the form

$$V(x) = \begin{cases} -35 \text{ MeV} & |x| < 10^{-12} \text{ cm} \\ \dfrac{2Ze^2}{|x|} & |x| > 10^{-12} \text{ cm} \end{cases}$$

where Z is the atomic number of the nucleus and e is the magnitude of the electron charge.

(a) Adjust the parameters V_0 and x_0 in Equation 2.15 in order to have the potential in that equation duplicate $V(x)$ given above as closely as possible.

(b) A typical speed observed for alpha particles that have escaped nuclei is 2×10^7 m/s. Is the total energy corresponding to this speed sufficient to permit an alpha particle that starts from the point $x = 0$ to get out of the potential well described by Equation 2.15 with the parameters as determined in (a)?

12. The potential function

$$V(x) = \frac{2V_B}{1 + \exp(x/x_0)} \qquad (x \geq 0)$$

where x_0 is a positive constant, describes a potential barrier of "height" V_B.

(a) Sketch a graph of $V(x)$.

(b) Expand $V(x)$ to second order in a Maclaurin series and introduce the result into the Second Law. Calculate the motion, given the initial conditions $x(0) = 0$, $v(0) = 0$.

13. Show that the period of oscillation between the points x_1 and x_2 for a particle in a potential well is given by

$$T = 2 \int_{x_1}^{x_2} \left[\frac{m/2}{V(x_1) - V(x)} \right]^{1/2} dx$$

where $V(x)$ describes the potential well. [Hint: Consider Equation 2.10.]

14. Verify the result derived in Problem 13 for the harmonic oscillator potential

$$V_H(x) = \tfrac{1}{2} k x^2$$

15. An equation describing the data in Figure 2.4a is

$$F(\xi) = -k\xi + b\xi^2$$

where b is a positive constant.

(a) Calculate and sketch the graph of the corresponding potential function, $V(\xi)$.

(b) Despite the fact that the object described by $F(\xi)$ is not an harmonic oscillator, it will vibrate (instead of stretch out the spring to which it is attached) if it starts its motion at $\xi = 0$ with a speed less than a certain value v_{crit}. Calculate this critical value.

16. A diatomic molecule may be described by Equation 2.24 if $\xi(t)$ is taken to be the displacement of one atom in the molecule away from the equilibrium position and the single-particle mass m in Equation 2.25 is replaced by

$$\mu = \frac{m_A m_B}{m_A + m_B}$$

the two-particle reduced mass, where A and B refer to the two atoms (μ is discussed in Chapter 4). Consider a series of diatomic molecules wherein atom A is the same always but atom B varies. Deduce a relation between the frequencies of vibration of the molecules in the series, assuming that the force constant remains the same in every case. Check this relation with the frequency data on the pairs CH, CD; HCl, DCl; HF, DF; and OH, OD in Table 2.1.

17. Calculate the motion of the harmonic oscillator, using the Energy Theorem as an alternative to the Second Law, according to the procedures outlined in Section 2.1. Take $\xi(0) = A > 0$, $v(0) = 0$, with $v(t) < 0$.

18. Show that the result (2.40) can be derived from the Energy Theorem without any particular initial conditions being prescribed.

19. Demonstrate, using Equations 2.43, that the phase constant for the harmonic oscillator is determined solely by the ratio of initial kinetic energy to initial potential energy. Why *should* this be so?

20. The phase constant for the harmonic oscillator is related to the time t_m, at which the first turning point occurs, and to the period of oscillation, T, by

$$\frac{\delta}{2\pi} = \frac{t_m}{T}$$

where δ is taken as positive. Prove this statement and demonstrate how the remaining turning points may be similarly expressed.

21. A man of mass 100 kg steps onto a bathroom scale with a platform supported by a spring with $k = 5 \times 10^5$ N/m. Calculate the resulting deflection of the platform. (Ignore friction and buoyancy in doing the calculation.)

22. A cylindrical buoy 1 m in length is observed to oscillate vertically in the still waters of a lake. The period of oscillation is observed to be 1.25 s.

(a) Derive an expression for the period of oscillation in terms of the density of the buoy, that of the water, and other relevant parameters.

(b) Calculate the density of the buoy.

23. Consider a parcel of lightly polluted air at a point z above the surface of the Earth in the atmosphere. If the effects of viscosity are ignored, the total force on the parcel is simply due to gravity and buoyancy: $F = -mg + \rho_a Vg$, where m is the mass and V is the volume of the parcel and ρ_a is the local density of the atmosphere.

(a) Write down the Second Law for the parcel in a form that does not show a dependence on the parcel volume V.

(b) Assume the air in the parcel and that in the atmosphere to be ideal gases at the same pressure (and molecular weight). Show that the Second Law then may be expressed in terms of the difference in local temperature of the two gases.

(c) Finally, use the fact that $T(z) = T(0) - \Gamma z$ for both the parcel and the atmosphere (but with generally different values of the constant Γ) to show that, if $T(0) \gg \Gamma_{atm} z$ and $\Gamma_{parcel} > \Gamma_{atm}$, the Second Law for the parcel describes an harmonic oscillator of angular frequency $[(\Gamma_{parcel} - \Gamma_{atm})g/T(0)]^{1/2}$. The constant Γ is known in meteorology as the *lapse rate*. The oscillatory motion of the parcel corresponds to what is called a condition of *stablity* in the atmosphere. It is evident that upward pollutant dispersal would be difficult in a stable atmosphere.

24. The plane pendulum is a particle constrained to move only along a vertical circle in a uniform gravitational field. Although this motion is not linear, it is one dimensional in that it can be described fully by the angle of deviation, measured along a radius from the center of the circle, of the pendulum from its lowest position. The potential energy of the plane pendulum is

$$V(\theta) = mgl(1 - \cos \theta)$$

where l is the fixed radius of the circle and $\theta(t)$ is the angle of deviation. Show that, for very small values of $\theta(t)$, the Second Law for the plane pendulum describes an harmonic oscillator of period

$$T_0 = 2\pi \left(\frac{l}{g}\right)^{1/2}$$

25. Use the Energy Theorem, a little trigonometry, and the expression for $V(\theta)$ given in Problem 24 to show that

$$t = -\frac{T_0}{4\pi} \csc \theta(0) \int_{\theta(0)}^{\theta(t)} \left[1 - \left(\frac{\sin (\theta/2)}{\sin (\theta(0)/2)}\right)^2\right]^{-1/2} d\theta$$

for the plane pendulum subject to the conditions:

$$\theta(0) > 0 \qquad v(0) = l\left(\frac{d\theta}{dt}\right)_{t=0} = 0 \qquad v(t) < 0$$

The rather complicated integral that appears here is known as an *elliptic integral of the first kind*. It has been extensively studied and tabulated.

26. Using the result of Problem 25, show that the period of the plane pendulum may be expressed in the form

$$T = \frac{2T_0}{\pi} \int_0^1 \frac{(1 - k^2 z^2)^{-1/2}}{(1 - z^2)^{1/2}} \, dz$$

where

$$z = \frac{\sin(\theta/2)}{\sin(\theta(0)/2)} \equiv \frac{1}{k} \sin(\theta/2)$$

Expand $(1 - k^2 z^2)^{-1/2}$ in a Maclaurin series to terms of order $(kz)^4$ and approximate

$$k \simeq \frac{\theta(0)}{2} - \frac{\theta^3(0)}{48}$$

to show that

$$T \simeq T_0 \left[1 + \left(\frac{\theta(0)}{4} \right)^2 + \frac{11}{12} \left(\frac{\theta(0)}{4} \right)^4 \right]$$

27. By direct computation verify that

$$\frac{dT}{dt} = F(\xi(t)) v(t)$$

for the linear harmonic oscillator, where T is the kinetic energy.

28. The *time-average* value of any periodic function $f(t)$ is defined by

$$\langle f \rangle \equiv \frac{1}{T_0} \int_0^{T_0} f(t) \, dt$$

where T_0 is the period of $f(t)$.

(a) Calculate $\langle T \rangle$ for the harmonic oscillator. For simplicity, assume the initial conditions $\xi(0) > 0$, $v(0) = 0$.

(b) Verify the Virial Theorem, which states that

$$\langle T \rangle = \frac{n + 1}{2} \langle V \rangle$$

for any motion governed by a force of the form $F(x) = cx^n$, where c is a constant of proportionality.

29. The *variance* of any periodic function $f(t)$ is defined by

$$(\Delta f)^2 \equiv \langle (f - \langle f \rangle)^2 \rangle$$

and is a measure of the mean deviation of the function from its average value.

(a) Show that, in general,

$$(\Delta f)^2 = \langle f^2 \rangle - \langle f \rangle^2$$

(b) For the case of an harmonic oscillator show that

$$(\Delta \xi)^2 = \langle \xi^2 \rangle$$

$$(\Delta v)^2 = \langle v^2 \rangle$$

(c) Finally, use Equation 2.33 to verify the relation

$$\Delta \xi \, \Delta p = \frac{E}{\omega_0}$$

where $p(t) = mv(t)$ is the linear momentum of the oscillator. This equation, unlike most of the theory of the harmonic oscillator in classical mechanics, is valid in quantum mechanics as well. It is known generally as an *uncertainty relation* because it prescribes a condition on the root-mean-square deviations of a pair of measurable quantities from their average values.

30. Explain how the variance of some physical quantity during periodic motion could be used as a criterion in determining whether that quantity were conserved.

Note.   *The following problems depend on the material discussed in Special Topic 2.*

31. Derive the following relations for the Fourier coefficients of a periodic but otherwise arbitrary position function $\xi(t)$:

(a) $a_0 = 2\langle \xi \rangle$

(b) $\dfrac{a_0^2}{2} + \displaystyle\sum_{n=1}^{\infty} (a_n^2 + b_n^2) = 2\langle \xi^2 \rangle$

The second expression is a form of what is called *Parseval's equation* for a Fourier series.

32. Show how Equations S.9 and S.10 can be derived very simply by assuming that both the force and the position can be written as Fourier series.

33. A particle moves under the gaussian potential function

$$V(x) = -V_0 \exp\left[-\left(\frac{x}{x_0}\right)^2 \right] \qquad (-\infty < x < \infty)$$

where V_0 and x_0 are positive, constant parameters.

   (a) Write $V(x)$ as a Maclaurin series in $[(x/x_0)^2]$ to terms of second order [i.e., fourth order in (x/x_0)].

   (b) Use the method of Special Topic 2 to calculate the motion, given the initial conditions $x(0) = 1/4 \, x_0$, $v(0) = 0$.

34. Consider a particle moving according to the Morse potential function, introduced in Problem 7.

   (a) Expand the potential function in a Taylor series about x_e to terms of order ξ^3.

   (b) Use the method of Special Topic 2 to calculate the motion subject to the initial conditions $\xi(0) = 1/10 \, x_e$, $v(0) = 0$. You may assume that the Fourier sine coefficients all vanish for this motion. Notice that, under the given conditions, *aperiodic motion* is, in general, possible according to the approximate potential function although it is not possible according to the true potential function. This fact illustrates one of the common difficulties associated with the theory of nonlinear oscillations.

35. Use the method of Special Topic 2 to show that the period of the plane pendulum is given by

$$T \simeq T_0 \left[1 + \left(\frac{\theta(0)}{4}\right)^2 \right]$$

in the second approximation. You may assume the initial conditions

$$\theta(0) > 0, \, v(0) = l\left(\frac{d\theta}{dt}\right)_{t=0} = 0$$

SPECIAL TOPIC 2
Fourier Series and the Potential Well

We have seen in Chapter 2 that, when it is possible to expand $V(x)$ about a point of minimum value in a Taylor series, we can make the first approximation

$$V(\xi) \simeq V_H(\xi) = \tfrac{1}{2}k\xi^2 \tag{S.1}$$

where ξ is the position, measured relative to the minimum point. To this degree of accuracy we have thereby reduced the problem to one capable of exact solution, with the general result

$$\xi(t) = c_1 \cos \omega_0 t + c_2 \sin \omega_0 t \tag{S.2}$$

Thus we find the motion of a particle under $V_H(\xi)$ to be periodic for all values of the total energy. This result we shall now use as a guide in setting up the remainder of a general approximation scheme for computing $\xi(t)$ from the Second Law. In particular we shall henceforth assume that the motion we are interested in is *always* periodic and that it will always involve periodic functions of the time such as $\cos(\omega_0 t)$ and $\sin(\omega_0 t)$. In terms of the physics, this amounts to saying that we shall limit our investigation to particles whose total energy is low enough to keep them down inside the potential well portion of $V(\xi)$. This restriction is broad enough to include the large number of problems involving "bound-state motion," but it does exclude scattering processes. (However, the perceptive reader may have seen already how at least the first step in our procedure can be modified to give a description of a particle striking a potential barrier.)

If we assume that $\xi(t)$ is a continuous periodic function of the time with period $2\pi/\omega$ and that the integral

$$\int_0^{2\pi} |\xi(\omega t)|^2 \, d(\omega t)$$

exists, that the branch of mathematics known as *harmonic analysis* provides us with the theorem[1] that $\xi(\omega t)$ may be represented by

$$\xi(\omega t) = \frac{a_0}{2} + \sum_{n=1}^{\infty} \left[a_n \cos(n\omega t) + b_n \sin(n\omega t)\right] \tag{S.3}$$

Equation S.3 is called the *Fourier series* for $\xi(\omega t)$. The Fourier coefficients a_0, a_n,

---

[1] Fourier series are discussed in most books on mathematical physics or the theory of functions. One good place to look is E. Butkov, *Mathematical Physics*, Addison-Wesley, Reading, Mass., 1968, Chapter 4. For a detailed, but very readable, mathematical discussion, see C. Lanczos, *Discourse on Fourier Series*, Hafner, New York, 1966.

and b_n are easily calculated in terms of $\xi(\omega t)$ as follows. First, we integrate both sides of (S.3) with respect to the dimensionless quantity ωt:

$$\int_0^{2\pi} \xi(\omega t)\, d(\omega t) = \frac{a_0}{2} \int_0^{2\pi} d(\omega t)$$

$$+ \sum_{n=1}^{\infty} \left[a_n \int_0^{2\pi} \cos (n\omega t)\, d(\omega t) + b_n \int_0^{2\pi} \sin (n\omega t)\, d(\omega t) \right]$$

$$= \frac{a_0}{2} \cdot 2\pi = a_0 \pi$$

since all of the integrals in the sum over n vanish. Therefore

$$a_0 = \frac{1}{\pi} \int_0^{2\pi} \xi(\omega t)\, d(\omega t) \qquad (S.4)$$

gives us the first Fourier coefficient. Next, we multiply both sides of Equation S.3 by $\cos (m\omega t)$ $(1 \leqslant m \leqslant \infty)$ and integrate as before. Then we obtain the cosine coefficient

$$a_m = \frac{1}{\pi} \int_0^{2\pi} \xi(\omega t) \cos (m\omega t)\, d(\omega t) \qquad (m \geqslant 1) \qquad (S.5)$$

because

$$\int_0^{2\pi} \cos (n\omega t) \cos (m\omega t)\, d(\omega t) = \begin{cases} 0 & n \neq m \\ \pi & n = m \end{cases}$$

$$\int_0^{2\pi} \sin (n\omega t) \cos (m\omega t)\, d(\omega t) = 0$$

If we repeat the procedure with $\sin (m\omega t)$ we get the sine coefficient

$$b_m = \frac{1}{\pi} \int_0^{2\pi} \xi(\omega t) \sin (m\omega t)\, d(\omega t) \qquad (m \geqslant 1) \qquad (S.6)$$

Equations S.4 to S.6 permit the computation of the Fourier series for any explicitly known displacement function. However, they are purely mathematical results without a dynamical meaning. What we must do now is relate them to the Second Law,

$$\frac{d^2\xi}{dt^2} = \frac{1}{m} F(\xi) \qquad (S.7)$$

This we do as follows. The coefficient a_0 may be expressed in terms of the initial displacement $\xi(0)$:

$$a_0 = 2 \left[\xi(0) - \sum_{n=1}^{\infty} a_n \right] \qquad (S.8)$$

The coefficients a_n and b_n $(1 \leqslant n \leqslant \infty)$ are found in terms of the force $F(\xi)$ after differentiating Equation S.3 twice with respect to the time, substituting from

Equation S.7, and employing the procedures used to obtain Equations S.5 and S.6. The results are:

$$a_n = \frac{-(n\omega)^{-2}}{\pi m} \int_0^{2\pi} F(\xi) \cos(n\omega t)\, d(\omega t) \qquad (n \geqslant 1) \qquad (S.9)$$

$$b_n = \frac{-(n\omega)^{-2}}{\pi m} \int_0^{2\pi} F(\xi) \sin(n\omega t)\, d(\omega t) \qquad (n \geqslant 1) \qquad (S.10)$$

These equations for the Fourier coefficients have a direct physical significance although, in general, they would be very difficult to compute because $F(\xi)$ is only an implicit function of the time. It should be clearly understood, on the other hand, that all of our results so far are *exact*. No approximations have been made in developing the Fourier series expansion for $\xi(t)$.

To get some feeling for the use of Equations S.8, S.9, and S.10, let us consider again the linear harmonic oscillator. In that case we have

$$F(\xi) = -k\xi$$

and Equation S.10 becomes

$$\begin{aligned}
a_n &= \frac{(n\omega)^{-2}k}{\pi m} \int_0^{2\pi} \xi(t) \cos(n\omega t)\, d(\omega t) \\
&= \frac{(n\omega)^{-2}}{m} k a_n \qquad (n \geqslant 1)
\end{aligned}$$

according to Equation S.5. This equality can be satisfied only if

$$\begin{aligned}
a_n &\equiv 0 & n &> 1 \\
1 &= \frac{\omega^{-2}k}{m} & n &= 1
\end{aligned} \qquad (S.11)$$

with the coefficient a_1 left as arbitrary. In a similar way we find

$$b_n \equiv 0 \qquad n > 1$$

and we see that Equation S.3 has the form

$$\xi(t) = \frac{a_0}{2} + a_1 \cos(\omega_0 t) + b_1 \sin(\omega_0 t) \qquad (S.12)$$

where here

$$\omega^2 = \frac{k}{m} = \omega_0^2$$

as stipulated in Equation S.11. If we set $\xi(0) = a_1$, then Equation S.8 tells us that a_0 is zero and that

$$\xi(t) = a_1 \cos(\omega_0 t) + b_1 \sin(\omega_0 t) \qquad (S.13)$$

This result in complete agreement with our previous expressions for the harmonic oscillator.

In the general situation, where $F(\xi)$ is an arbitrary function of the position, we must differentiate Equation 2.20 and write

$$F(\xi) = -k\xi - \cdots - \frac{1}{(n-1)!}\left(\frac{d^n V}{d\xi^n}\right)_{\xi=0} \xi^{n-1} - \cdots.$$

$$\equiv -k\xi + \tilde{F}(\xi) \tag{S.14}$$

Equation S.14 is then introduced into Equations S.9 and S.10, with the results

$$a_n = \frac{-(n^2\omega^2 - \omega_0^2)^{-1}}{\pi m} \int_0^{2\pi} \tilde{F}(\xi) \cos(n\omega t)\, d(\omega t) \qquad (n \geqslant 1) \tag{S.15}$$

$$b_n = \frac{-(n^2\omega^2 - \omega_0^2)^{-1}}{\pi m} \int_0^{2\pi} \tilde{F}(\xi) \cos(n\omega t)\, d(\omega t) \qquad (n \geqslant 1) \tag{S.16}$$

Equations S.15 and S.16 are the working equations in the approximation scheme, which develops now as follows.

1. We represent $\tilde{F}(\xi)$ by the first nonvanishing term after $F_H(\xi)$ in the Maclaurin series (S.14) for the force. The assumption here is that (S.14) converges with reasonable rapidity for any ξ within the potential well being dealt with.

2. In agreement with this approximation, we truncate the Fourier series (S.3) at the first nonvanishing terms after $a_1 \cos(\omega t)$ and b_1 (sin ωt). (Just which of the Fourier coefficients are nonvanishing can often be determined from symmetry arguments, as will be seen presently.) The approximation in this step is based on the assumption (or hope) that the Fourier coefficients a_n and b_n $(n > 1)$ decrease in absolute magnitude rapidly as the index n increases.

3. The truncated $\xi(\omega t)$ we put into Equations S.15 and S.16. The integrals are then computed, using trigonometric identities when appropriate, to produce a set of simultaneous equations for the Fourier coefficients. The solutions of these equations generate $\xi(\omega t)$.

4. The calculation is increased in accuracy by either keeping more terms in the Fourier series for $\xi(\omega t)$ when we estimate the Fourier coefficients, or by going to the next nonvanishing term in the Taylor series for $F(\xi)$.

To see how all this works out in practice, we shall investigate further the motion of a particle in the potential well of $V(x)$ given by Equation 2.15. The first approximation to this potential function was presented in Equation 2.22. The second approximation is

$$V_2(x) = 4V_0\left(\frac{x}{x_0}\right)^2 - 6V_0\left(\frac{x}{x_0}\right)^4 \tag{S.17}$$

and is plotted for comparison with $V(x)$ in Figure 2.8. (Recall that $\xi(t) = x(t)$ for this potential function because the minimum point is at the origin.) Only even

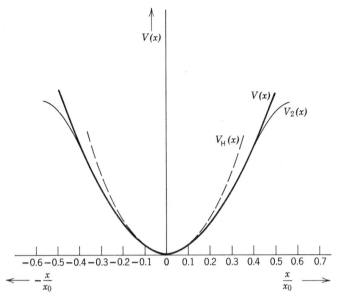

FIGURE 2.8. A comparison of $V(x)$, given in Equation 2.15, with the approximate potential function $V_2(x)$, given in Equation S.17.

powers of x appear in the Taylor series for $V(x)$ because it is invariant under the replacement of $-x$ for x. In this instance, moreover, the coefficients of the powers of the dimensionless ratio (x/x_0) are not decreasing in magnitude in successive terms. This fact indicates that, unfortunately, the contribution from the second term in (S.17) to $x(t)$ may be as important as that from the first term.

Now that we have a potential function we can compute $\tilde{F}(x)$. This quantity is

$$\tilde{F}(x) = \frac{24V_0}{x_0}\left(\frac{x}{x_0}\right)^3$$

and must be put into (S.15) and (S.16) to calculate the Fourier coefficients. Thus

$$a_n = \frac{-(3/\pi x_0^2)}{[(n\omega/\omega_0)^2 - 1]} \int_0^{2\pi} x^3(\omega t) \cos(n\omega t)\, d(\omega t) \qquad (n \geq 1) \qquad \text{(S.18)}$$

$$b_n = \frac{-(3/\pi x_0^2)}{[(n\omega/\omega_0)^2 - 1]} \int_0^{2\pi} x^3(\omega t) \sin(n\omega t)\, d(\omega t) \qquad (n \geq 1) \qquad \text{(S.19)}$$

where we have noted that, according to Equation 2.22,

$$\omega_0^2 = \frac{8V_0}{mx_0^2}$$

In order to proceed farther we must come up with a second approximation for $x(t)$ through Equation S.3. At this point a knowledge of the initial conditions and the symmetry properties of the problem can be most valuable. Suppose, for the sake of this example, we choose the initial conditions $x(0) = A > 0$ and $v(0) = 0$. Since the potential function $V_2(x)$ is invariant under a reflection through the origin, we know that $x(t)$ will be a comparably symmetric function and, therefore, that it must divide its time equally among positive and negative values. Moreover, the initial conditions we have chosen in conjunction with the symmetry and periodicity of $x(t)$ indicate that the position at the midpoint of each cycle of oscillation cannot be at the origin. For, a symmetric, periodic, position function may not start at a turning point and come a half cycle later to the point of stable equilibrium. (It must, in fact, come to another turning point then.) These arguments are enough to lead to the conclusion that the Fourier coefficients a_0, b_n ($n \geqslant 1$), and a_n ($n = 2, 4, 6, \ldots$) all *vanish identically* in Equation S.3. The coefficient a_0 is zero because the integral in Equation S.4 receives equal and opposite contributions from the two half cycles that make up its range of integration. The sine coefficients b_n disappear because sin $(n\omega t)$, unlike $x(t)$, does take on a zero value in the middle of the full cycle of oscillation and is thus an odd function about that midpoint. It follows that every value of the integrand in Equation S.6 that comes during the first π radians is canceled by the corresponding value of the integrand during the second π radians. The Fourier cosine coefficients a_n vanish whenever n is an even integer because they are even functions about the midpoint of the first *half* cycle while $x(t)$, as mentioned above, is an odd function about that midpoint. Thus every value of the integrand in Equation S.5 during the first $\pi/2$ radians is canceled by the corresponding value of the integrand during the following $\pi/2$ radians. What remain, then, are the odd Fourier cosine coefficients, which do not vanish simply because cos (ωt), cos $(3\omega t)$, and so on, take on one of their extremum values when $\omega t = \pi$ and therefore are of the same symmetry character as $x(t)$. It follows that we may write

$$x(t) \simeq a_1 \cos (\omega t) + a_3 \cos (3\omega t) \tag{S.20}$$

as the second approximation for $x(t)$. Upon cubing $x(t)$ and applying the trigonometric identities

$$\cos^3 (\omega t) \equiv \tfrac{3}{4} \cos (\omega t) + \tfrac{1}{4} \cos (3\omega t)$$

$$\cos^2 (\omega t) \cos (3\omega t) \equiv \tfrac{1}{4} \cos (\omega t) + \tfrac{1}{2} \cos (3\omega t) + \tfrac{1}{4} \cos (5\omega t)$$

$$\cos (\omega t) \cos^2 (3\omega t) \equiv \tfrac{1}{2} \cos (\omega t) + \tfrac{1}{4} \cos (5\omega t) + \tfrac{1}{4} \cos (2\omega t)$$

$$\cos^3 (3\omega t) \equiv \tfrac{3}{4} \cos (3\omega t) + \tfrac{1}{4} \cos (9\omega t)$$

we get

$$\begin{aligned}
x_3(t) = \ &\tfrac{3}{4}(a_1{}^3 + a_1{}^2 a_3 + 2a_1 a_3{}^2) \cos (\omega t) + \tfrac{1}{4}(a_1{}^3 + 6a_1{}^2 a_3 \\
&+ 3a_3{}^3) \cos (3\omega t) + \tfrac{3}{4}(a_1 a_3{}^2 + a_1 a_3{}^2) \cos (5\omega t) \\
&+ \tfrac{3}{4} a_1 a_3{}^2 \cos (7\omega t) + \tfrac{1}{4} a_3{}^3 \cos (9\omega t)
\end{aligned}$$

We should not retain the last three terms in this expression because they go beyond Equation S.20 and are likely to be inaccurate. Moreover, only the first term will contribute to an equation for a_1 and only the second term will remain in that for a_3 because of the integral properties of the cosine function mentioned in conjunction with Equation S.5. The Fourier coefficients are, then, according to Equation S.18,

$$a_1 = \frac{9/4x_0^2}{[1 - (\omega/\omega_0)^2]} (a_1^3 + a_1^2 a_3 + 2a_1 a_3^2) \tag{S.21}$$

$$a_3 = \frac{3/4x_0^2}{[1 - 9(\omega/\omega_0)^2]} (a_1^3 + 6a_1^2 a_3 + 3a_3^3) \tag{S.22}$$

This is a pair of rather complicated simultaneous equations for a_1 and a_3. They are typical of what commonly develops in the kind of approximation method we are using and can be solved by straightforward if tedious means. But we would do best to see first if physical insight can be employed to reduce their complexity. One thing we may notice, for example, is that the equations can both be rewritten in dimensionless form as

$$1 = \frac{9/4}{[1 - (\omega/\omega_0)^2]} \left[\left(\frac{a_1}{x_0}\right)^2 + \left(\frac{a_1}{x_0}\right)\left(\frac{a_3}{x_0}\right) + 2\left(\frac{a_3}{x_0}\right)^2 \right] \tag{S.23}$$

$$\frac{a_3}{a_1} = \frac{3/4}{[1 - 9(\omega/\omega_0)^2]} \left[\left(\frac{a_1}{x_0}\right)^2 + 6\left(\frac{a_1}{x_0}\right)\left(\frac{a_3}{x_0}\right) + 3\left(\frac{a_3}{x_0}\right)^2 \left(\frac{a_3}{a_1}\right) \right] \tag{S.24}$$

respectively. Equation S.23 leads immediately to an expression for the angular frequency:

$$\omega^2 = \omega_0^2 \left\{ 1 - \frac{9}{4} \left[\left(\frac{a_1}{x_0}\right)^2 + \left(\frac{a_1}{x_0}\right)\left(\frac{a_3}{x_0}\right) + 2\left(\frac{a_3}{x_0}\right)^2 \right] \right\} \tag{S.25}$$

The angular frequency is apparently decreased from what it was in the harmonic oscillator approximation, which is consistent with the observation that $V_2(x) \leqslant V_H(x)$ for all values of the position. Moreover, *the angular frequency now depends on the amplitude of oscillation*, resulting from the appearance of the Fourier coefficients in Equation S.25. This amplitude dependence is evidence of the complicated nature of motion under an arbitrary conservative force and is a frequently encountered property of nonlinear oscillations.[2] Equation S.24 can provide the relation of the Fourier coefficients to one another. A look at Figures 2.1 and 2.8 suggests that this relation will be one of dominance for the coefficient a_1 if we do not permit the total energy of the oscillator to have any but low values. This conclusion is not as clear-cut as we might wish because of the lack of rapid decrease in the coefficients of the powers of (x/x_0) in $V_2(x)$. However, for low total energies

---

[2] For a complete introductory discussion of this rather difficult subject, see N. Minorsky, *Non-Linear Oscillations*, D. Van Nostrand, Princeton, 1962.

we have low values of (x/x_0), since x_0 is a point at which $V(x)$ achieves a *maximum* value, and this implies a small amplitude of oscillation dominated by a_1. With this physical requirement in mind we can write Equation S.23 approximately as

$$\omega^2 \simeq \omega_0^2 \left[1 - \frac{9}{4} \left(\frac{a_1}{x_0} \right)^2 \right]$$

This, then, leads to

$$\frac{a_3}{a_1} \simeq -\frac{1/27}{[(32/81) - (a_1/x_0)^2]} \left(\frac{a_1}{x_0} \right)^2$$

For small enough chosen values of (a_1/x_0) these expressions should be relatively accurate. As an example, suppose we put $a_1 = 0.4x_0$, which corresponds to a point where $V_2(x)$ is still an accurate representation of $V(x)$, although $V_H(x)$ is not. Then we find

$$\omega \simeq 0.8\omega_0 \qquad \frac{a_3}{a_1} \simeq -\frac{1}{39.7}$$

If we now introduce the second of these approximate equations into Equation S.25 we get the more accurate result

$$\omega \simeq 0.804\omega_0$$

that fully substantiates our first estimate. At this point we can go on to a more complete discussion of the motion, especially at higher total energies, by either augmenting Equation S.20 with $a_5 \cos(5\omega t)$ or carrying out the Taylor series for $V(x)$ to terms of order $(x/x_0)^6$. The procedure then is the same as we have just gone through. It should be evident from our example that we can calculate the motion this way to as great a degree of accuracy as we wish, although the complexity of equations like (S.23) and (S.24) would make elaborate approximation methods and a digital computer necessary before long.

3

The Motion of a Particle in One Dimension

II. NONCONSERVATIVE FORCES

3.1 DISSIPATIVE FORCES

It is a familiar experience that, in the description of the motion of *macroscopic* particles, the analysis we have carried through in the preceding chapter is, to some extent, irrelevant because the total mechanical energy usually is not conserved. A ball thrown upward rapidly, for example, does not quite achieve the height predicted by Equation 2.7, which its initial kinetic energy would seem to warrant, even if we take into account its spinning. A weight hung on a spring and set into motion by a downward tug does not in fact oscillate forever, as Equation 2.50 would have us believe, but instead comes to a standstill after a time. The reason for these observations and many like them is that frictional forces are acting on the particles under consideration. These forces generally oppose the action of the other, conservative forces and, since they persist throughout the motion, tend to bring about a condition of equilibrium characterized by a loss of total energy.[1] The non-conservative forces that bring about this dissipation of total energy we shall call, accordingly, *dissipative forces*.

The precise mathematical form of a dissipative force is a matter dependent on the circumstances under which the force arises and must be determined on the basis of experiment. For a wide variety of situations, however, it has been observed that the dissipative force may be expressed as a function of the velocity of the particle on which it acts and that, for a given, fixed direction of the velocity, it may be written as a Maclaurin expansion in powers of that quantity:

$$F(v) = a_0 + a_1 v + a_2 v^2 + \cdots. \tag{3.1}$$

Quite often, all but one of the expansion coefficients a_n can be neglected over the range of velocity under investigation and the dissipative force will be simply proportional to some power of v. In those cases the Second Law usually can be solved exactly (although it may be technically a nonlinear differential equation). This fact has brought about the common practice of breaking up the observed $F(v)$ into segments that can be fit to single powers of the velocity and studied analytically. We shall follow this practice here, avoiding any consideration of complex forms for $F(v)$, and examine only the behavior of a particle subject to the power-law forces generally employed to describe dissipative motion. In order to get the clearest possible physical picture of the difference among these forces, we shall restrict our discussion initially to free particles undergoing dissipative motion, or *damping*, as it is more familiarly known. More complicated situations will appear in the sections to follow.

1. *Coulomb Damping*. This kind of damping occurs in the classical example of a solid object sliding in a straight line along a surface capable of supporting its weight. The dissipative force is expressed by

$$F_C(v) = a_0 \equiv -\mu m g \qquad (v > 0) \tag{3.2}$$

where μ is a positive constant called the *coefficient of kinetic friction*, mg is the weight of the object, and the direction of the velocity is chosen positive. The factor μ in Equation 3.2 depends on the character of the interface between the object and the surface on which it moves, but does not depend on the area of contact or the magnitude of the velocity. Tables of μ appear in many handbooks on the properties of materials. Typical values range between 0.04, for teflon on teflon, and 0.6, for steel on steel.

---

[1] The Energy Theorem can be recovered, however, if we broaden its meaning to include thermal energy as well as kinetic and potential energy. The conservation principle then states that the sum of mechanical energy transfer (*work done*) and thermal energy transfer (*heat flux*) equals the change in total energy during the motion of a macroscopic particle. In this form the Energy Theorem involves nonmechanical quantities that, despite their very great importance, are not directly related to the discussion here. Thus we suggest that the interested reader consult any textbook on thermodynamics for more information about the Energy Theorem as generalized to include dissipative forces.

When $F_C(v)$ is introduced into the Second Law we find the differential equation

$$\frac{dv}{dt} = -\mu g \qquad (v(t) > 0) \tag{3.3}$$

which is very similar to Equation 1.13. The motion of the particle accordingly may be calculated using the method of the integrating factor as presented in Section 1.2. The results are

$$v(t) = v(0) - \mu g t$$
$$x(t) = x(0) + v(0)t - \tfrac{1}{2}\mu g t^2 \tag{3.4}$$

in terms of the initial conditions. Equations 3.4 predict that the particle starting with velocity $v(0) > 0$ will come to rest in the time

$$t_{stop} = \frac{v(0)}{\mu g}$$

and will travel the distance

$$x_{stop} = x(0) + \frac{v^2(0)}{2\mu g}$$

before doing so. A quantity of physical interest here (and for all damping processes) is the *half-life* of the motion, which is defined as the time, $t_{1/2}$, required for the velocity to reduce to one half its initial value. Thus, according to the first of Equations 3.4,

$$v(t_{1/2}) \equiv \tfrac{1}{2}v(0)$$
$$= v(0) - \mu g t_{1/2}$$

and

$$t_{1/2} = \frac{v(0)}{2\mu g} \qquad \text{(Coulomb half-life)} \tag{3.5}$$

Equation 3.5 contains parameters that are readily measured and so provides a means for empirically checking the theory of Coulomb damping. We note in particular that it predicts the half-life of the motion to be independent of the mass of the particle but to be directly proportional to its initial speed.

2. *Stokes Damping.* This kind of damping takes place when a particle travels through a viscous medium, such as air, water, or oil and, more generally, when its motion involves the deformation of a substance or an object in contact with it. The damping force is given by

$$F_S(v) = a_1 v \equiv -\beta v \tag{3.6}$$

where β is a positive constant whose value depends on the nature of the medium or the object deformed by the motion. We have noted by the minus sign that $F_S(v)$ opposes the motion of the particle. If the particle has the shape of a perfect sphere and travels slowly through a fluid, Equation 3.6 represents exactly the frictional force acting upon it and is known as *Stokes' Law*. In that case

$$\beta = 6\pi\eta r$$

where η is called the *coefficient of viscosity* of the fluid and r is the radius of the particle. With the methods of hydrodynamics the value of β can actually be computed for a slowly moving ellipsoid of arbitrary dimensions.

The Second Law, with Equation 3.6 providing the force, is now written

$$\frac{dv}{dt} + \frac{1}{\tau} v(t) = 0 \tag{3.7}$$

where

$$\tau \equiv \frac{m}{\beta} \qquad \text{(Stokes time constant)} \tag{3.8}$$

is a parameter having the dimensions of time and is called, appropriately, the *time constant* for the damping process. Equation 3.7 has the same mathematical form as Equation 1.14 if we set $p(t) = 1/\tau$ and $q(t) = 0$. The method of the integrating factor then gives the solution

$$v(t) = v(0)e^{-t/\tau} \tag{3.9}$$

The velocity of a free particle under Stokes damping decreases *exponentially* with the time (see Figure 3.1). The half-life follows directly as

$$t_{1/2} = (\ln 2)\tau \simeq 0.693\tau \qquad \text{(Stokes half-life)} \tag{3.10}$$

In this instance the half-life does depend on the mass of the particle through τ (Equation 3.8), but does not depend on the initial velocity. This mathematical behavior clearly distinguishes the two classes of damping we have considered so far and, therefore, could be used to distinguish between them experimentally. The position of the particle as a function of the time may be found in the usual way by solving

$$\frac{dx}{dt} = v(0)e^{-t/\tau}$$

with the method of Section 1.2 to get

$$x(t) = x(0) + v(0)\tau(1 - e^{-t/\tau}) \tag{3.11}$$

One peculiar thing we notice about Equations 3.9 and 3.11 is that, although the particle does not stop moving until an infinite amount of time has passed, during this time it travels the finite distance $x(0) + v(0)\tau$. This result comes about directly from Equation 3.6, which tells us that the damping force decreases in strength as it slows down the particle and thus becomes less effective as time goes by. The diminishing of its strength causes an infinite amount of time to pass before it can stop the particle, but the distance traveled is finite because the particle is moving so slowly

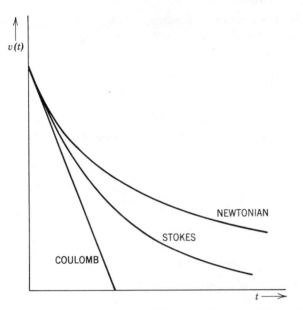

FIGURE 3.1. The velocity of a free particle under Coulomb, Stokes, and Newtonian damping. In each case the initial velocity and damping force and the mass of the particle have been given unit absolute value.

so much of the time. The prediction of an infinite stopping time, of course, cannot agree with experiment and represents a weakness in the theory. In practice this difficulty is avoided by defining the stopping time to be that which makes $v(t)$ small beyond the sensitivity of any velocity measurement being made on the particle. Since $v(t)$ in Equation 3.9 is quite "flat" for all times much larger than about five time constants, this procedure usually has little ambiguity associated with it.

3. *Newtonian Damping.* This kind of damping is also known to occur when a particle travels through a viscous medium, a particular example being a rocket sent through the atmosphere. The damping force is expressed mathematically by

$$F_N(v) = a_2 v^2 \equiv -\alpha v^2 \qquad (v > 0) \tag{3.12}$$

where the direction of the velocity is taken to be positive. The Second Law for Newtonian damping is

$$\frac{dv}{dt} = -\left(\frac{\alpha}{m}\right) v^2(t) \qquad (v(t) > 0) \tag{3.13}$$

which is an example of a nonlinear differential equation.

As has been suggested before, nonlinear differential equations are not ordinarily easy to solve. Equation 3.13, fortunately, is an exception to the rule in that it is a particular case of the nonlinear equation

$$\frac{dy}{dt} + ay(t) = by^n(t) \qquad (n > 1) \tag{3.14}$$

where n is any real number greater than one and a and b are arbitrary constants. Equation 3.14 may be transformed[2] into the integrable expression

$$dt = \frac{dy/y}{(by^{n-1} - a)} \qquad (n > 1) \tag{3.15}$$

A table of integrals will then lead to the results

$$y(t) = \begin{cases} \left[\dfrac{b}{a}(1 - C^{-1}e^{a(n-1)t})\right]^{1/(1-n)} & (a \neq 0) \\[2ex] [b(1 - n)(C + t)]^{1/(1-n)} & (a = 0) \end{cases} \tag{3.16}$$

where C is a constant of integration.

Now, Equation 3.13 corresponds to Equation 3.14 if we set $a = 0$, $b = -(\alpha/m)$, and $n = 2$. It follows that

$$v(t) = \left[\left(\frac{\alpha}{m}\right)(C + t)\right]^{-1}$$

or

$$v(t) = \frac{v(0)}{[1 + (\alpha v(0)t/m)]} \tag{3.17}$$

upon evaluating the constant of integration C in terms of the initial condition on $v(t)$. Equation 3.17 shows that a free particle subjected to Newtonian damping, similarly to what occurs under Stokes damping, will not come to rest until an infinite amount of time has elapsed. The physical reason for this is the same in both cases. But observe that, in general, Newtonian damping is even less effective than Stokes damping (see again Figure 3.1). The entire cause of the relative difference here is mathematical: v^2 is much smaller than v when the velocity is near zero. The half-life for Newtonian damping follows directly from Equation 3.17:

$$t_{1/2} = \frac{m}{\alpha v(0)} \qquad \text{(Newtonian half-life)} \tag{3.18}$$

Here we have the same dependence of $t_{1/2}$ on the mass of the particle as we had for Stokes damping, but also we have a dependence on the initial velocity that is inverse to that for Coulomb damping. (Notice, however, that in all three cases

---

[2] For a technical discussion of Equation 3.14, see, for example, E. D. Rainville and P. E. Bedient, *Elementary Differential Equations*, The Macmillan Co., New York, 1969, §89.

$t_{1/2}$ has the *same* dependence on the parameter that measures the strength of the damping force: μ, β, or α.) Because both Newtonian and Stokes damping are caused by forces that are independent of the weight of the particle, the half-life corresponding to them is lengthened (the damping is less effective) as the mass of the particle—its inertia—is increased. These forces do not compensate for the mass and, accordingly, are less effective as the mass gets larger. The force that causes Coulomb damping, on the other hand, does increase proportionally with the mass of the particle and yields a half-life that is not sensitive to that parameter. It is sensitive to the initial velocity, however, because it does not compensate for any increase in that quantity, as do $F_S(v)$ and $F_N(v)$. Therefore, the Coulomb half-life is longer the greater is the initial speed. The half-life for Newtonian damping is, on the contrary, shorter, the greater is $v(0)$, because $F_N(v)$ compensates for a large velocity even better than does $F_S(v)$. These comments should make it evident that the important physical parameters in damping do include the initial velocity of the particle, despite its absence from the very familiar Equation 3.10.

3.2 PARTICLE-SIZE FRACTIONATION BY GRAVITY

Now we shall consider the dynamics of a very important geophysical phenomenon: the settling by gravity of particles suspended in a viscous fluid. Examples of this phenomenon occur when solid or liquid air pollutants drift to the ground after being emitted by a tall stack or when suspended sediments drop out of the water in a river and cause siltation of the stream bed. The theory of settling by gravity predicts that the asymptotic value of the settling velocity (called the *terminal velocity*) depends on the size of the particle. Therefore, the theory also has an important application to the laboratory process of particle-size analysis of powders and to the investigation of turbidity in reservoirs used for water supply.

We shall consider the case of Stokes damping, which is almost universally employed to represent the frictional force for particles falling at relatively low speeds in fluids. The total force on the settling particle is

$$F = -m_{\text{eff}}g - \beta v \qquad (3.19)$$

where, according to Archimedes' Principle,

$$m_{\text{eff}} = (\rho - \rho_f)V$$

is an effective mass which accounts for the influence of buoyancy in the fluid. The density of the particle is ρ, ρ_f is that of the fluid, and V is the particle volume. The Second Law comes from Equation 3.19 as

$$\frac{dv}{dt} + \frac{1}{\tau}v(t) = -g_e \qquad (3.20)$$

where $g_e \equiv m_{\text{eff}}g/m$. Note that the parameter g_e may be positive or negative, depending on whether the particle density is greater or less than the fluid density, respectively. We shall assume $g_e > 0$ in order to describe the process of settling. Equation 3.20 is readily seen to be an inhomogeneous form of Equation 3.7. Because the inhomogeneous term is just a constant we can deal with it as we dealt with the comparable term in Equation 2.45, where a mathematical problem of this kind made its first appearance. Thus we define

$$u(t) \equiv v(t) - v_T \tag{3.21}$$

where v_T is a constant parameter that will be used to remove g_e from Equation 3.20. After introducing (3.21) into (3.20) and setting

$$v_T = -g_e \tau \qquad \text{(Stokes terminal velocity)} \tag{3.22}$$

we get the homogeneous differential equation

$$\frac{du}{dt} + \frac{1}{\tau} u(t) = 0 \tag{3.23}$$

The solution of Equation 3.23 is, of course,

$$u(t) = u(0)e^{-t/\tau}$$

and, through Equation 3.21, it gives us the result

$$v(t) = v_T + (v(0) - v_T)e^{-t/\tau} \tag{3.24}$$

Equation 3.24 is plotted in Figure 3.2. There we see that the velocity of the particle

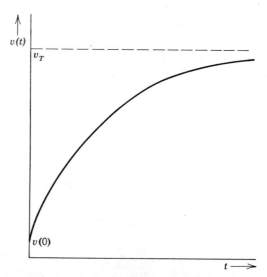

FIGURE 3.2. The velocity of a particle in a uniform gravitational field under Stokes damping.

changes gradually from $v(0)$ to v_T, approaching the latter value asymptotically. For this reason v_T is called the *terminal velocity* of the particle. That a terminal velocity exists for the total force given by Equation 3.19 is very easy to see on purely physical grounds. Suppose that the force of Stokes friction opposes the effect of the gravitational field with that of damping. Eventually the two effects must nullify one another because of the dependence of the damping on the velocity of the particle. When this cancellation occurs the total force vanishes and Equation 3.19 immediately yields the constant velocity given by Equation 3.22. We note also that v_T depends on the particle volume through m_{eff}. This fact is the basis for particle-size fractionation by settling.

The position of the particle is the solution of

$$\frac{dx}{dt} = v_T + (v(0) - v_T)e^{-t/\tau}$$

which has the form of Equation 1.14 if we set $p(t) = 0$ and $q(t) = v_T + (v(0) - v_T)e^{-t/\tau}$. The solution for $x(t)$ then follows directly after applying Equation 1.16:

$$x(t) = x(0) + v_T t + (v(0)\tau - v_T\tau)(1 - e^{-t/\tau}) \tag{3.25}$$

in terms of the initial conditions. It is worthwhile to look at the limiting forms of Equation 3.25 to see the physical picture they imply. When the ratio (t/τ) is very small we can safely expand the exponential in (3.25) in a Maclaurin series to second order to get

$$x(t) \simeq x(0) + v_T t + (v(0)\tau - v_T\tau)\left[\left(\frac{t}{\tau}\right) - \frac{1}{2}\left(\frac{t}{\tau}\right)^2\right]$$

$$= x(0) + v(0)t - \frac{1}{2}g_e\left(1 - \frac{v(0)}{v_T}\right)t^2$$

which, if we can neglect $v(0)$ in relation to v_T, is precisely of the form of Equation 2.4. This result states, reasonably, that for small elapsed times (or large values of the time constant) the particle behaves as if it were moving in the absence of a dissipative force. A small elapsed time means that there has been little chance for the velocity to change because of the force due to the gravitational field and, therefore, that there is only a small effect of the damping force. Similarly, a large value of τ necessarily means light damping, by virtue of Equation 3.8. When the ratio (t/τ) is at the other extreme and is large enough to reduce the exponential in (3.25) to a value very nearly equal to zero, we have

$$x(t) \sim x(0) + (v(0) - v_T)\tau + v_T t \tag{3.26}$$

which describes a particle subject to no force and traveling with the terminal velocity. This result agrees with what we have already encountered in connection with Equation 3.24. Notice also that $x(t)$ does not extrapolate to $x(0)$ for zero time. This is simply because the slope of the line to which $x(t)$ conforms asymptotically does not in general equal the slope of the line to which it conforms initially and, therefore, a different value of the intercept at time zero is required.

One other item that we must investigate is the location in space and time of the turning point of the motion if the initial velocity is directed upward. This turning point, it will be recalled, can occur only if $v(0)$ and $-g_e$ have the opposite signs necessary to make $v(t)$ vanish for some value of the time variable. When this is the case Equation 3.24 yields in the usual way

$$t_m = \tau \ln \left(1 - \frac{v(0)}{v_T} \right) \tag{3.27}$$

as the modified expression for the time at which the particle turns around. If, in practice, the ratio $|v(0)/v_T|$ is very small compared with unity, we can expand the logarithm in (3.27) to first order in a Maclaurin series to get

$$t_m \simeq \tau \left(\frac{-v(0)}{v_T} \right) = \frac{v(0)}{g_e}$$

which is the result to be found in the absence of frictional forces! After putting (3.27) into (3.25) we find, for the turning point,

$$x_m(t_m) = x(0) + v(0)\tau + v_T\tau \ln \left(1 - \frac{v(0)}{v_T} \right) \tag{3.28}$$

This rather complicated expression may be compared with Equation 2.7, which

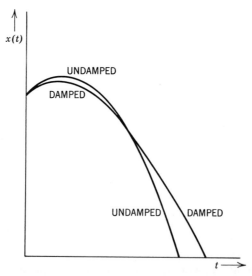

FIGURE 3.3. The position of a particle thrown upward in a uniform gravitational field. The case of Stokes damping is compared with that of no damping.

describes the friction-free case, if we expand the logarithm to third order in powers of $(v(0)/v_T)$:

$$x_m(t_m) \simeq x(0) - \tfrac{1}{2}v_T\tau \left(\frac{v(0)}{v_T}\right)^2 - \tfrac{1}{3}v_T\tau \left(\frac{v(0)}{v_T}\right)^3$$

$$= x(0) + \tfrac{1}{2}\frac{v^2(0)}{g_e}\left[1 + \tfrac{1}{6}\left(\frac{v(0)}{v_T}\right)\right] \tag{3.29}$$

Because $(v(0)/v_T)$ is a negative number, we see that the turning point comes *nearer* to $x(0)$ than it did when only the uniform field was present. The effect of the dissipative force, as expected, is to reduce the velocity to zero sooner than it would be otherwise (see Figure 3.3).

3.3 FRICTIONAL SLOWING OF A PROJECTILE

Generally the frictional force on a projectile moving through the atmosphere may be represented by the Newtonian expression, Equation 3.12. This expression also could apply to the downward motion of a parachutist or other massive object and to the descent of a space vehicle returning from a moon flight. The total force on any of these objects would therefore be given by

$$F = -m_{\text{eff}}g + \alpha v^2 \qquad (v < 0) \tag{3.30}$$

The Second Law is then

$$\frac{dv}{dt} - \frac{\alpha}{m}v^2(t) = -g_e \qquad (v(t) < 0) \tag{3.31}$$

which is a generalized form of Equation 3.13. Proceeding as before, we define a function $u(t)$ by Equation 3.21, but now with

$$v_T = \left(\frac{mg_e}{\alpha}\right)^{1/2} \qquad \text{(Newtonian terminal } speed\text{)} \tag{3.32}$$

Equation 3.31 then becomes the differential equation

$$\frac{du}{dt} - \frac{1}{\tau}u(t) = \frac{1}{2v_T\tau}u^2(t) \tag{3.33}$$

where the time constant τ is now defined by

$$\tau \equiv \tfrac{1}{2}\frac{m}{\alpha v_T} \qquad \text{(Newtonian time constant)} \tag{3.34}$$

Equation 3.33 has the mathematical form of Equation 3.14 if we put

$$a = -\frac{1}{\tau} \qquad b = \frac{1}{2v_T\tau} \qquad n = 2$$

According to the first of Equations 3.16, we may write

$$u(t) = \left[-\frac{1}{2v_T}(1 - C^{-1}e^{-t/\tau}) \right]^{-1}$$

subject to $u(0) = v(0) - v_T$. After evaluating the arbitrary constant C in the usual way we find

and

$$u(t) = \left[-\frac{1}{2v_T}\left(1 - \left(1 + \frac{2v_T}{u(0)}\right)e^{-t/\tau}\right) \right]^{-1}$$

$$v(t) = v_T - 2v_T\left[1 + \frac{v_T + v(0)}{v_T - v(0)}e^{-t/\tau}\right]^{-1}$$

$$= \frac{v(0) - v_T \tanh(t/2\tau)}{1 - (v(0)/v_T)\tanh(t/2\tau)} \tag{3.35}$$

The second step in (3.35) comes after some tedious but straightforward algebra. To understand the physical consequences of Equation 3.35 we consider the limit $t \to \infty$, for which the hyperbolic tangent becomes equal to 1. This shows us that the velocity becomes asymptotically equal to $-v_T$, as expected. When $(t/2\tau)$ is a very small quantity, $\tanh(t/2\tau) \simeq t/2\tau \ll 1$ and

$$v(t) \simeq v(0) - \frac{v_T t}{2\tau} = v(0) - g_e t$$

according to Equations 3.32 and 3.34. [We have assumed that the absolute value of the ratio $(v(0)/v_T)$ is small enough to justify setting the denominator equal to unity in (3.35).] Once again the result for short elapsed times is independent of the influence of the frictional force.

The position of the particle as a function of time is determined from Equation 3.35 in the usual way. The result is

$$x(t) = x(0) + v_T t - 2v_T\left\{t + \tau\ln\left[1 + \frac{v_T + v(0)}{v_T - v(0)}e^{-t/\tau}\right] - \tau\ln\left(\frac{2v_T}{v_T - v(0)}\right)\right\}$$

$$= x(0) - 2v_T\tau\ln\left[\cosh\left(\frac{t}{2\tau}\right) - \frac{v(0)}{v_T}\sinh\left(\frac{t}{2\tau}\right)\right] \tag{3.36}$$

An interesting application of this equation occurs when the rapid downward motion of a spherical object is investigated. If the fall of the body is from rest, we have

$$x(t) = x(0) - 2v_T\tau\ln\cosh\left(\frac{t}{2\tau}\right) \tag{3.37}$$

as its position at the time t. In air under normal conditions we have $\alpha \simeq 1.02 \times 10^{-3}R^2$, where R is the radius of the body in centimeters. Therefore, by Equations 3.32 and 3.34, $2\tau \simeq \sqrt{m}/R$ (τ in seconds). An iron sphere of radius 10 cm has a mass of about 3.18×10^4 g. This leads to $2\tau \simeq 18$ s, which is quite large in comparison with the usual elapsed time in a falling-body experiment. For a wooden sphere of the same radius the mass is about 6.6 percent as great and we find $2\tau \simeq 4.6$ s, which is a little on the small side, yet is still about five times as large as the time required for a sphere to drop from the ceiling to the floor of an average laboratory. It follows that we can, in practice, expand Equation 3.37 to, for example, fourth order in powers of $(t/2\tau)$ and lose little accuracy:

$$x(t) \simeq x(0) - v_T\tau \left(\frac{t}{2\tau}\right)^2 + \tfrac{1}{6}v_T\tau \left(\frac{t}{2\tau}\right)^4$$

$$= x(0) - \tfrac{1}{2}g_e t^2 \left[1 - \tfrac{1}{6}\left(\frac{t}{2\tau}\right)^2\right] \tag{3.38}$$

Equation 3.38 says that all bodies do *not* fall at the same rate in air regardless of their masses. Instead, those of larger mass (corresponding to larger τ) of a given size will achieve a given distance more quickly. For an iron sphere of the size mentioned above, the correction to $x(t)$ because of dissipative forces amounts to about 0.2 percent in a falling time of 2 s, while for the comparable wooden sphere the correction is 3.2 percent in the same falling time.

3.4 DYNAMICS OF THE PENDULUM SEISMOGRAPH: THE DAMPED LINEAR OSCILLATOR

The pendulum seismograph is an instrument in which a suspended mass is used to provide a reference point for measuring the movements of the ground caused by earthquakes. In its essential parts, the seismograph consists of a boom held in stable equilibrium by means of a spring, a massive object attached to the boom, and a damping device to resist the oscillations of the boom (see Figure 3.4). A seismograph usually produces its final record by moving a stylus or a light point by an amount proportional to the ground displacement. The constant of proportionality is called the magnification of the instrument. The magnification can range in value between 1 and 10^6 and ordinarily will depend on the frequency of oscillation of the boom.

Since the seismograph is a kind of pendulum, its response is measured in terms of the angular displacement θ of the boom. The total force on the boom in the absence of earth movement is then

$$F = -K\theta - B\frac{d\theta}{dt} \tag{3.39}$$

where K is a force constant for the restraining spring and B is a constant that measures the strength of damping. The form of this expression for the total force suggests that the seismograph may be considered as a special case of a linear harmonic oscillator subject to Stokes damping, with θ playing the role of the usual

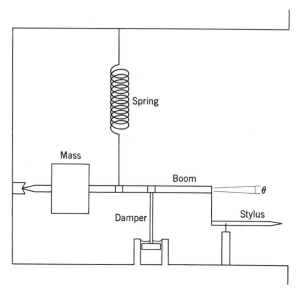

FIGURE 3.4. The vertical pendulum seismograph.

linear displacement ξ. With this idea in mind we may write the Second Law for the pendulum seismograph as follows:

$$\frac{d^2\theta}{dt^2} + \frac{1}{\tau}\frac{d\theta}{dt} + {\omega_0}^2\theta(t) = 0 \tag{3.40}$$

where $\tau \equiv M/B$, M is, technically, the moment of inertia of the boom about its hinge to the frame, and ${\omega_0}^2 \equiv K/M$. The angular frequency ω_0 is known in the present context as the *natural angular frequency* of the seismograph. (The reason for this name will become clear soon.)

Equation 3.40 is a linear, homogeneous differential equation of the same mathematical form as Equation 2.26 and, therefore, it can be solved readily by the method of the auxiliary equation. The auxiliary equation for (3.40) is, in fact,

$$m^2 + \frac{1}{\tau}m + {\omega_0}^2 = 0$$

and possesses the two general solutions

$$m_1 = -\frac{1}{2\tau} + \omega_0\left(\frac{1}{4Q^2} - 1\right)^{1/2} \qquad m_2 = -\frac{1}{2\tau} - \omega_0\left(\frac{1}{4Q^2} - 1\right)^{1/2} \tag{3.41}$$

In Equations 3.41 we have combined the natural angular frequency and the Stokes time constant τ of the oscillator to define the *quality factor* Q:

$$Q \equiv \omega_0\tau \tag{3.42}$$

The quality factor provides a convenient physical measure of the damping, a *large* value of Q corresponding to *small* damping. Moreover, it is the numerical value of Q that distinguishes the three general classes of solution of the auxiliary equation. These are enumerated as follows.

1. $Q > \frac{1}{2}$. When the quality factor is larger than 0.5, the oscillator is said to be *underdamped* and the solutions of the auxiliary equation become the complex-conjugate pairs

$$m_1 = -\frac{1}{2\tau} + i\omega_d \qquad m_2 = -\frac{1}{2\tau} - i\omega_d$$

where

$$\omega_d = \omega_0 \left(1 - \frac{1}{4Q^2}\right)^{1/2} \tag{3.43}$$

is called the *angular frequency of the damped oscillator*. The solution of Equation 3.40, according to Equation 2.31, is then

$$\theta(t) = e^{-t/2\tau}(c_1 \cos \omega_d t + c_2 \sin \omega_d t)$$

where c_1 and c_2 are arbitrary constants of integration. The physical significance of this result is more easily appreciated if we use Equations 2.41 to rewrite it as

$$\theta(t) = Ce^{-t/2\tau} \cos(\omega_d t - \delta) \tag{3.44}$$

where C and δ are related to the initial conditions by

$$C^2 = \theta^2(0)\left[1 + \left(\frac{\dot{\theta}(0)}{\omega_d\theta(0)} + \frac{1}{2\omega_d\tau}\right)^2\right] \qquad \tan \delta = \frac{\dot{\theta}(0)}{\omega_d\theta(0)} + \frac{1}{2\omega_d\tau}$$

where $\dot{\theta}(0)$ means $(d\theta/dt)_{t=0}$. Equation 3.44 is plotted in Figure 3.5 for the case $C = \theta(0)$ and $\delta = 0$. We see that the motion of the underdamped oscillator is still periodic, but that the period of oscillation has been shifted by the damping to

$$T_d \equiv \frac{2\pi}{\omega_d} = \frac{T}{[1 - (1/4Q^2)]^{1/2}} \tag{3.45}$$

in terms of the period T of the undamped oscillator, given by Equation 2.38. The dissipative force retards the motion of the oscillator throughout its cycle and thereby lengthens the period.

The amplitude of the oscillator is by definition the positive value of the displacement for all values of the time variable that satisfy

$$\left(\frac{d\theta}{dt}\right)_{t=t_m} = -Ce^{-t_m/2\tau} \cos(\omega_d t_m - \delta)\left[\frac{1}{2\tau} + \omega_d \tan(\omega_d t_m - \delta)\right] = 0$$

which means that t_m is the solution of

$$\tan(\omega_d t_m - \delta) = -\frac{1}{2\omega_d\tau} \tag{3.46}$$

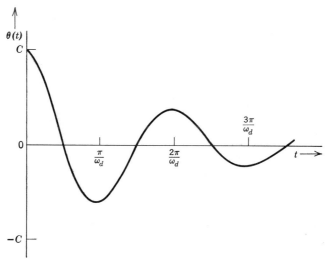

FIGURE 3.5. The angular displacement of the boom of an underdamped pendulum seismograph.

If we combine (3.46) with the well-known trigonometric identity relating the squares of the sine and cosine we find

$$\cos(\omega_d t_{max} - \delta) = \frac{2\omega_d \tau}{[1 + 4(\omega_d \tau)^2]^{1/2}}$$

where t_{max} is any value of t_m leading to a maximum displacement, and

$$\text{amplitude}(t_{max}) = \frac{2C\omega_d \tau}{[1 + 4(\omega_d \tau)^2]^{1/2}} e^{-t_{max}/2\tau} \qquad (3.47)$$

Equation 3.47 states that the amplitude decreases *exponentially* during each succeeding cycle of oscillation and suggests that the natural logarithm of the amplitude ratio for consecutive cycles will be a constant. To check this prediction we write

$$\ln\left[\frac{\text{amplitude}(t_{max}(n))}{\text{amplitude}(t_{max}(n+1))}\right] = \ln\left[\frac{e^{-t_{max}(n)/2\tau}}{e^{-t_{max}(n+1)/2\tau}}\right] = \frac{t_{max}(n+1) - t_{max}(n)}{2\tau}$$

according to Equation 3.47, where $t_{max}(n)$ is the time at which the displacement achieves its maximum value during the nth period of oscillation. The logarithm of the amplitude ratio given above is called the *logarithmic decrement* and may be given the symbol Δ. If we note that the quantity $[t_{max}(n+1) - t_{max}(n)]$ is just $2\pi/\omega_d$, the period of damped oscillation, we have

$$\Delta = \frac{\pi}{\omega_d \tau} = \left(\frac{\omega_0}{\omega_d}\right)\frac{\pi}{Q} \qquad (3.48)$$

which verifies the implication of Equation 3.47. Equation 3.48 provides a useful relation between the logarithmic decrement and the quality factor. In the case of very light damping we have

$$\omega_d \tau \simeq \omega_0 \tau = Q \gg 1$$

and Equations 3.47 and 3.48 become, respectively,

$$\text{amplitude}\,(t_{\max}) \simeq Ce^{-t_{\max}/2\tau}$$

$$\Delta \simeq \frac{\pi}{Q}$$

where, according to (3.46),

$$t_{\max} \simeq \frac{\delta}{\omega_0}, \frac{2\pi\delta}{\omega_0}, \frac{4\pi\delta}{\omega_0}, \dots$$

in complete analogy with the undamped oscillator.

2. $Q = \frac{1}{2}$. When the quality factor takes on the special value of 0.5, the oscillator is said to be *critically damped* and Equations 3.41 reduce to

$$m_1 = m_2 \equiv m_0 = -\frac{1}{2\tau}$$

The displacement of the oscillator then is specified by Equation 2.30 to be

$$\theta(t) = (c_1 + c_2 t)e^{-t/2\tau} \tag{3.49}$$

Equation 3.49 is graphed in Figure 3.6. There we see that, for the critically damped oscillator, the quality factor has become small enough for the dissipative force to prevent any periodic motion at all and the oscillator returns to its equilibrium position after going but once through a turning point. Thus we may characterize critical damping by the statement that it occurs just when the

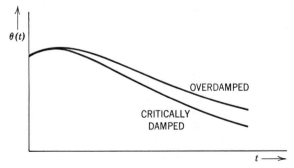

FIGURE 3.6. The angular displacement of the boom of critically damped and overdamped pendulum seismographs.

quality factor Q has the largest value possible for which the displacement of the oscillator is purely exponential in the time.

3. $Q < \frac{1}{2}$. When the quality factor falls below 0.5 the oscillator is said to be *overdamped* and the roots of the auxiliary equation become

$$m_1 = -\frac{1}{2\tau} + \frac{1}{2\tau_0} \qquad m_2 = -\frac{1}{2\tau} - \frac{1}{2\tau_0}$$

where

$$\tau_0 \equiv \frac{\tau}{(1 - 4Q^2)^{1/2}} \tag{3.50}$$

defines a second characteristic time constant. The displacement of the oscillator is now prescribed by Equation 2.29 to be

$$\theta(t) = e^{-t/2\tau}(c_1 e^{t/2\tau_0} + c_2 e^{-t/2\tau_0}) \tag{3.51}$$

Equation 3.51 is compared with Equation 3.49 in Figure 3.6. It is evident from the graphs that, for given initial conditions, the overdamped oscillator is much slower in its approach to equilibrium than is the critically damped oscillator. (Hence the adjective *over*damped.) Indeed, the asymptotic form of (3.51) is

$$\theta(t) \sim c_1 \exp\left[-\left(\frac{1}{2\tau} - \frac{1}{2\tau_0}\right)t\right]$$

which is always larger in absolute value than is $\exp(-t/2\tau)$. The critical damping of an oscillator results in the most rapid return to equilibrium, should the particle be set into motion. This fact is of great importance in the design of seismographs and other macroscopic oscillators such as ballistic galvanometers and analytical balances. Underdamping is, of course, considered to be a serious defect for a seismograph because it leads to slowly decaying responses and sharp seismometer resonances (to be discussed in Section 3.5) that give earth waves with periods close to $2\pi/\omega_0$ an undue emphasis on the chart record. Heavily overdamped seismographs are more undesirable, with the result that the value of Q is usually set between 0.5 and 1.0.

3.5 THE DRIVEN, DAMPED OSCILLATOR

An harmonic oscillator subject to Stokes damping and to an arbitrary external force that is an *explicit* function of the time is known as a driven, damped oscillator. The Second Law as applied to this type of particle has been found to provide a reasonably faithful picture of the behavior of a large number of physical systems that range in character from the recording pen of the seismograph, driven by the translatory motion of the ground, to the optically active electron in a classical atom driven by electromagnetic radiation. In its most general form the total force on the driven, damped oscillator is

$$F = -k\xi - \beta v + F(t)$$

which gives rise to the differential equation

$$\frac{d^2\xi}{dt^2} + \frac{1}{\tau}\frac{d\xi}{dt} + \omega_0{}^2\xi(t) = a(t) \qquad (3.52)$$

where τ is the Stokes time constant, ω_0 is the natural angular frequency of the oscillator, and $a(t) = F(t)/m$, $F(t)$ is the arbitrary "driving force." Equation 3.52 is an inhomogeneous form of Equation 3.40. Its mathematical type has been investigated extensively and there exists a number of results that is of great value in obtaining its solutions for a chosen $a(t)$. We shall state some of these results now, without proofs, in a way that will be of direct use to us in doing applications.[3] A summary is presented in Table 3.1.

TABLE 3.1 Some Possible Forms of the Steady-State Response of a Driven, Damped Oscillator

Form of $a(t)$	Form of $\xi_S(t)$
c (constant)	b_1 (constant)
t^m	$b_1 t^m + b_2 t^{m-1} + \cdots + b_m$
e^{bt}	$b_1 e^{bt}$
$\sin \omega t$	$b_1 \sin \omega t + b_2 \cos \omega t$
$\cos \omega t$	$b_1 \cos \omega t + b_2 \sin \omega t$

Note. If $a(t)$ has a term equal to t^n ($n \geqslant 0$) times a term in $\xi_T(t)$, $\xi_S(t)$ contains t^{n+1} times the duplicated term and its nonredundant derivatives, *in addition to the functions shown above.*

1. The displacement of a driven, damped oscillator is generally the sum of any solution of the Second Law for the damped oscillator (Equation 3.40) and any particular solution of the Second Law for the driven, damped oscillator (Equation 3.52). Thus

$$\xi(t) = \xi_T(t) + \xi_S(t)$$

is the general form of the displacement, where $\xi_T(t)$—called in this context the *complementary function* of Equation 3.52—is a solution of (3.40) and $\xi_S(t)$ is some determined particular solution of (3.52). Because the complementary function always dies out exponentially in time it has been given the alternate name *transient solution.* The displacement $\xi_S(t)$ may not decrease monotonically as

[3] For proofs of these statements and an excellent discussion, see M. Tenebaum and H. Pollard, *Ordinary Differential Equations,* Harper & Row, New York, 1963.

time passes but, instead, may persist as the response of the oscillator to a continually applied driving force $F(t)$. For this reason, when the term is appropriate, $\xi_S(t)$ is given the name *steady-state solution*. As a general rule the transient solution is the part of $\xi(t)$ containing the initial conditions on the oscillator while the steady-state solution is independent of these but contains parameters reflecting the nature of the damping and driving forces.

2. If $a(t)$ in Equation 3.52 is equal to any one of the functions c, t^m, exp (bt), sin ωt, cos ωt, or to combinations of them, c, m, b, and ω being constants, and if *no* term in $a(t)$ is the same as one in $\xi_T(t)$, then $\xi_S(t)$ will, in general, be a sum of all the functions in $a(t)$ and all of the derivatives of these functions that are not simply linear combinations of them. As an example, suppose

$$a(t) = 2t^2 + e^{3t}$$

and $\xi_T(t)$ were given by Equation 3.49. A comparison shows that $a(t)$ has no term the same as or proportional to any in $\xi_T(t)$ and we may write down immediately the particular solution

$$\xi_S(t) = b_1 t^2 + b_2 e^{3t} + b_3 t + b_4$$

The coefficients in this expression are then determined by substituting it into Equation 3.52 and collecting the terms containing like functions of t. Note that the derivatives of e^{3t} do not appear in $\xi_S(t)$ because they happen to be simply proportional to the original function.

3. If $a(t)$ is equal to any one of the functions listed above in (2) and contains one or more terms, each of which is t^n times a term in $\xi_T(t)$, where n is a nonnegative integer, then $\xi_S(t)$ will be a sum of all the terms in $a(t)$, and their nonredundant derivatives that are not in $\xi_T(t)$, plus a sum of t^{n+1} times each duplicated term of $\xi_T(t)$ along with its nonredundant derivatives. For example, let

$$a(t) = e^{-t} + t$$

and consider $\xi_T(t)$ to be given by Equation 3.49 with $\tau = 0.5$ s. Then $a(t)$ is just t^0 times the first term in $\xi_T(t)$ plus a term not in the complementary function, and we have

$$\xi_S(t) = b_1 t + b_2 t e^{-t} + b_3 e^{-t} + b_4$$

3.6 PERIODIC, IMPULSIVE, AND EXPONENTIAL DRIVING FORCES

Now we shall consider some specific examples of driven, damped oscillators which are of practical interest. We shall begin by discussing the response to

$$F(t) = F_0 \cos \omega t \qquad (F_0 > 0) \qquad (3.53)$$

which represents a periodic driving force. The Second Law (3.52) then becomes

$$\frac{d^2\xi}{dt^2} + \frac{1}{\tau}\frac{d\xi}{dt} + \omega_0^2 \xi(t) = a_0 \cos \omega t \qquad (3.54)$$

where $a_0 = F_0/m$ is simply a constant. According to our previous discussion, the solution of Equation 3.54 is a sum containing a solution of Equation 3.40 plus a linear combination of cos ωt and sin ωt. We shall choose the transient solution to represent a critically damped oscillator. Then we have

$$\xi(t) = (c_1 + c_2 t)e^{-t/2\tau} + b_1 \cos \omega t + b_2 \sin \omega t \qquad (3.55)$$

The coefficients b_1 and b_2 are found by substituting $\xi(t)$ from (3.55) into Equation 3.54. The result is

$$b_1 = \frac{a_0(\omega_0{}^2 - \omega^2)}{[(\omega_0{}^2 - \omega^2)^2 + (\omega/\tau)^2]} \qquad b_2 = \frac{a_0\omega/\tau}{[(\omega_0{}^2 - \omega^2)^2 + (\omega/\tau)^2]} \qquad (3.56)$$

In order to see conveniently the effect of the transient response, we shall set $\omega = \omega_0$, $\xi(0) = A > 0$, and $v(0) = 0$. Then Equation 3.55 becomes

$$\xi(t) = \left[A + \left(\frac{A}{2\tau} - a_0\tau\right)t\right]e^{-t/2\tau} + \frac{a_0\tau}{\omega_0}\sin \omega_0 t \qquad (3.57)$$

Equation (3.57) is graphed in Figure 3.7. There we can see that the initial conditions and the damping force at first combine to distort the periodicity of the displacement, but that the transient contribution becomes insignificant after a time equal to about 10 time constants, when the response "settles down" to become a sine function. From that point on the steady-state contribution—which shows no effect of the initial conditions—is the dominant term in the displacement. This behavior is typical of any damped oscillator driven by a periodic force.

In order to get a more complete idea of the character of the steady-state part of $\xi(t)$, we shall study its dependence on the frequency of the driving force and the

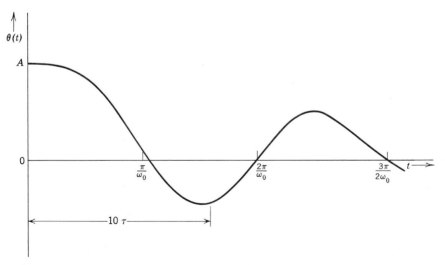

FIGURE 3.7. The displacement of a critically damped oscillator driven by $F(t) = F_0 \cos \omega_0 t$.

quality factor. Our investigation is made simpler by defining

$$\sin \delta \equiv \frac{\omega/\omega_0 Q}{[(1 - (\omega/\omega_0)^2)^2 + (1/Q^2)(\omega/\omega_0)^2]^{1/2}} \tag{3.58}$$

$$\cos \delta \equiv \frac{1 - (\omega/\omega_0)^2}{[(1 - (\omega/\omega_0)^2)^2 + (1/Q^2)(\omega/\omega_0)^2]^{1/2}} \tag{3.59}$$

With these equations put into it, the steady-state solution of Equation 3.54 is

$$\xi_S(t) = A(\omega) \cos (\omega t - \delta) \tag{3.60}$$

where

$$A(\omega) = \frac{a_0/\omega_0{}^2}{[(1 - (\omega/\omega_0)^2)^2 + (1/Q^2)(\omega/\omega_0)^2]^{1/2}} \tag{3.61}$$

$$\tan \delta = \frac{\omega/\omega_0 Q}{1 - (\omega/\omega_0)^2} \tag{3.62}$$

The amplitude of the steady-state oscillation, $A(\omega)$, is portayed in Figure 3.8, for several values of Q, as a function of the frequency of the driving force. When the oscillator is critically damped or overdamped ($Q \leqslant 1/2$), the amplitude is largely insensitive to changes in the rate of driving and decreases slowly to zero with increasing ω, starting from a low maximum at zero frequency. On the other hand, an underdamped oscillator ($Q > 1/2$) exhibits an amplitude that has a pronounced

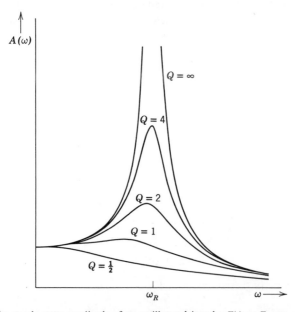

FIGURE 3.8. The steady-state amplitude of an oscillator driven by $F(t) = F_0 \cos \omega t$.

maximum at the angular frequency ω_R defined implicitly by the condition

$$\left(\frac{dA}{d\omega}\right)_{\omega=\omega_R} = 0 \tag{3.63}$$

According to Equation 3.61, this frequency is

$$\omega_R = \omega_0\left(1 - \frac{1}{2Q^2}\right)^{1/2} \tag{3.64}$$

and the corresponding peak amplitude is

$$A(\omega_R) = \frac{a_0 Q}{\omega_0{}^2[1 - (1/4Q^2)]^{1/2}} = \frac{a_0 Q}{\omega_0 \omega_d} \tag{3.65}$$

where ω_d is the angular frequency of the underdamped oscillator. As the damping is decreased (i.e., as Q increases) the frequency ω_R approaches the natural angular frequency of the oscillator and the height of the maximum in $A(\omega)$ becomes infinitely large. Regardless of the amount by which Q is greater than one half, when the frequency of the driving force has been adjusted to produce a maximum in the steady-state amplitude, the oscillator is said to be in a condition of maximum response or *resonance*. The resonance frequency is, of course, given in Equation 3.64 and, for quality factors equal to 10 or more, it is negligibly different in value from the natural angular frequency of the oscillator.

In Figure 3.9 graphs of the phase constant δ are plotted as a function of the driving frequency for fixed values of the quality factor. We see that, when the quality factor is large, the phase constant remains nearly equal to zero until resonance, at which it suddenly takes on the value $\pi/2$ radians, then shoots up almost to π radians and makes a gradual ascent to that limiting value. A look at Equations 3.53 and 3.60 tells us that this behavior corresponds to a displacement that is in phase with the driving force for low-driving frequencies, is one quarter of a period behind it at resonance, and is about half a period behind it thereafter. At low frequencies, then, the oscillator is merely responding in a natural way to a force which would set it into motion. The driving force pulls and pushes on the oscillator with a frequency ω and it responds quickly (because the damping is so light), through its own restoring force, to vibrate in accordance with this disturbance. As the frequency of the driving force increases, the oscillator is less able to follow it and the displacement begins to lag behind the driving force. When the oscillator happens to lag by exactly one quarter of a period, it is in the rare situation of having its *velocity* in phase with the driving force, since $v(t)$ always leads the displacement by $\pi/2$ radians. In this situation the driving force continuously acts on the oscillator in a direction parallel with the direction of motion. This efficient use of the force then causes the vibrations of the particle to have their greatest possible amplitude. As the frequency of the driving force increases further, the oscillator lags further behind and, since its velocity no longer is in a critical phase

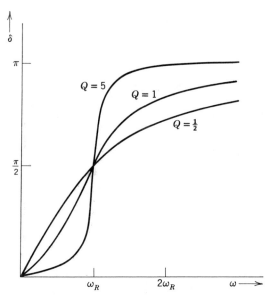

FIGURE 3.9. The steady-state phase constant for an oscillator driven by $F(t) = F_0 \cos \omega t$.

relationship with $F(t)$, the oscillations soon drop sharply in amplitude. In effect what has happened is that the oscillator progressively is unable to follow the rapid variations of the driving force and so becomes almost oblivious to the disturbance.

When the quality factor is near the critical value of 0.5, much of what has just been said is still pertinent, with the exception that the changes in the phase factor as the driving frequency increases are more gradual because of the stronger damping force. The oscillator is now more retarded in its response to the driving force. This sluggishness results in a less sudden approach to resonance and a pronounced lowering of the peak amplitude, as shown in Figures 3.8 and 3.9.

Another physically interesting driving force is the impulsive force, expressed by

$$F(t) = \begin{cases} F_0 > 0 & (0 \leqslant t < t_0) \\ 0 & (t_0 \leqslant t) \end{cases} \tag{3.66}$$

where it is to be understood that t_0 is not much greater than zero. A mass-on-a-spring oscillator driven by an impulsive force would appear to be one subjected to a tug of short duration that sets it into motion. We shall look for a verification of this intuitive idea by investigating the behavior of an initially quiescent, underdamped oscillator acted on by $F(t)$. Because the driving force here is a discontinuous function of the time, we must write the Second Law in the form

$$\frac{d^2\xi}{dt^2} + \frac{1}{\tau}\frac{d\xi}{dt} + \omega_0^2\xi(t) = a_0 \qquad (0 \leqslant t < t_0) \tag{3.67a}$$

$$\frac{d^2\xi}{dt^2} + \frac{1}{\tau}\frac{d\xi}{dt} + \omega_0{}^2\xi(t) = 0 \qquad (t_0 \leqslant t) \tag{3.67b}$$

The conditions on $\xi(t)$ are

$$\xi(0) = 0 \qquad v(0) = 0$$

and the continuity requirement, that $\xi(t_0)$ and $v(t_0)$ computed from Equations 3.67a and 3.67b be the same in each case. Equation 3.67a can be solved either by noting that, since its complementary function, given by Equation 3.44, has no constant term, the particular solution $\xi_S(t)$ is just a constant like $a(t)$ itself, or by shifting $\xi(t)$ to remove the inhomogeneous term a_0, as we have done before on several occasions. Whichever method we choose, the result for the displacement is

$$\xi(t) = \frac{a_0}{\omega_0{}^2} + Ce^{-t/2\tau}\cos(\omega_d t - \delta) \qquad (0 \leqslant t < t_0)$$

Upon applying the initial conditions we find that this general solution becomes

$$\xi(t) = \frac{a_0}{\omega_0{}^2}\left[1 - \frac{(1 + 4(\omega_d\tau)^2)^{1/2}}{2\omega_d\tau}e^{-t/2\tau}\cos(\omega_d t - \delta)\right]$$

where

$$\tan\delta = \frac{1}{2\omega_d\tau}$$

We shall assume, for purposes of illustration, that the quality factor is very large, so that $\omega_d\tau \simeq \omega_0\tau = Q \gg 1$ and $\delta \simeq 0$. Then our solution reduces to

$$\xi(t) \simeq \frac{a_0}{\omega_0{}^2}[1 - e^{-t/2\tau}\cos\omega_0 t] \qquad (0 \leqslant t < t_0) \tag{3.68}$$

Now, the solution of Equation (3.67b) is just

$$\xi(t) = Ce^{-t/2\tau}\cos(\omega_d t - \delta) \qquad (t_0 \leqslant t) \tag{3.69}$$

where the constants C and δ are to be determined from the continuity requirement derived from Equation 3.68 at $t = t_0$. This is

$$\xi(t_0) = \frac{a_0}{\omega_0{}^2}[1 - e^{-t_0/2\tau}\cos\omega_0 t_0]$$

$$v(t_0) = \frac{a_0}{\omega_0{}^2}\left[\frac{1}{2\tau}\cos\omega_0 t_0 + \omega_0\sin\omega_0 t_0\right]e^{-t_0/2\tau}$$

Again we shall simplify the calculation. This time we shall assume that the impulsive force is applied for exactly one quarter of a period. This makes $t_0 = \pi/2\omega_0$ and reduces the continuity condition to

$$\xi(t_0) = \frac{a_0}{\omega_0{}^2} \qquad v(t_0) = \frac{a_0}{\omega_0}e^{-\pi/4\omega_0\tau} \simeq \frac{a_0}{\omega_0}$$

the last step coming from the condition of light damping. With these two equations imposed on (3.68), we find

$$\frac{a_0}{\omega_0{}^2} = Ce^{-\pi/4\omega_0\tau} \cos\left(\frac{\pi}{2} - \delta\right) \simeq C \sin\delta$$

$$-\frac{a_0}{\omega_0} = C\left(\omega_0 \cos\delta - \frac{1}{2\tau}\sin\delta\right)e^{-\pi/4\omega_0\tau} \simeq C\omega_0 \cos\delta$$

which gives us $\delta \simeq (\pi/4)$ and $C \simeq \sqrt{2}a_0/\omega_0{}^2$. Therefore,

$$\xi(t) \simeq \frac{\sqrt{2}a_0}{\omega_0{}^2}e^{-t/2\tau}\cos\left(\omega_0 t - \frac{\pi}{4}\right) \qquad \left(\frac{\pi}{2\omega_0} \leqslant t\right) \qquad (3.70)$$

is the approximate solution of Equation 3.67b, subject to $Q \gg \pi/4$. Equations 3.68 and 3.70 show that the impulsive force brings about a positive displacement of the underdamped oscillator initially in equilibrium and that this displacement dies out in the usual way after the force has ceased to act. This behavior is precisely what was expected on physical grounds (see Figure 3.10).

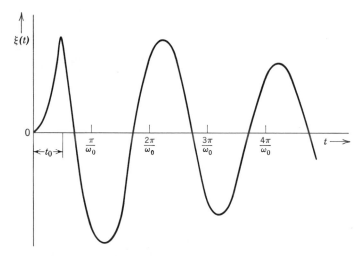

FIGURE 3.10. The displacement of an underdamped oscillator subjected to an impulsive force at time zero. The quality factor has been set equal to 10.

Finally, let us consider an underdamped oscillator subjected to the driving force

$$F(t) = F_0 e^{-t/2\tau} \qquad (F_0 > 0)$$

where τ is the Stokes time constant for the particle. This force acts to keep the oscillator from returning to its equilibrium position and does so with a resistance that diminishes as the oscillations themselves do. The Second Law in this case is

$$\frac{d^2\xi}{dt^2} + \frac{1}{\tau}\frac{d\xi}{dt} + \omega_0{}^2\xi(t) = a_0 e^{-t/2\tau} \qquad (3.71)$$

The complementary function for this equation has no terms in common with $a(t)$. The particular solution of (3.71) therefore must have the form

$$\xi_s(t) = b e^{-t/2\tau}$$

where b is a constant. [No derivatives of $a(t)$ appear in $\xi_s(t)$ because they are proportional to $\exp(-t/2\tau)$.] Upon introducing $\xi_s(t)$ into (3.71) we find

$$b = \frac{4a_0\tau^2}{4Q^2 - 1} \qquad (Q > \tfrac{1}{2})$$

It follows that

$$\xi(t) = \left[\frac{4a_0\tau^2}{4Q^2 - 1} + C\cos(\omega_d t - \delta)\right]e^{-t/2\tau}$$

is the general solution of Equation 3.71. To bring out dramatically the character of the driving force, we shall choose the initial conditions

$$\xi(0) = \frac{8a_0\tau^2}{4Q^2 - 1} \qquad v(0) = \frac{-4a_0\tau}{4Q^2 - 1}$$

These transform the general solution into

$$\xi(t) = \frac{4a_0\tau^2}{4Q^2 - 1}(1 + \cos\omega_d t)e^{-t/2\tau}$$

$$= \frac{8a_0\tau^2}{4Q^2 - 1}e^{-t/2\tau}\cos^2\left(\frac{\omega_d t}{2}\right) \qquad (3.72)$$

We see that the driving force is just strong enough, under the given conditions at zero time, to prevent the oscillator from ever getting past its equilibrium position. Instead the particle turns around near the equilibrium position and continues to vibrate with an exponentially-decreasing amplitude, as shown in Figure 3.11.

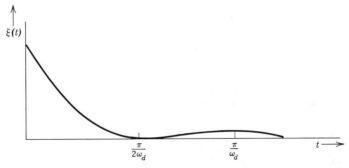

FIGURE 3.11. The displacement of an underdamped oscillator driven by an exponentially decreasing force whose time constant is equal to that for the damping force acting on the oscillator.

FOR FURTHER READING

T. C. Bradbury
Theoretical Mechanics, Wiley, New York, 1968. Chapter 4 contains an extensive discussion of dissipative forces, damped and driven oscillators, and electrical oscillations.

K. R. Symon
Mechanics, Addison-Wesley, Reading, Mass., 1971. Chapter 2 gives a solid introductory discussion of damped and driven oscillators.

G. K. Batchelor
An Introduction to Fluid Dynamics, Cambridge University Press, London, 1970. Sections 4.9 and 4.10 comprise a rather complete introductory discussion of fluid friction. In particular, a derivation of the Stokes dissipative force is given.

S. J. Williamson
Fundamentals of Air Pollution, Addison-Wesley, Reading, Mass., 1973. This fine textbook contains a discussion on particle size fractionation by gravity in Chapter 11.

G. F. Feinberg
"Fall of Bodies near the Earth," *American Journal of Physics* 33: 501–502 (1965). This short article deals with the problem of indirectly observing the mass-independence of g for objects not falling in a vacuum.

P. L. Willmore
"The Detection of Earth Movements" in *Methods and Techniques of Geophysics* (ed. by S. K. Runcorn), Wiley, New York, 1960, Volume I, pp. 230–276. This is a quite readable introduction to seismographs and the physics of their design.

J. R. Barker
Mechanical and Electrical Vibrations, Barnes and Noble, New York, 1964. Chapters 2 through 5 of this short book provide a very fine introduction to damped, driven oscillators of all kinds: mechanical vibrators, rotating flywheels, pendula, electric circuits, and the like.

A. A. Andronov,
A. A. Vitt, and
S. E. Khaikin
Theory of Oscillators, Addison-Wesley, Reading, Mass., 1966. Chapters I, III, and V give an extensive discussion of damped and driven oscillators, with special emphasis on the electric circuit analogies.

PROBLEMS

1. The dissipative force on a sphere of radius r that moves through a fluid of density ρ and viscosity η is expressed by

$$F = -6\pi\eta rv\left(1 + \frac{3\rho rv}{8\eta}\right) \quad (v > 0)$$

Calculate the velocity of a particle subject to this force, then compute the half-life of the motion and compare your result with Equation 3.10.

2. Calculate the velocity of a particle subject to the general dissipative force

$$F = -\alpha_n v^n \quad (v > 0)$$

where n is any positive integer greater than one. Compute the half-life of the motion and sketch a graph of it as a function of n.

3. An experiment with fluid friction has indicated that the dissipative force may have the form

$$F = -F_0 e^{\beta v} \qquad (v > 0)$$

where F_0 and β are positive constants.
  (a) Calculate the velocity of a particle subject to this force.
[*Hint.* Set $\beta v(t) = \ln u(t)$ in the Second Law.]
  (b) Expand the force in a Maclaurin series to first order in the velocity and compute $v(t)$ once again. Compare and interpret your result in relation to what you found in (a).
4. Compute the position of a particle subject to Newtonian damping. Investigate your result in the limit $t \to \infty$ and compare your findings with Equation 3.11 considered in the same limit.
5. Show that, for Newtonian damping, the velocity of a particle decreases exponentially with the distance traveled. How can this result be used to estimate the Newtonian friction coefficient α?
6. The local office of the State Highway Patrol wishes to have a chart showing how many times longer is the distance an automobile continuously braked will travel before stopping on wet pavement as compared with dry pavement. You respond to the query by stating that you can provide figures for the chart for any initial speed if you are told only the coefficients of kinetic friction for rubber tires on wet and dry pavements. How do you know this?
7. A glass of beer of total mass 0.8 kg is slid by a bartender toward a customer 1.5 m down the bar. What initial velocity should the bartender give the glass in order that it stop right in front of the customer? The value of μ for glass on polished wood is 0.5.
8. In the list below are some data taken in order to determine the combined effect of water and air resistance on the gliding to rest of a model boat. Use these data to make an hypothesis about the velocity-dependence of the effective frictional force.

Time (s)	Velocity (m/s)			
	Trial 1	Trial 2	Trial 3	Trial 4
0	0.60	0.90	1.50	2.40
1	0.50	0.69	1.00	1.32
2	0.43	0.56	0.75	0.92
3	0.38	0.47	0.60	0.71
4	0.33	0.41	0.50	0.57
5	0.30	0.36	0.43	0.48

9. Calculate the decrease in velocity, in m/s, of a 0.1-kg rider that moves the length of a 2-m linear air track after being given a gentle push. Take $\beta = 240\,\eta$ in this case, where η is the coefficient of viscosity of air in cgs units. ($\eta \simeq 1.8 \times 10^{-4}$ g/cm-s.)
10. An automobile braking to a stop is subject to both Coulomb and Stokes damping. The total frictional force is thus

$$F = -\mu mg - \beta v \qquad (v > 0)$$

in terms of the paràmeters discussed in Section 3.1.
  (a) Show that an automobile subject to this force will come to rest in the time

$$t_{stop} = \tau \ln\left(1 + \frac{t_{stop}^c}{\tau}\right)$$

where t_{stop}^c is the "stopping time" for purely Coulomb damping and τ is the Stokes time constant.

(b) Show that the degree to which t_{stop} approximates t_{stop}^c is determined completely by the ratio of the strength of the Stokes friction force to that of the Coulomb friction force at the instant the brakes are applied.

11. Show that the terminal speed of a sphere falling slowly in a viscous medium is proportional to the cross-sectional area of the sphere.

12. The Reynolds number is a quantity that measures the degree of turbulence in a flow of fluid. It is defined by

$$N_R \equiv \frac{vd}{\eta/\rho_f}$$

where v is the average flow speed, d is an appropriate linear dimension for the flow (e.g., a pipe diameter), η is the viscosity, and ρ_f is the density of the fluid.

(a) For a sphere falling in a viscous fluid, v is the terminal speed and d is the diameter of the sphere. (In the frame of reference of the sphere, the fluid flows upward with speed v_T past an obstacle of diameter d.) Show that, for Stokes friction,

$$N_R = \frac{gd^3}{18(\eta/\rho_f)^2}\left(\frac{\rho}{\rho_f} - 1\right)$$

where ρ is the density of the sphere

(b) It is observed experimentally that the expression for the terminal speed derived from the Stokes frictional force does not work for particles falling in air if $N_R > 1$. Calculate the largest raindrop diameter for which the expression is accurate. Take $\rho/\rho_f \simeq 780$ and $\eta/\rho_f \simeq 0.15$ cm^2/s in your computation.

13. Repeat the calculation in Problem 12b for mineral particles falling in water. Take $\rho/\rho_f \simeq 2.7$ and $\eta/\rho_f \simeq 0.01$ cm^2/s.

14. How quickly will a clay-sized particle ($d \leqslant .002$ mm) achieve terminal speed falling in water? Assume $v_T \simeq v(t = 5\tau)$ and take $\rho \simeq 2.8$ g/cm^3, $\eta \simeq 10^{-2}$ cgs units (poise).

15. The turbidity of lake and stream waters—often a serious environmental problem—depends on the concentration of solid particles suspended in them. To understand the persistence of small particles in suspension, it is instructive to calculate the time it takes a silt-sized particle (.002 mm $< d <$ .05 mm) to settle 1 m at terminal speed. Do this, using the data given in Problem 14. Assume the particles to be spherical, of course.

16. In a famous remark about the rate of fall of objects toward the Earth's surface, Galileo states that two iron balls, one of weight 100 lb and one of weight 1 lb, will not be separated by more than "two finger-breadths" after falling 200 ft. Check this prediction for objects falling in air by the following method.

(a) Calculate the friction coefficient α for each iron ball, given $\alpha \simeq 10^{-3} R^2$, where R is the ball radius in cm, and the fact that the density of iron is 7.6 g/cm^3.

(b) Compute the Newtonian time constant and use the result to simplify Equation 3.36 applied to an object falling from rest.

(c) Derive an equation for the vertical separation between the balls at any instant, given that they are dropped from the same height at the same time. Assume the separation is small compared with the total distance fallen.

(d) Did Galileo do the experiment?

17. In the following list are some data on the free fall from rest of a sky diver. Assume Newtonian damping and compute the terminal speed of the diver. The combined mass of the diver and his equipment is 118.7 kg.

$-[x(t) - x(0)]$	t
0.189 km	9.9 s
0.341	14.5
0.777	19.1
1.128	23.7
1.600	28.3
2.073	32.9

18. Solve the differential equation (3.31) by quadrature without using the shift method employed in Section 3.3.

19. Presumably the Newtonian frictional force provides a better description of the settling of large particles in air than does the Stokes frictional force. Calculate the terminal speed of a grain of fine sand ($d = 0.1$ mm) according to both types of force and compare your results. Take $\rho/\rho_f \simeq 2100$, $\rho_f \simeq 1.29$ g/m^3, $\eta/\rho_f \simeq 0.15$ cm^2/s, and $\alpha = 2.55 \times 10^{-4} d^2$ (d in cm).

20. After the electric current to be detected by a ballistic galvanometer has ceased, the angular displacement ϕ of the galvanometer coil from its position of equilibrium satisfies the differential equation

$$J \frac{d\omega}{dt} + B\omega(t) + K\phi(t) = 0$$

J is the moment of inertia of the moving coil about an axis through its center, $\omega(t) = d\phi/dt$ is the angular velocity of the coil, B is a parameter that represents the retarding force on the coil produced by air resistance and by motionally induced electromotive force, and K is the force constant for the torsion suspension that restores the coil to its equilibrium position after a displacement occurs.

(a) Calculate τ, ω_0, and Q for the galvanometer.

(b) Calculate the frequency of underdamped oscillations. What is the condition for critical damping in terms of J, B, and the like?

21. A pendulum seismograph must be calibrated in order to interpret unambiguously its response to an earthquake. Often this is done by observing the movement of the seismometer boom after applying a specific type of disturbance that has been arranged to provide a measured value of one of the parameters ω_0 or τ. For example, with the damping system removed, the free oscillations of the boom are timed to obtain ω_0. However, it may happen that the oscillations will still be damped by air resistance and residual friction in this case, and that a period T_d will be observed instead of the desired period T. Show that the true natural period then can be found from the expression

$$\frac{1}{T^2} = \frac{1}{T_d^2}\left(1 + \frac{\Delta^2}{4\pi^2}\right)$$

where Δ is the observed logarithmic decrement of the swinging boom.

22. An interesting kind of damped harmonic oscillator is a heavy ball bearing of radius r that oscillates under the force of gravity near the bottom of a shallow, spherical dish of radius

R. In the absence of friction and slipping, the equation of motion for the bearing has the form

$$\frac{d^2\theta}{dt^2} + \omega_0^2\theta(t) = 0$$

where $\omega_0^2 = 5\ g/7(R - r)$ and $\theta(t)$ is the (small) angle between a radial line drawn from the center of curvature of the dish to its bottom and one drawn from the center of curvature to the center of the bearing. (Thus the ball bearing is a special case of a plane pendulum.) Coulomb friction will damp the oscillations of the bearing.

(a) Write down the equation of motion for the bearing subject to Coulomb damping. You may consider $d\theta/dt$ to be positive when the bearing rolls up to the right in the dish.

(b) Write down a general solution for $\theta(t)$ that will be valid for either possible sign of $d\theta/dt$. Indicate how the arbitrary constants in this solution will be determined using the initial conditions (e.g., $\theta(0) = \theta_0 > 0$, $(d\theta/dt)_{t=0} = 0$) and the requirement that $\theta(t)$ be a continuous function of the time.

(c) Sketch a graph of $\theta(t)$ as you would expect it to be. It is not necessary to solve the equation of motion completely beforehand.

23. In the case of simple harmonic ground motion the Second Law for the seismograph becomes

$$\frac{d^2\theta}{dt^2} + \frac{1}{\tau}\frac{d\theta}{dt} + \omega_0^2\theta(t) = \omega^2 A \cos \omega t$$

where A is a positive constant related to the mechanical properties of the instrument and to the amplitude of ground motion. ω is the angular frequency of ground oscillation. Show that, for very small values of ω (very slow ground oscillations) and after transient effects have died out, $\theta(t)$ is proportional to the ground acceleration.

24. Determine the particular solution $\theta_S(t)$ for a critically damped seismograph driven by the ground acceleration

$$a(t) = a_0 e^{-t/2\tau_E} \cos \omega t$$

where a_0 is a positive constant and τ_E is a time constant which expresses the rate of attenuation of ground motion.

25. When an electric current flows through a ballistic galvanometer, the "driving force" $A\ i(t)$ must be added to the right-hand side of the equation of motion given in Problem 20. A is a constant whose value depends on the nature of the coil and the static magnetic field in which it is moving. Calculate $\phi(t)$ for a critically damped galvanometer through which a capacitor has discharged:

$$i(t) = i(0)e^{-t/\tau_D}$$

where $i(0) > 0$ and τ_D is the time constant for the discharge. The initial conditions are: $\phi(0) = 0$, $\omega(0) = 0$.

26. A critically damped ballistic galvanometer is subjected to the current pulse

$$i(t) = \begin{cases} i_0 & 0 < t < t_0 \\ 0 & t_0 \leqslant t \end{cases}$$

where t_0 is very small. Calculate $\phi(t)$ given the initial conditions $\phi(0) = 0$, $\omega(0) = 0$.

27. In the classical atom an electron is considered to be elastically bound to an equilibrium position and to be subject to a dissipative force arising from the radiation it emits when

accelerated. The equation of motion of the electron when the atom is exposed to monochromatic light of angular frequency ω is

$$\frac{d^2\xi}{dt^2} + \frac{1}{\tau}\frac{d\xi}{dt} + \omega_0{}^2\xi(t) = \frac{e}{m}E_0 \cos \omega t$$

where e/m is the charge-to-mass ratio of the electron and E_0 is the electric field intensity created by the light.

(a) Show that the steady-state response of the electron can be separated into a part in phase with the driving force (elastic response) and a part $\pi/2$ radians out of phase with the driving force (absorptive response).

(b) Show that the elastic response disappears and the absorptive response is maximum when the frequency of the impinging light and the natural frequency ω_0 are equal and the quality factor is very large. Under this condition there is no radiation from the electron at the frequency of the incident light (there are no in-phase oscillations) and a medium composed of these classical atoms will be opaque.

28. Antivibration mountings are springs set between the base of a machine or instrument and the floor. In the case of a rotating device that is slightly unbalanced the mountings receive a periodic application of force $F = F_0 \sin \omega t$, where F_0 is a positive constant and ω is the angular speed of rotation.

(a) Sketch a plot of the steady-state amplitude of vibration of the mounting as a function of ω. Assume that the mounting is underdamped.

(b) In what range of frequency is the mounting a satisfactory antivibration device? What recommendation would you make regarding the magnitude of ω_0?

29. A simple spring-platform scale is to be constructed so that a 200-kg load will cause a downward deflection of 2 cm. Assume that the scale is critically damped and calculate the parameters k and β for the system. The mass of the platform is 5 kg.

30. A brick wall near a busy intersection is fitted with a well-anchored spring snubber that compresses upon head-on impact by an automobile. If the spring constant is 2×10^4 kg/s^2, calculate the value of β for critical damping upon impact by a freely rolling automobile moving with a speed of 2 m/s. The mass of the auto is 1500 kg. How important is critical damping in this application?

31. A cylindrical object, moving vertically along the direction of its axis of symmetry in a viscous liquid, is a type of damped harmonic oscillator. (See Problem 22 in Chapter 2.) The natural frequency of oscillation is given by $\omega_0 = (\rho_f g/\rho l)^{1/2}$, where ρ is the density of the object, l is its length, and ρ_f is the density of the liquid. The damping force coefficient β is equal to $4\pi\eta R$, where R is the cylinder radius. Calculate the "critical mass" of the cylinder that will lead to critical damping of the oscillations in water ($\eta \simeq 10^{-2}$ g/cm-s) and thereby show that any macroscopic cylinder will, in fact, be underdamped.

32. Prove the following statement: *the only effect of a uniform gravitational field on an harmonic oscillator subject to arbitrary dissipative and driving forces is the lowering of the equilibrium position of the oscillator.*

33. The Froude pendulum is a type of plane pendulum where the swinging bob is attached to a rigid rod that, by means of a ring at its end, can be made to move by rotating a shaft whose symmetry axis points perpendicularly to the plane of swing of the pendulum. This arrangement requires that the force of friction of the shaft acting on the ring from which the pendulum is suspended be taken into account in the equation of motion. This new force acts as a driving

mechanism for the pendulum and is, in a first approximation, proportional to the angular velocity of the pendulum. The Second Law for this system, then, has the form

$$ml^2 \frac{d^2\theta}{dt^2} + B\frac{d\theta}{dt} + mgl \sin\theta(t) = F\left(\omega - \frac{d\theta}{dt}\right) \simeq F(\omega) - F'(\omega)\frac{d\theta}{dt}$$

where l is the length of the rigid rod, B is a frictional coefficient, m is the mass of the bob, ω is the angular speed of the shaft, and $F'(\omega) \equiv (\partial F/\partial x)_{x=\omega} < 0$, $x \equiv \omega - (d\theta/dt)$. The driving force F is assumed to decrease as the difference between ω and $d\theta/dt$ increases.

(a) Transform the equation of motion using the approximation $\theta(t) = \theta_E + \xi(t), |\xi(t)| \ll \theta_E$, where θ_E is determined by the equilibrium condition $mgl \sin\theta_E = F(\omega)$. The result should be a linear differential equation for $\xi(t)$.

(b) Calculate and sketch a graph of $\theta(t)$ in each of the two possible cases: $B > |F'(\omega)|$ and $B < |F'(\omega)|$. Physically interpret your results. Take the initial conditions to be $\xi(0) = 0$, $(d\xi/dt)_{t=0} = \omega_0 > 0$.

34. The power absorption of a driven, damped oscillator is defined by

$$P(t) \equiv F(t)v_S(t)$$

where $F(t)$ is the driving force and $v_S(t)$ is the velocity of the steady-state response. Show that the time-averaged value of $P(t)$ is given by

$$\langle P \rangle = \tfrac{1}{2}\frac{F_0 a_0 \omega^2 / \tau}{[(\omega_0{}^2 - \omega^2)^2 + (\omega/\tau)^2]}$$

when $F(t) = F_0 \cos\omega t$. (Recall that

$$\langle P \rangle \equiv \frac{\omega}{2\pi}\int_0^{2\pi/\omega} P(t)\,dt)$$

35. Show that only the *absorptive* response of an electron in a classical atom (see Problem 27) contributes to the mean power absorption $\langle P \rangle$.

36. With the result of Problem 34, show that
(a) the maximum value of $\langle P \rangle$ occurs when $\omega = \omega_0$,
(b) the two values of ω at which $\langle P \rangle$ equals half its maximum value are

$$\omega_+ = \omega_0\left[\left(1 + \frac{1}{4Q^2}\right)^{1/2} + \frac{1}{2Q}\right]$$

$$\omega_- = \omega_0\left[\left(1 + \frac{1}{4Q^2}\right)^{1/2} - \frac{1}{2Q}\right] \qquad \text{and}$$

(c) the relations

$$\Delta\omega = \omega_+ - \omega_- = \frac{1}{\tau} \qquad Q = \frac{\omega_0}{\Delta\omega}$$

are valid. The expressions in (c) are of great practical importance. The first one provides a means of deducing the time constant from a measurement of $\langle P \rangle$ as a function of ω. The second one shows that the quality factor is a direct measure of the sharpness of the power absorption resonance.

37. Calculate the time-averaged value of

$$E_S(t) = \tfrac{1}{2}mv_S{}^2(t) + \tfrac{1}{2}k\xi_S{}^2(t)$$

which is the "total energy" of a driven, damped oscillator in the steady-state, for the driving force $F(t) = F_0 \cos \omega t$. Show that

$$Q = \omega_0 \frac{\langle E_S \rangle}{\langle P \rangle}$$

and interpret this result.

Note. *The following problems make use of the material presented in Special Topic 3.*

38. An *LRC* circuit is to be designed such that an applied potential difference of 100 volts will cause a charge of 0.05 coulomb to build up on the capacitor. Assume that the circuit is critically damped and calculate the parameters C and R for the system. The inductance L is 1.25 millihenry.

39. An inductor and a resistor are arranged in series with a battery as shown in Figure 3.13 (but without the capacitor). With a constant current flowing in the circuit the switch S is moved suddenly from point 2 to point 1. Write down Kirchoff's law for this situation and calculate $i(t)$. The initial condition is $i(0) > 0$.

40. Electric currents in the liquid core of the Earth are believed to be responsible for the terrestrial magnetic field, and the ohmic decay of these currents is presumed to be primarily responsible for the observed decrease to zero (followed by a reversal) of the magnetic field at the rate of about 0.05 percent/year. A simple picture of the "electric circuit" in the liquid core has it that electrons spiral around the core, which has both an inner and outer boundary, like wire wound around on a torus. Thus the "circuit" is equivalent to an *LR* circuit, as discussed in Problem 39, and the decay time constant may be computed readily. Given the estimates $R \simeq 1.19 \times 10^{-10}$ ohm and $L \simeq 3.32$ henry, calculate the time constant for the core currents and compare your result with the observed rate of decay.

SPECIAL TOPIC 3
Oscillations in an Electric Circuit

The ideas developed in this chapter concerning damped and driven oscillators have a remarkable applicability to the observed behavior of certain types of electric circuit. A careful observation of those circuits that contain as "lumped" entities (i.e., as localized devices situated at given points in the circuit) capacitors, inductors, resistors, and sources of electromotive force indicates that the mathematical structure of Newtonian dynamics may be superimposed on circuit theory once a certain analogy is made. This analogy says that *we may associate logically the force acting on a particle at a given point in space and time with the potential difference, or voltage, in a circuit at a given point in space and time.* With this statement as a guiding principle and with an understanding of the character of circuit devices[1] it is possible to construct Table 3.2. In each case listed there, the electrical quantity is analogous to the mechanical quantity in that both display the same *logical* relation to the physical system of which they are a part. Thus we are *not* saying, for example, that voltage is some kind of force or that it has the same

---

[1] A good discussion of the basic circuit elements is found in W. T. Scott, *The Physics of Electricity and Magnetism*, Wiley, New York, 1966.

TABLE 3.2 A Table of Logical Correspondence Between Certain Electrical and Mechanical Quantities

Mechanical Quantity		Electrical Quantity	
Force	$F(x, v, t)$	Voltage	$V(q, i, t)$
Mass	m	Self-inductance	L
Position	$x(t)$	Charge	$q(t)$
Velocity	$v(t)$	Current	$i(t)$
Kinetic energy	$\frac{1}{2}mv^2$	Magnetic energy	$\frac{1}{2}Li^2$
Oscillator potential energy	$\frac{1}{2}kx^2$	Capacitor electric energy	$\frac{1}{2}q^2/C$
Power	Fv	Power	iV

dimensions as force. Instead, we are saying that self-inductance is to voltage as mass is to force, and so on. This fact should become more obvious as we develop the analogies through examples.

Since we know now that the force on a particle is equivalent to the voltage in a circuit, we can go on to create a dynamics for circuits by correspondence with that for particles. To accomplish that we write the Second Law in the so-called Bernoulli-D'Alembert form:

$$\sum_{i=1}^{N} F_i(x, v, t) - ma(t) = 0 \qquad (S.1)$$

This equation states that the algebraic sum of the $N + 1$ forces on a particle, with ma considered to be a force, equals zero. The Bernoulli-D'Alembert principle in itself, of course, tells us nothing new about mechanics; but when we transcribe it into the language of circuit theory using Table 3.2 it becomes

$$\sum_{i=1}^{N+1} V_i(q, i, t) = 0 \qquad (S.2)$$

Equation S.2 says that the algebraic sum of potential differences at a given place and at a given instant in a circuit must equal zero.[2] This statement is indeed known to be accurate for circuits with lumped capacitance, inductance, and resistance, and it generally goes under the name *Kirchoff's voltage law*. The dynamics of these kinds of circuits, by analogy with the Newtonian program as outlined in Section 1.1, is investigated by first specifying the dependence on charge, current, and time of each voltage rise or drop contributing to the total voltage. (The *positive* sign is given to a voltage *rise*.) When this has been done, the results are put into Equation S.2, which then becomes a differential equation either for the charge or the current. The equation is then solved and the behavior of the circuit has been established analytically.

---

[2] By convention we do not single out the voltage drop across the inductor in Equation S.2 in the way that *ma* is singled out in the Bernoulli-D'Alembert principle.

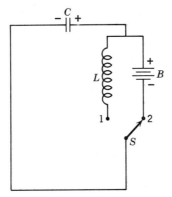

FIGURE 3.12. An *LC* circuit. With the switch *S* connected to point 1, the capacitor discharges through the inductor.

We can best get a feeling for this program by investigating some representative circuits. The first one we shall consider is diagrammed in Figure 3.12. It comprises a capacitor and an inductor arranged in series and is generally known as an *LC* circuit. When the switch *S* connects to point 2, a charge is placed on the capacitor until the voltage drop across its plates cancels the voltage rise due to the battery *B*. From that time on, everything remains quiescent until the switch connects to point 1, whereupon the capacitor discharges through the inductor. For this process, Equation S.2 takes on the form[1]

$$-L\frac{di}{dt} - \frac{q(t)}{C} = 0 \qquad (S.3)$$

where *L* is the self-inductance and *C* is the capacitance. (The commonly employed units for these quantities are listed in Table 3.3.) The first term in (S.3) represents the voltage drop across the inductor. If we take note of the fundamental definition

$$i(t) \equiv \frac{dq}{dt} \qquad (S.4)$$

TABLE 3.3   The Units of Certain Electrical Quantities

Quantity	Units	
	MKS	CGS
Charge	Coulomb (C)	Statcoulomb (esu)
Current	Ampere (amp)	Statampere
Potential difference	Volt (V)	Statvolt
Self-inductance	Henry (h)	s²/cm
Capacitance	Farad (F)	cm
Resistance	Ohm (Ω)	s/cm

Equation S.3 may be rewritten entirely in terms of the charge $q(t)$:

$$\frac{d^2q}{dt^2} + \frac{1}{LC}\, q(t) = 0 \tag{S.5}$$

Equation S.5 has precisely the same mathematical form as Equation 2.24, the Second Law for an harmonic oscillator, if we define

$$\omega_0 \equiv (LC)^{-1/2} \tag{S.6}$$

This suggests *immediately* that the charge and the current in the LC circuit follow

$$\left. \begin{aligned} q(t) &= A \cos{(\omega_0 t - \delta)} \\ i(t) &= -A\omega_0 \sin{(\omega_0 t - \delta)} \end{aligned} \right\} \tag{S.7}$$

where, by direct analogy with Equations 2.43,

$$A^2 = q^2(0) + LCi^2(0) \qquad \tan \delta = \frac{(LC)^{1/2}i(0)}{q(0)} \tag{S.8}$$

The mechanical-electrical correspondence thus predicts that the charge at a point in an LC circuit is a periodic function of the time with angular frequency $(LC)^{-1/2}$. This means that positive charge will alternately build up and deplete forever on the plates of the capacitor. Such a prediction makes sense when it is realized that we are assuming that the total electromagnetic energy is conserved (no dissipation in the form of thermal energy) and that the capacitor is acting toward the charge as the agent of the restoring force does toward the displacement of the mechanical oscillator. When a large positive charge builds up on one plate of the capacitor, a large voltage drop is induced that tends to promote discharge through the inductor. This voltage drop, for a given amount of charge, is larger, the smaller is the capacitance, and, therefore, it is $1/C$ which is analogous to the force constant k, as indicated in Table 3.2. The discharge of the capacitor is governed not only by the size of the voltage drop across its plates, but also by the degree of opposition of the inductor to any change in the current. The inductor creates a voltage drop that is proportional to the first derivative of the current with respect to time just as the harmonic oscillator provides a force (in the Bernoulli-D'Alembert sense) that is proportional to the first derivative of the velocity with respect to time, that is, ma. It follows that the self-inductance L is analogous to the mass m as already shown in Table 3.2. The charge in an LC circuit, then, is continually shuttled back and forth between the plates of the capacitor through the combined influence of the voltage drop across it and the "inertial reaction" of the inductor. For the circuit shown in Figure 3.13, the appropriate initial conditions are

$$q(0) > 0 \qquad i(0) = 0 \tag{S.9}$$

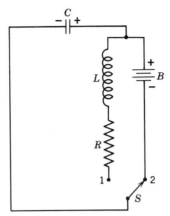

FIGURE 3.13. An LRC circuit. With the switch S connected to point 1, the capacitor discharges through the inductor and the resistor.

With these we easily find that the total electromagnetic energy

$$E = \tfrac{1}{2}Li^2(t) + \frac{1}{2C} q^2(t) = \tfrac{1}{2}\frac{q^2(0)}{C} \qquad (S.10)$$

is a constant of the motion.

Just as there are no harmonic oscillators without damping there are no electric circuits without resistance. We can get a more realistic picture of the circuit we have just described if we include a lumped resistance in series with the inductor and capacitor, as diagrammed in Figure 3.13. When the capacitor is discharging, Kirchoff's law states[1]

$$-L\frac{di}{dt} - i(t)R - \frac{1}{C} q(t) = 0$$

where the second term in the equation represents the voltage drop across the resistor (Ohm's law) and R is the value of the resistance. With the definition of current (S.4) inserted, this expression becomes

$$\frac{d^2q}{dt^2} + \left(\frac{R}{L}\right)\frac{dq}{dt} + \omega_0{}^2 q(t) = 0 \qquad (S.11)$$

where ω_0 is given by Equation S.6. This differential equation is mathematically identical with the Second Law for an harmonic oscillator subject to Stokes damping (Equation 3.40) if we define

$$\tau \equiv L/R \qquad (S.12)$$

to be the time constant for the circuit. As we know from our analysis of the damped oscillator, Equation S.11 has three classes of solution whose forms depend on the value of the quality factor, given in the present context by

$$Q = \omega_0 \tau = (L/R^2 C)^{1/2} \tag{S.13}$$

Therefore, if $L/R \leqslant \frac{1}{4} RC$ the circuit is overdamped or critically damped, according to our previous discussion. In those cases the charge on the capacitor simply decays to zero exponentially as time passes and the electrical energy in the capacitor is converted to thermal energy by the resistor. On the other hand, if $L/R > \frac{1}{4} RC$ the circuit is underdamped and the charge on the capacitor will oscillate to zero with a rapidity controlled by the value of the time constant L/R. In the overdamped or critically damped case the resistance is large enough to quickly dissipate all of the electrical energy in spite of the presence of the inductor. In the opposite case, the resistance is too small to bring about complete dissipation of the electromagnetic energy before the charge has been transferred a few times from one plate of the capacitor to the other.

We can exploit the mechanical-electrical analogy further by considering a sine-wave generator in series with an LRC circuit, as shown in Figure 3.14. Kirchoff's law applied to this circuit reads

$$-L \frac{di}{dt} - i(t)R - \frac{q(t)}{C} + V_0 \sin \omega t = 0 \tag{S.14}$$

where V_0 is the peak voltage of the generator. This expression can be rewritten as

$$\frac{d^2 q}{dt^2} + \frac{1}{\tau} \frac{dq}{dt} + \omega_0{}^2 q(t) = \frac{V_0}{L} \sin \omega t \tag{S.15}$$

Equation S.15 has the mathematical form of the Second Law for an harmonic oscillator subject to Stokes damping and driven by a force proportional to $\sin \omega t$.

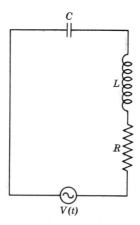

FIGURE 3.14. An LRC circuit driven by a sine-wave generator.

Thus it is in every way comparable with Equation 3.54 if we put

$$a_0 \equiv \frac{V_0}{L} \qquad \text{(S.16)}$$

The solution of (S.16) is found in exactly the same manner as was that of (3.54). The short-lived transient solution is of the form of one of Equations 3.44, 3.49, or 3.51 (depending on the magnitude of the quality factor), with $q(t)$ replacing $\xi(t)$. Since the driving voltage is a sine function we can use Table 3.1 and Equation S.15 to establish the steady-state solution

$$q\,(t) = \frac{a_0(\omega_0{}^2 - \omega^2)}{[(\omega_0{}^2 - \omega^2)^2 + (\omega/\tau)^2]} \sin \omega t - \frac{a_0\omega/\tau}{[(\omega_0{}^2 - \omega^2)^2 + (\omega/\tau)^2]} \cos \omega t \qquad \text{(S.17)}$$

where, of course, a_0, ω_0, and τ are given by Equations S.16, 2.6, and S.12, respectively. As done in Section 3.6, we can make the character of (S.17) more transparent by defining

$$\sin \delta \equiv \frac{\omega/\tau}{[(\omega_0{}^2 - \omega^2)^2 + (\omega/\tau)^2]^{1/2}} = \frac{R\omega/L}{\{[(1/LC) - \omega^2]^2 + (R\omega/L)^2\}^{1/2}}$$

$$\equiv \frac{R}{[(X_C - X_L)^2 + R^2]^{1/2}} \equiv \frac{R}{z(\omega)} \qquad \text{(S.18)}$$

$$\cos \delta \equiv \frac{X_C - X_L}{z(\omega)} \qquad \text{(S.19)}$$

where

$$X_C = \frac{1}{C\omega} \qquad \text{(S.20)}$$

is called the *capacitative reactance*,

$$X_L = L\omega \qquad \text{(S.21)}$$

is called the *inductive reactance*, and

$$z(\omega) = [(X_C - X_L)^2 + R^2]^{1/2} \qquad \text{(S.22)}$$

is called the *impedance* of the circuit. With these definitions employed, the steady-state solution becomes

$$q_s(t) = \frac{V_0}{\omega z(\omega)} \sin (\omega t - \delta) \qquad \text{(S.23)}$$

where

$$\tan \delta = \frac{R}{X_C - X_L} \qquad \text{(S.24)}$$

If the transient solution represents an underdamped circuit there can be a resonance in the steady-state solution. According to Equation 3.64 this will occur at the angular frequency

$$
\begin{aligned}
\omega_R &= \omega_0 \left(1 - \frac{1}{2Q^2}\right)^{1/2} \\
&= (LC)^{-1/2} \left(1 - \frac{R^2 C}{2L}\right)^{1/2} \\
&= \frac{1}{L}\left(X_L X_C - \frac{R^2}{2}\right)^{1/2}
\end{aligned}
\qquad \text{(S.25)}
$$

The resonance amplitude of $q_s(t)$ is, according to Equation 3.65,

$$
A(\omega_R) = \frac{V_0}{\omega_R z(\omega_R)} = \frac{V_0 Q/L}{\omega_0^2 [1 - (1/4Q^2)]^{1/2}} = \frac{V_0 L/R}{[X_L X_C - (R^2/4)]^{1/2}} \qquad \text{(S.26)}
$$

which increases without limit as the resistance in the circuit approaches zero.

4

Bounded Motion in
Three Dimensions

4.1 CLASSICAL DYNAMICS IN THREE-DIMENSIONAL SPACE

In order to describe the motion of a single particle in three-dimensional space we must extend the mathematical language we have employed up to this point. For the case of a particle coursing along a straight line it has been possible to represent the dynamical variables, position and velocity, by single coordinates whose numerical values are dependent on the time. Now we have to enlarge the number of coordinates to *three* and write, for example,

$$\mathbf{r} = \{x_1, x_2, x_3\}$$

for a position in three-dimensional space. This extension of our concepts immediately raises the question of an algebra. For example, how do we prescribe the sum of two positions, or their difference? How do we multiply these quantities? How do we relate position unambiguously to velocity? The answer to these queries is by no means obvious. It is clear that, say, the addition of two positions defined as is **r** above cannot be carried through in just the same way as was that of two values of x, because **r** is associated with a *set* of three numbers, not just a single coordinate. As is explained in detail in Section A.1 of the Appendix, and as the reader presumably knows, the solution of our problem lies with the concept of a *vector*. Very briefly, a vector is a mathematical object that is a member of a set that satisfies four axioms. These axioms permit the definition of vector addition, which is given the symbol **+** to distinguish it from +, the addition of ordinary numbers; of multiplication of a vector by a number (scalar); of the zero vector **0**, and of the inverse of a vector. Besides these defining axioms we give to a set of vectors the property of multiplication, denoted by the centered dot symbol ·. This multiplication of two vectors produces a scalar rather than a vector and makes possible the necessary geometric interpretation of the position vector **r**. Again, these basic ideas are discussed in detail in the Appendix and will be assumed to be understood from now on. The reader who wishes additional review should read Section A.1 before continuing.

Given the concept of a vector and its geometric picture as a directed line segment, we shall begin our discussion by presenting the definitions of the kinematical and dynamical quantities that were introduced as scalars in Chapter 1.

Position   The position vector of a particle is specified by three continuous and twice-differentiable functions of the time called position coordinates. Position may be represented in a purely algebraic manner as the set $\{x_1(t), x_2(t), x_3(t)\}$ or, in a more useful geometric fashion, as the vector sum

$$\mathbf{r}(t) = x_1(t)\hat{\mathbf{x}}_1 + x_2(t)\hat{\mathbf{x}}_2 + x_3(t)\hat{\mathbf{x}}_3 \tag{4.1}$$

where $\{\hat{\mathbf{x}}_1, \hat{\mathbf{x}}_2, \hat{\mathbf{x}}_3\}$ is a set of three orthonormal, cartesian basis vectors and **+** means addition of directed line segments, as discussed in the Appendix (see Figure 4.1).

Velocity   The velocity vector of a particle is specified by the set of three coordinates $\{v_1(t), v_2(t), v_3(t)\}$, where

$$v_i(t) = \frac{dx_i}{dt} \quad (i = 1, 2, 3) \tag{4.2}$$

The velocity can be represented geometrically by the vector sum

$$\mathbf{v}(t) = v_1(t)\hat{\mathbf{x}}_1 + v_2(t)\hat{\mathbf{x}}_2 + v_3(t)\hat{\mathbf{x}}_3 \tag{4.3}$$

where $v_i(t)\hat{\mathbf{x}}_i$ is the ith component of $\mathbf{v}(t)$. Thus, whenever the vector expression

$$\mathbf{v}(t) = \frac{d\mathbf{r}}{dt} \tag{4.4}$$

appears, it is understood to summarize Equation 4.3, by the three Equations 4.2.

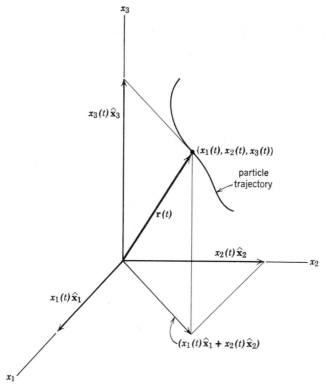

FIGURE 4.1. The position vector as a directed line segment.

Acceleration The acceleration vector of a particle is represented either by the set $\{a_1(t), a_2(t), a_3(t)\}$ of first derivatives of the velocity coordinates with respect to the time, or by the vector sum

$$\mathbf{a}(t) = a_1(t)\hat{\mathbf{x}}_1 + a_2(t)\hat{\mathbf{x}}_2 + a_3(t)\hat{\mathbf{x}}_3 \tag{4.5}$$

The vector equation

$$\mathbf{a}(t) = \frac{d\mathbf{v}}{dt} \tag{4.6}$$

is understood accordingly to be a shorthand notation for the three equations

$$a_i(t) = \frac{dv_i}{dt} \quad (i = 1, 2, 3) \tag{4.7}$$

Force The force on a particle is a vector function of the position, velocity, and the time. By this is meant

$$\mathbf{F}(\mathbf{r}, \mathbf{v}, t) = F_1(\mathbf{r}, \mathbf{v}, t)\hat{\mathbf{x}}_1 + F_2(\mathbf{r}, \mathbf{v}, t)\hat{\mathbf{x}}_2 + F_3(\mathbf{r}, \mathbf{v}, t)\hat{\mathbf{x}}_3 \tag{4.8}$$

where

$$F_i(\mathbf{r}, \mathbf{v}, t) \equiv F_i(x_1, x_2, x_3, v_1, v_2, v_3, t)$$

The Second Law in the language of vectors is, therefore,

$$\mathbf{F}(\mathbf{r}, \mathbf{v}, t) = m\mathbf{a}(t) \tag{4.9}$$

Alternatively, it is the set of three differential equations summarized in the expression

$$\mathbf{F}(\mathbf{r}, \mathbf{v}, t) = m\frac{d\mathbf{v}}{dt} \tag{4.10}$$

Four things of importance should be mentioned now in connection with the concept of force. (1) Notice that Equations 4.5, 4.8 and 4.9 tell us that, in general, the ith acceleration coordinate for a particle will be dependent explicitly on *all* of the position and velocity coordinates, not just the ith position and velocity coordinates. This is an intrinsic aspect of motion in three-dimensional space that stands as a primary reason for why the concept of force is more subtle than for motion along a straight line. (2) Note also that the constant of proportionality in Equation 4.9 remains simply a number, as it was for the case of one-dimensional motion. Thus *the mass of a particle is a scalar quantity.* (3) Note that, because Equation 4.10 means

$$F_i(\mathbf{r}, \mathbf{v}, t) = m\frac{dv_i}{dt} \qquad (i = 1, 2, 3) \tag{4.11}$$

we must expect to solve *three* differential equations for either $\mathbf{r}(t)$ or $\mathbf{v}(t)$ in order to carry through the Newtonian Program in three-dimensional space. If we use (4.11) to find the velocity, then Equations 4.2 must be solved to produce the position. Moreover, since each force coordinate $F_i(\mathbf{r}, \mathbf{v}, t)$ $(i = 1, 2, 3)$ depends on seven variables in general, Equations 4.11 will be a set of coupled differential equations whose solutions may be far from simple to obtain. (4) Finally, note that the assumption made in Chapter 1 concerning two or more forces acting at the same time on a particle must now be broadened to the extent that the term "sum" is understood in the sense of *vector* addition. Because + has precisely the same algebraic properties of commutation and association as does +, we need say no more about this generalization in order to apply it to dynamical problems.

Linear Momentum The linear momentum of a particle is the vector that results from multiplying the velocity of the particle by its mass. Therefore

$$\mathbf{p}(t) = m\mathbf{v}(t) \tag{4.12}$$

in vector notation.

Energy The kinetic energy of a particle moving in three dimensions is defined by a straightforward extension of Equation 1.26:

$$T(\mathbf{v}) \equiv \tfrac{1}{2}m(\mathbf{v} \cdot \mathbf{v}) = \tfrac{1}{2}mv^2 \tag{4.13}$$

where $v \equiv \|\mathbf{v}\|$ is called the *speed* of the particle and is a nonnegative, scalar function of the time. The equality between the scalar product $(\mathbf{v} \cdot \mathbf{v})$ and v^2 is, of course, a direct consequence of the definition of the length of a vector. Equation 4.13 states that the kinetic energy is a scalar function. This property is in agreement with the fact that no experiment has shown the need to associate $T(\mathbf{v})$ with any directional characteristic. It remains simply a number.

The potential energy of a particle is defined by the integral

$$V(\mathbf{r}_2) - V(\mathbf{r}_1) \equiv -\int_{\mathbf{r}_1}^{\mathbf{r}_2} \mathbf{F}(\mathbf{r}) \cdot d\mathbf{s} \qquad (4.14)$$

The integral on the right-hand side of Equation 4.14 sums up the projections of the force $\mathbf{F}(\mathbf{r})$ along the path between $\mathbf{r}_1$ and $\mathbf{r}_2$. (See Figure 4.2.) It is a generalization, to three dimensions, of the integral appearing in Equation 1.28. The differential $d\mathbf{s}$ is expressed by

$$d\mathbf{s} = dx_1\hat{\mathbf{x}}_1 + dx_2\hat{\mathbf{x}}_2 + dx_3\hat{\mathbf{x}}_3 \qquad (4.15)$$

and is to be regarded as an infinitesimal element of arc length pointing in a positive sense when it is directed along the path of integration from $\mathbf{r}_1$ to $\mathbf{r}_2$. In Figure 4.2, then, the contribution to the line integral will be positive at point A, but will be negative at point B, since the angle between $d\mathbf{s}$ and $\mathbf{F}(\mathbf{r})$ is greater than $\pi/2$ at the latter point and is less than $\pi/2$ at the former.

Because the path of integration in Equation 4.14 is now a curve in three-dimensional space, instead of a line, and because each of the coordinates of $\mathbf{F}(\mathbf{r})$ is a function of all three position coordinates, the definition of $V(\mathbf{r})$ is not unambiguous without the specification of further conditions on the force besides its

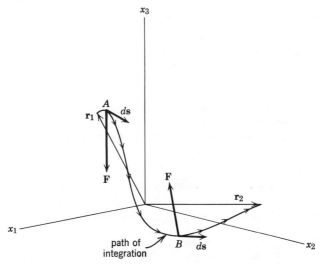

FIGURE 4.2. An element in the line integral given by Equation 4.14.

lack of time and velocity dependence. In particular, it is not automatic that the difference $V(\mathbf{r}_2) - V(\mathbf{r}_1)$ will result correctly from the integration of $\mathbf{F} \cdot d\mathbf{s}$ along *any* curve between $\mathbf{r}_1$ and $\mathbf{r}_2$. For this condition to obtain and, accordingly, for the relation

$$\mathbf{F}(\mathbf{r}) = -\nabla V \tag{4.16}$$

to hold, where

$$\nabla V = \frac{\partial V}{\partial x_1} \hat{\mathbf{x}}_1 + \frac{\partial V}{\partial x_2} \hat{\mathbf{x}}_2 + \frac{\partial V}{\partial x_3} \hat{\mathbf{x}}_3$$

is the *gradient* of $V(\mathbf{r})$, the force must satisfy

$$(\nabla \times \mathbf{F})_1 \equiv \frac{\partial F_3}{\partial x_2} - \frac{\partial F_2}{\partial x_3} \equiv 0$$

$$(\nabla \times \mathbf{F})_2 \equiv \frac{\partial F_1}{\partial x_3} - \frac{\partial F_3}{\partial x_1} \equiv 0 \tag{4.17a}$$

$$(\nabla \times \mathbf{F})_3 \equiv \frac{\partial F_2}{\partial x_1} - \frac{\partial F_1}{\partial x_2} \equiv 0$$

To appreciate that Equations 4.17a imply Equation 4.16 we need only cite the identity $\nabla \times \nabla \varphi \equiv \mathbf{0}$, where $\varphi(\mathbf{r})$ is any scalar function. (See Problem 4 at the end of this chapter.) The demonstration that (4.17a) also imply that the integral in (4.14) may be computed over an *arbitrary* path between $\mathbf{r}_1$ and $\mathbf{r}_2$ is more complicated[1].

The operator represented by the symbol ∇ is called the *del* or *gradient operator*. (See Section A.2 of the Appendix.) It may be expressed in the "hungry" form

$$\nabla \equiv \frac{\partial}{\partial x_1} \hat{\mathbf{x}}_1 + \frac{\partial}{\partial x_2} \hat{\mathbf{x}}_2 + \frac{\partial}{\partial x_3} \hat{\mathbf{x}}_3 \tag{4.18}$$

in order to make sense of Equations 4.17a written in the vector notation

$$\nabla \times \mathbf{F} \equiv \mathbf{0} \tag{4.17b}$$

$\nabla \times \mathbf{F}$ is thus an axial vector, as discussed in Section A.1 of the Appendix. Although it is possible to manipulate ∇ algebraically in this way, as one would an ordinary vector, ultimately the operator must act on some function. It must not be left "hungry"!

This completes the specification of the quantities in Newtonian dynamics in vector form. Before we go on to discuss concrete examples of solutions of the Second Law, however, we must pause to consider a very important, particular aspect of three-dimensional motion: that of its geometric symmetry.

---

[1] A careful, informative discussion of the condition (4.17) is presented in Chapter III of H. M. Schey, *Div, Grad, Curl, and All That*, W. W. Norton, New York, 1973.

4.2 SYMMETRY AND THE CONCEPT OF ANGULAR MOMENTUM

The study of motion in three-dimensional space often confronts us with the difficulty that the differential equations which follow from Equation 4.10 are rather complicated. Although the cause of this mathematical unruliness can be subtle, usually it is the result of the motion under consideration exhibiting a natural symmetry character that is not rectangular and, therefore, is not expressed well in cartesian coordinates. As an example of this situation we might imagine any motion for which the force acting on the particle depends only on the distance from the origin of the reference frame to the point at which the particle is located:

$$\mathbf{F} = -\frac{k}{(x_1{}^2 + x_2{}^2 + x_3{}^2)^{3/2}} (x_1\hat{\mathbf{x}}_1 + x_2\hat{\mathbf{x}}_2 + x_3\hat{\mathbf{x}}_3) \qquad \text{(cartesian)}$$

$$\mathbf{F} = -\frac{k}{r^2}\hat{\mathbf{r}} \qquad \text{(spherical polar)}$$

The force then has a relatively complicated dependence on the cartesian coordinates of the particle, but can be simplified greatly if spherical polar coordinates are used instead to express it mathematically.

It should be clear enough from these few remarks that we might look into the effect on the Second Law of transforming from the cartesian coordinates $\{x_1, x_2, x_3\}$ into, for example, the sets $\{r, \theta, \varphi\}$ and $\{\rho, \theta, z\}$. (Spherical polar and cylindrical polar coordinates are discussed in Section A.3 of the Appendix.) This effect turns out to be rather complicated because the expressions relating the cartesian basis vectors $(\hat{\mathbf{x}}_1, \hat{\mathbf{x}}_2, \hat{\mathbf{x}}_3)$ to $(\hat{\mathbf{r}}, \hat{\boldsymbol{\theta}}, \hat{\boldsymbol{\varphi}})$ or $(\hat{\boldsymbol{\rho}}, \hat{\boldsymbol{\theta}}, \hat{\mathbf{z}})$ are not particularly simple. The net result of the coordinate transformation, in fact, is rather unfortunate; for it develops that the "standard form" of the Second Law,

$$m\frac{d\dot{q}_i}{dt} = F_i \qquad (i = 1, 2, 3) \tag{4.19}$$

where $\dot{q}_i = dq/dt$ and q_i is a noncartesian coordinate, *is not generally valid*. Thus, the apparent simplification of the function $\mathbf{F}$ brought on by transforming to coordinates that reflect natural geometric symmetries is canceled out by the loss of the general form for the differential equations that represent the Second Law.

A better approach, and the one we shall take now, is to consider the coordinate transformations with respect to the toal energy. This quantity is a scalar function and, therefore, cannot change in value under such transformations. We shall begin by taking up the case of spherical symmetry. This kind of symmetry is said to exist if $V(x_1, x_2, x_3) = V(r)$, where $r = (x_1{}^2 + x_2{}^2 + x_3{}^2)^{1/2}$. Potentials of this form are called *central-field potentials* because the forces to which they correspond are directed toward or away from the origin—the center of the frame of reference.

Thus, spherical symmetry also may be defined by the geometric condition: $\mathbf{r} \times \mathbf{F} = \mathbf{0}$. The natural set of coordinates to which the cartesian $\{x_1, x_2, x_3\}$ should be transformed is $\{r, \theta, \varphi\}$, the spherical polar coordinates (see Figure 4.3). These coordinates are defined by:

$$x_1 = r \sin \theta \cos \varphi$$
$$x_2 = r \sin \theta \sin \varphi \qquad (4.20)$$
$$x_3 = r \cos \theta$$

The "velocity coordinates" corresponding to r, θ, and φ are evidently $\dot{r}$, $\dot{\theta}$, and $\dot{\varphi}$, respectively, where the dot refers to differentiation with respect to the time. These velocities are related to the cartesian velocities by:

$$v_1 = \dot{r} \sin \theta \cos \varphi + r \cos \theta \cos \varphi \, \dot{\theta} - r \sin \theta \sin \varphi \, \dot{\varphi}$$
$$v_2 = \dot{r} \sin \theta \sin \varphi + r \cos \theta \sin \varphi \, \dot{\theta} + r \sin \theta \cos \varphi \, \dot{\varphi} \qquad (4.21)$$
$$v_3 = \dot{r} \cos \theta - r \sin \theta \, \dot{\theta}$$

Equations 4.21 follow from differentiating both sides of Equations 4.20. With these expressions we may write the total energy in terms of spherical polar coordinates. Thus, for a central-field potential,

$$E = \tfrac{1}{2}m(v_1{}^2 + v_2{}^2 + v_3{}^2) + V(x_1, x_2, x_3) \qquad (4.22)$$

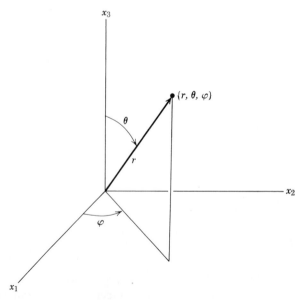

FIGURE 4.3. Spherical polar coordinates.

becomes

$$E = \tfrac{1}{2}m(\dot{r}^2 + r^2\dot{\theta}^2 + r^2 \sin^2 \theta\, \dot{\phi}^2) + V(r) \tag{4.23}$$

after repeated application of the familiar trigonometric identity involving the sums of squares of the sine and cosine.

The kinetic-energy part of Equation 4.23 looks rather different from that of Equation 4.22. It appears that some care will be required to interpret physically the three terms containing the squares of the "velocities" $\dot{r}$, $\dot{\theta}$, and $\dot{\phi}$. With $\dot{r}$ there is really no problem since this quantity has the dimensions of velocity and is obviously the rate of change of distance along a radial line between the particle and the origin. With $\dot{\theta}$ and $\dot{\phi}$, however, we do not have velocity dimensions. Perhaps we can proceed by analogy with what we know for a linear coordinate such as x_1, x_2, x_3, or r. For example, the linear momenta mv_1, mv_2, mv_3, and $m\dot{r}$, which play important roles in the conservation theorems and in the Second Law, are each the partial derivative of the kinetic energy with respect to the appropriate velocity coordinate. What do we have in the case of the angular coordinates $\dot{\theta}$ and $\dot{\phi}$? Evidently

$$p_\theta = \frac{\partial T}{\partial \dot{\theta}} = mr^2\dot{\theta} \qquad p_\phi = \frac{\partial T}{\partial \dot{\phi}} = mr^2 \sin^2 \theta\, \dot{\phi} \tag{4.24}$$

These equations we shall take to define the "momentum coordinates" p_θ and p_ϕ (although they do not have the dimensions of linear momentum). Consider now the coordinate p_ϕ. By the chain rule for partial derivatives, we have the relation

$$p_\phi = \frac{\partial T}{\partial \dot{\phi}} = \sum_{i=1}^{3} \frac{\partial T}{\partial v_i} \frac{\partial v_i}{\partial \dot{\phi}} = \sum_{i=1}^{3} p_i \frac{\partial v_i}{\partial \dot{\phi}}$$

$$= -p_1 r \sin \theta \sin \varphi + p_2 r \sin \theta \cos \varphi \qquad \text{(by Equations 4.21)}$$

$$= x_1 p_2 - x_2 p_1 \qquad \text{(by Equations 4.20)} \tag{4.25}$$

Equation 4.25 tells us that p_φ is a rather curious combination of cartesian position and momentum coordinates lying only in the x_1-x_2 plane. Even so, this combination is in fact *a constant of the motion* for the central force we assume here:

$$\frac{dp_\varphi}{dt} = \frac{d}{dt}(x_1 p_2 - x_2 p_1) = (v_1 p_2 - v_2 p_1) + (x_1 F_2 - x_2 F_1) \equiv 0 \tag{4.26}$$

since $v_1 p_2 = mv_1 v_2 = v_2 p_1$ and $x_1 F_2 - x_2 F_1 = 0$ follows directly from the condition that a force be central, $\mathbf{r} \times \mathbf{F} = 0$. Therefore, p_φ is a conserved quantity! Moreover, we see that p_φ has the combination of coordinates that occurs in an axial vector, as discussed in Section A.1 of the Appendix. It appears that p_φ is the "vertical coordinate" of an axial vector constructed from $\mathbf{r}$ and $\mathbf{p}$. This axial vector, which we shall call $\mathbf{J}$, would be written

$$\mathbf{J} = \mathbf{r} \times \mathbf{p} \tag{4.27}$$

and would have the vertical cartesian coordinate $J_3 = x_1p_2 - x_2p_1 = p_\varphi$. What about J_1 and J_2? Evidently

$$J_1 = x_2p_3 - x_3p_2 = -\sin\varphi\, p_\theta - \cot\theta\cos\varphi\, p_\varphi \qquad (4.28a)$$

$$J_2 = x_3p_1 - x_1p_3 = \cos\varphi\, p_\theta - \cot\theta\sin\varphi\, p_\varphi \qquad (4.28b)$$

according to Equations 4.20 and 4.21. The method of proof demonstrated in Equation 4.26 makes it clear that these two coordinates also will be constants of the motion given the crucial requirement $\mathbf{r} \times \mathbf{F} = \mathbf{0}$. Therefore, the vector $\mathbf{J}$ is itself a conserved quantity:

$$\frac{d\mathbf{J}}{dt} \equiv \mathbf{0} \qquad (\mathbf{r} \times \mathbf{F} = \mathbf{0}) \qquad (4.29)$$

The vector $\mathbf{J}$, according to Equations 4.27 and 4.28, is connected with the angular motion of the particle and not with the potential-energy-dependent radial motion. The fact that it is conserved is a reflection of the spherical symmetry of the problem as represented in the equivalent conditions $V(x_1, x_2, x_3) = V(r)$ and $\mathbf{r} \times \mathbf{F} = \mathbf{0}$. Equation 4.27 shows that $\mathbf{J}$ points perpendicularly to the plane in which $\mathbf{r}$ and $\mathbf{p}$ lie (see Figure 4.4). But this means that Equation 4.29 demonstrates that $\mathbf{r}$ and $\mathbf{p}$ lie in a *stationary* plane as the particle moves. Because these two vectors

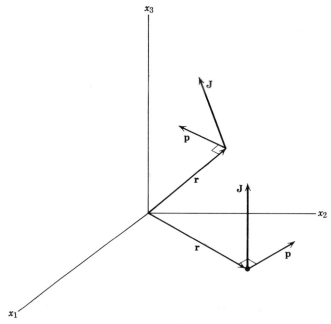

FIGURE 4.4. The angular momentum vector is perpendicular to the plane of $\mathbf{r}$ and $\mathbf{p}$.

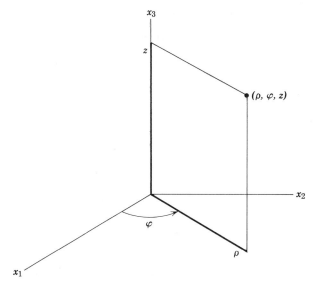

FIGURE 4.5. Cylindrical polar coordinates.

in fact constitute the motion of the particle, we are led to the important conclusion that *the motion of a particle under a central force always makes a trajectory that lies entirely in a fixed plane.*

Of course, the magnitude of the vector $\mathbf{J}$ is also a constant of the motion. By the relation $J_3 = p_\varphi$, along with Equations 4.28, the magnitude of $\mathbf{J}$ is

$$
\begin{aligned}
J = \|\mathbf{J}\| &= (J_1{}^2 + J_2{}^2 + J_3{}^2)^{1/2} \\
&= [(\sin \varphi \, p_\theta + \cot \theta \cos \varphi \, p_\varphi)^2 + (\cos \varphi \, p_\theta - \cot \theta \sin \varphi \, p_\varphi)^2 + p_\varphi{}^2]^{1/2} \\
&= \left(p_\theta{}^2 + \frac{p_\varphi{}^2}{\sin^2 \theta} \right)^{1/2}
\end{aligned}
\tag{4.30}
$$

where we have used the identity $1 + \cot^2 \theta \equiv \csc^2 \theta$. We see that J is related more or less directly to p_θ and p_φ. Because of this fact $\mathbf{J}$ has been called the *angular momentum* of a particle. For motion exhibiting spherical symmetry, Equation 4.29 is the natural generalization of the Momentum Theorem developed in Chapter 1. It is, then, a statement of the *conservation of angular momentum.*

The case of cylindrically symmetric motion is defined by $V(x_1, x_2, x_3) = V(\rho, z)$, where $\rho = (x_1{}^2 + x_2{}^2)^{1/2}$. The natural set of coordinates to which the cartesian $\{x_1, x_2, x_3\}$ should be transformed is $\{\rho, \theta, z\}$, the cylindrical polar coordinates (see Figure 4.5). These coordinates and their corresponding "velocities" are defined by:

$$
x_1 = \rho \cos \theta \qquad\qquad x_2 = \rho \sin \theta \qquad\qquad x_3 = z \tag{4.31}
$$

$$
v_1 = \dot{\rho} \cos \theta - \rho \sin \varphi \, \dot{\theta} \qquad v_2 = \dot{\rho} \sin \theta + \rho \cos \varphi \, \dot{\theta} \qquad v_3 = \dot{z} \tag{4.32}
$$

The total energy becomes, accordingly, the expression

$$E = \tfrac{1}{2}m(\dot{\rho}^2 + \rho^2\dot{\theta}^2 + \dot{z}^2) + V(\rho, z) \tag{4.33}$$

If we follow the line of thought developed for the case of spherical symmetry, we are led to write

$$J_3 = p_\varphi = m\rho^2\dot{\theta} \tag{4.34}$$

as the conserved, vertical coordinate of the angular momentum **J**. In fact

$$J_3 = \frac{\partial T}{\partial\dot{\theta}} = \sum_{i=1}^{3} p_i \frac{\partial v_i}{\partial\dot{\theta}} = -p_1\rho\sin\theta + p_2\rho\cos\theta$$

$$= x_1 p_2 - x_2 p_1$$

as before, according to the chain rule and Equations 4.31 and 4.32. The constancy of J_3 then follows from the condition for cylindrical symmetry, $V(x_1, x_2, x_3) = V(\rho, z)$, or its geometric equivalent, $(\mathbf{r} \times \mathbf{F})_3 = 0$. The coordinates J_1 and J_2 will not generally be constant. In the special case that $V(x_1, x_2, x_3) = V(\rho)$, however, the angular momentum vector will point along the the x_3 axis and the motion will be confined to the x_1-x_2 plane.

4.3   THE ISOTROPIC OSCILLATOR

In this section we shall consider a bounded particle motion where the natural geometric symmetry is not rectangular. The standard example to be taken up is the isotropic oscillator, a particle that moves under the potential function

$$V_H(r) = \tfrac{1}{2}kr^2 \tag{4.35}$$

where k is a positive force constant. We know already, because of the analysis in Section 4.2, that the motion of the isotropic oscillator should be described with spherical polar coordinates and that the angular momentum will be a constant of the motion. Thus the particle moves in a fixed plane. We lose nothing if we call this plane the x_1-x_2 plane and set $\theta = \pi/2$. This done, we can write Equation 4.23 for the total energy of the oscillator in terms of the *plane* polar coordinates $\{r, \varphi\}$:

$$E = \tfrac{1}{2}m(\dot{r}^2 + r^2\dot{\varphi}^2) + \tfrac{1}{2}kr^2$$

$$= \tfrac{1}{2}m\dot{r}^2 + \frac{J^2}{2mr^2} + \tfrac{1}{2}kr^2 \tag{4.36}$$

where the last line comes from applying Equation 4.30.

Our method for finding the motion will make use of the Energy Theorem in a manner very similar to what we did in Section 2.1. We begin by writing

$$\frac{m}{2}\left(\frac{dr}{dt}\right)^2 + \frac{J^2}{2mr^2} + \tfrac{1}{2}kr^2 = E$$

or

$$\frac{dr}{dt} = \left[\frac{2E}{m} - \frac{J^2}{(mr)^2} - \omega_0^2 r^2 \right]^{1/2} \qquad \left(\omega_0^2 = \frac{k}{m} \right)$$

which finally becomes the integral expression

$$t = \int_{r(0)}^{r(t)} \left[\frac{2E}{m} - \frac{J^2}{(mr)^2} - \omega_0^2 r^2 \right]^{-1/2} dr \qquad (4.37)$$

Equation 4.37 is the implicit solution of our problem. If we make the substitution $x = r^2$, the integral has the general form

$$\frac{1}{2} \int \left[\frac{2E}{m} x - \frac{J^2}{m^2} - \omega_0^2 x^2 \right]^{-1/2} dx = \frac{1}{2\omega_0} \sin^{-1} \left[\frac{m\omega_0^2 x - E}{(E^2 - J^2\omega_0^2)^{1/2}} \right]$$

according to the tables, and we get

$$2\omega_0 t = \sin^{-1} \left[\frac{m\omega_0^2 r^2(t) - E}{(E^2 - J^2\omega_0^2)^{1/2}} \right] - \sin^{-1} \left[\frac{m\omega_0^2 r^2(0) - E}{(E^2 - J^2\omega_0^2)^{1/2}} \right] \qquad (4.38)$$

We can simplify this equation if we choose dr/dt to vanish when $t = 0$. In Section 4.5 we shall find that this corresponds to choosing $r(0)$ as a turning point of the motion. Under that condition Equation 4.36 may be written

$$E = \frac{J^2}{2mr^2(0)} + \tfrac{1}{2}m\omega_0^2 r^2(0) \qquad (4.39)$$

and (4.38) becomes

$$2\omega_0 t = \sin^{-1} \left(\frac{m\omega_0^2 r^2(t) - E}{E - m\omega_0^2 r^2(0)} \right) - \sin^{-1} (-1)$$

or, since $\sin^{-1}(-1) = -\pi/2$,

$$r(t) = \left[\frac{E}{m\omega_0^2} - \left(\frac{E}{m\omega_0^2} - r^2(0) \right) \cos(2\omega_0 t) \right]^{1/2} \qquad (4.40)$$

upon solving for $r(t)$. The trajectory of the oscillator in the r-φ plane depends sensitively on the magnitude of the angular momentum $\mathbf{J}$. If, for example,

$$J = m\omega_0 r^2(0) \qquad (4.41)$$

then, according to Equations 4.39 and 4.40,

$$r(t) = \left(\frac{E}{m\omega_0^2} \right)^{1/2} = \left(\frac{J}{m\omega_0} \right)^{1/2} \qquad (4.42)$$

and the oscillator must move in a *circle* of radius $(E/m\omega_0^2)^{1/2} = r(0)$. If, on the other hand,

$$J > m\omega_0 r^2(0) \qquad (4.43)$$

We have $E > m\omega_0^2 r^2(0)$, according to (4.39), and the radial coordinate varies between

$$r_{max} = \left[\frac{2E}{m\omega_0^2} - r^2(0)\right]^{1/2} = \frac{J}{m\omega_0 r(0)} \tag{4.44a}$$

when $t = n\pi/2\omega_0$ $(n = 1, 3, 5, \ldots)$ and

$$r_{min} = r(0) \tag{4.44b}$$

when $t = m\pi/\omega_0$ $(m = 0, 1, 2, \ldots)$. This result suggests that the path of the oscillator in the r-φ plane could be an *ellipse* with the orbital period equal to π/ω_0. To verify the hypothesis, we may calculate the parametric equation of the orbit in cartesian coordinates. We begin by using the identity

$$\cos 2\theta \equiv \cos^2 \theta - \sin^2 \theta$$

along with the result

$$\frac{E}{m\omega_0^2} = \frac{1}{2}\left(\frac{J}{m\omega_0 r(0)}\right)^2 + \tfrac{1}{2}r^2(0) = \tfrac{1}{2}r^2(0)\left[\left(\frac{\dot\phi(0)}{\omega_0}\right)^2 + 1\right]$$

to write Equation 4.40 in the form

$$r^2(t) = r^2(0)\left[1 + \left(\frac{\dot\phi(0)}{\omega_0}\right)^2 \tan^2 \omega_0 t\right]\cos^2 \omega_0 t \tag{4.45}$$

The expression

$$\dot\phi(t) = \frac{J}{mr^2(t)}$$

also may be combined with Equation 4.40 and integrated to produce

$$\tan\left[\varphi(t) - \varphi(0)\right] = \frac{\dot\phi(0)}{\omega_0}\tan \omega_0 t$$

(See Problem 8 at the end of this chapter.) This equation then is decomposed into the relations

$$\sin^2\left[\varphi(t) - \varphi(0)\right] = \frac{(\dot\phi(0)/\omega_0)^2 \tan^2 \omega_0 t}{1 + (\dot\phi(0)/\omega_0)^2 \tan^2 \omega_0 t} \tag{4.46a}$$

$$\cos^2\left[\varphi(t) - \varphi(0)\right] = \frac{1}{1 + (\dot\phi(0)/\omega_0)^2 \tan^2 \omega_0 t} \tag{4.46b}$$

With Equations 4.45 and 4.46 we may write

$$\frac{r^2(t)\cos^2 \Delta\varphi(t)}{r^2(0)} + \frac{r^2(t)\sin^2 \Delta\varphi(t)}{[r(0)\dot\phi(0)/\omega_0]^2} = 1$$

or,

$$\frac{x_1^2(t)}{r^2(0)} + \frac{x_2^2(t)}{[r(0)\dot\phi(0)/\omega_0]^2} = 1 \tag{4.47}$$

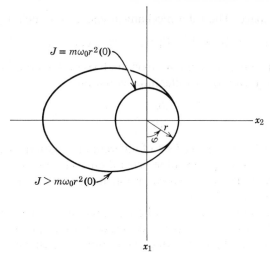

FIGURE 4.6. Trajectories of the isotropic oscillator.

Equation 4.47, it will be recalled,[2] is the equation for an ellipse in standard form. This conic section is characterized by two independent parameters, the length of the semimajor axis, a, and the eccentricity, e:

$$a \equiv \tfrac{1}{2}(r_{\text{max}} + r_{\text{min}}) \tag{4.48}$$

$$e \equiv \frac{r_{\text{max}} - r_{\text{min}}}{r_{\text{max}} + r_{\text{min}}} \tag{4.49}$$

In dynamical terms, these parameters are

$$a = \tfrac{1}{2}r(0)\left[1 + \frac{\dot{\phi}(0)}{\omega_0}\right] \qquad e = \frac{\dot{\phi}(0) - \omega_0}{\dot{\phi}(0) + \omega_0}$$

respectively, for the isotropic oscillator, according to Equations 4.44, 4.48, and 4.49.

The physical basis for the elliptical path can be understood directly on noting that the term $J^2/2mr^2(0)$ in Equation 4.39 is a contribution from the "rotational kinetic energy." The overall effect of increasing J is to enlarge this contribution and "stretch out" the oscillator a bit, causing it to deviate from a strictly circular path (see Figure 4.6).

4.4 THE PROBLEM OF TWO PARTICLES

One of the central problems of classical dynamics has to do with the motion of two particles whose total potential energy is a function only of their relative

---

[2] Those who do not recall should consult their textbooks on introductory calculus!

displacement in space. The total mechanical energy for such a pair will be

$$E_T = \tfrac{1}{2}m_A v_A{}^2 + \tfrac{1}{2}m_B v_B{}^2 - V(x_{A1} - x_{B1}, x_{A2} - x_{B2}, x_{A3} - x_{B3}) \qquad (4.50)$$

where A and B label the particles. A direct analysis of this situation in terms of the Second Law would lead to the differential equations

$$\frac{dp_{Ai}}{dt} = -\frac{\partial V}{\partial x_{Ai}} \qquad \frac{dp_{Bi}}{dt} = -\frac{\partial V}{\partial x_{Bi}} \qquad (i = 1, 2, 3) \qquad (4.51)$$

These expressions form a *coupled* set of differential equations because the partial derivatives such as $\partial V / \partial x_{Ai}$ depend in general on every one of the six particle coordinates. Therefore, the mathematical problem presented to us by Equations 4.51 usually will not be a simple one.

But this is not the first occasion where a direct attack would have produced mathematical difficulties instead of physical results. In the previous sections of this chapter we have seen that purely cartesian coordinates often may be too insensitive to the natural symmetries of an expected motion as suggested by the potential function. Whenever this lack of correspondence is the only source of complication in the equations of motion, experience has shown that great simplifications are possible by choosing an appropriate set of new coordinates. The motion of two particles apparently requires six cartesian position coordinates. Therefore, the line of thought we have been pursuing would suggest that six new coordinates derived from the character of the potential function in Equation 4.50 may be in order. A look at (4.50) tells us rather clearly that three of the generalized coordinates ought to be the *relative coordinates* $x_i \equiv (x_{Ai} - x_{Bi})$ $(i = 1, 2, 3)$. The other three coordinates we can choose almost at will; for example, they could be simply the position coordinates of particle B. We shall pin them down by asking that they be such as to preserve the total kinetic energy as a sum of independent terms just the way it is in (4.50). (The reason for choosing them so will become apparent quite soon.) Thus we shall define

$$X_i \equiv \frac{m_A x_{Ai} + m_B x_{Bi}}{M} \qquad (i = 1, 2, 3) \qquad (4.52)$$

$$x_i \equiv x_{Ai} - x_{Bi} \qquad (i = 1, 2, 3) \qquad (4.53)$$

to be our new set of coordinates, where M is the total mass of the particles. The geometric meaning of these definitions can be seen by rewriting them in vector notation:

$$\mathbf{R} \equiv \frac{m_A \mathbf{r}_A + m_B \mathbf{r}_B}{M} \qquad (4.54)$$

$$\mathbf{r} \equiv \mathbf{r}_A - \mathbf{r}_B \qquad (4.55)$$

(See Figure 4.7.) Because Equations 4.54 and 4.55 are not redundant, $\mathbf{R}$ and $\mathbf{r}$ represent independent coordinates and it must be possible to express $\mathbf{r}_A$ and $\mathbf{r}_B$ as

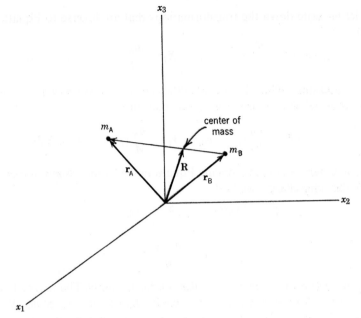

FIGURE 4.7. Center-of-mass and relative-position vectors for a two-particle system.

a linear combination of them. Since

$$\mathbf{r_A} - \mathbf{R} = \left(1 - \frac{m_A}{M}\right)\mathbf{r_A} - m_B\mathbf{r_B} = \frac{m_B}{M}\,\mathbf{r}$$

and

$$\mathbf{R} - \mathbf{r_B} = \frac{m_A}{M}\mathbf{r_A} + \left(\frac{m_B}{M} - 1\right)\mathbf{r_B} = \frac{m_A}{M}\,\mathbf{r}$$

are colinear vectors, the vector **R** always must refer to a point somewhere on the line of centers of the two particles, as shown in the figure. The exact point, according to Equation 4.54, will depend on the fraction of the total mass carried by each particle. For this reason the vector **R** is often considered to be a "weighted" average of $\mathbf{r_A}$ and $\mathbf{r_B}$ and is called the *center-of-mass vector*. Its physical meaning becomes apparent once we note that

$$\mathbf{P} = M\dot{\mathbf{R}} \equiv M\mathbf{V} = m_A\mathbf{v_A} + m_B\mathbf{v_B} \tag{4.56}$$

which is to say that **V** is the velocity of a point on the line of centers and $M\mathbf{V}$ is the total linear momentum of the two particles. Evidently the coordinates $\{X_1, X_2, X_3\}$ describe the motion of the pair "taken as a body" and the coordinates $\{x_1, x_2, x_3\}$ describe the internal, relative motion.

Now let us write down the transformations that are inverse to Equations 4.52 and 4.53:

$$x_{Ai} = X_i + \frac{m_B}{M} x_i \qquad x_{Bi} = X_i - \frac{m_A}{M} x_i \qquad (i = 1, 2, 3) \qquad (4.57)$$

These six equations define what is called the *center-of-mass coordinate system*. The corresponding velocity coordinate transformations are

$$v_{Ai} = V_i + \frac{m_B}{M} v_i \qquad v_{Bi} = V_i - \frac{m_A}{M} v_i \qquad (i = 1, 2, 3) \qquad (4.58)$$

where $v_i = \dot{x}_i$ here. Upon introducing Equations 4.57 and 4.58 into the expression (4.50) for the total energy, we find

$$E_T = \tfrac{1}{2}MV^2 + \tfrac{1}{2}\mu v^2 + V(x_1, x_2, x_3) \qquad (4.59)$$

where

$$\mu = \frac{m_A m_B}{M} \qquad (4.60)$$

is a parameter known as the *reduced mass* of the assembly. The reduced mass has the dimensions of mass but is always smaller than either m_A or m_B. We see in Equation 4.59 that the kinetic energy portion of E has remained as the sum of two independent terms, as proposed. The motion of the center of mass is described in terms of

$$E_{CM} = \tfrac{1}{2}MV^2 \qquad (4.61)$$

which represents the kinetic energy of a "particle" of mass M moving with the constant velocity V. This motion carries with it the total linear momentum $P = MV$. The relative motion of the two particles is described by

$$E = \tfrac{1}{2}\mu v^2 + V(x_1, x_2, x_3) \qquad (4.62)$$

which represents the energy of a "particle" of mass μ subject to the potential function $V(x_1, x_2, x_3)$. The energy E is most certainly not that of either one of the particles A or B because μ is not the mass of either one of them. Instead, E describes a property of *the assembly*—its internal or relative motion—which is complementary with that described by E_{CM}. Since Equations 4.61 and 4.62 each refer to a single "particle," we can deal with them *mathematically* in the same way as we would any single-particle problem. The beauty and utility of the center-of-mass transformation, then, comes from its permitting the decomposition of a two-particle problem into two one-particle problems whose solutions, in principle, we already know how to obtain.

4.5 ORBITAL MOTION AND THE EFFECTIVE POTENTIAL

An important example of the relative motion of two particles occurs when the total potential energy of the pair depends only on their distance of separation. In this case the natural symmetry is spherical and Equation 4.62 may be expressed conveniently in plane polar coordinates:

$$E = \tfrac{1}{2}\mu\dot{r}^2 + \frac{J^2}{2\mu r^2} + V(r) \tag{4.63}$$

By virtue of our experience with the isotropic oscillator, we know that Equation 4.63 can lead to a closed trajectory or *orbit* for the relative motion if $V(r)$ has the appropriate mathematical form and the initial conditions on the motion are right. In Section 4.3 we discussed the influence of these factors in a special case ($V(r) = \tfrac{1}{2}kr^2$) through a direct solution by quadrature for $r(t)$. This procedure would certainly suffice, but, in fact, it is not necessary if we choose to rely a little more upon physical arguments than we have thus far. To be specific, we might point out that the right-hand side of Equation 4.63 is dependent solely on the radial coordinate r and in so being is quite similar to the expression for the total energy in *one-dimensional* motion. Indeed, if we set

$$U(r) \equiv \frac{J^2}{2\mu r^2} + V(r) \tag{4.64}$$

we can write

$$E = \tfrac{1}{2}\mu\dot{r}^2 + U(r) \tag{4.65}$$

which has exactly the appearance of the total energy in a one-dimensional dynamical problem expressed in *cartesian* coordinates. (Note that $\mu\dot{r}$ is a linear momentum.) Now, in Chapter 2 we saw that a great deal of qualitative information about the motion of a single particle could be gained by analyzing the potential function under the requirement of energy conservation. In the present case it appears that the same kind of physical insight can be had if we regard $U(r)$ as a kind of effective potential to be analyzed similarly.

If one is to see orbital motion coming out of a solution of Equation 4.65, the effective potential energy must display a minimum at some finite value of r. Otherwise, the radial kinetic energy in (4.65) could not achieve the maximum value necessary to induce periodic motion. The term $J^2/2\mu r^2$ in $U(r)$ is in reality a contribution from the rotational part of the relative kinetic energy, which tends to make the orbiting pair move farther apart as J is increased.[3] Thus it behaves as a potential function describing an essentially *repulsive* interaction between two particles. In order that there be a minimum in $U(r)$, then, $V(r)$—the true relative potential—must contain either a term opposite in sign to the rotational term or one exhibiting a strong minimum in its own right. Relative potentials that are positive valued, and monotonically decreasing with an increase in the radial coordinate cannot lead to a minimum in the effective potential and, therefore,

---

[3] This "centrifugal effect" of increasing angular momentum should be well known from introductory mechanics. Indeed, the derivative of $J^2/2\mu r^2$ is just the rotationally induced *centrifugal force* $\mu r\dot{\phi}^2$. A discussion of this force will be given in Chapter 7. For now we regard it simply as an effective repulsive force generated naturally by the relative motion.

will not produce bounded motion of any kind. We shall encounter these potential functions in Chapter 5.

To get a more physical viewpoint on the significance of the effective potential, let us consider the isotropic oscillator as a problem in relative motion. We can accomplish this by imagining, say, two point atoms rushing through space as a bound unit and oscillating about their center of mass. Equation 4.64 then takes the form

$$U(r) = \frac{J^2}{2\mu r^2} + \tfrac{1}{2}\mu\omega_0^2 r^2 \tag{4.66}$$

where now $\omega_0^2 = k/\mu$, in terms of the force constant and reduced mass. Equation 4.66 is plotted in Figure 4.8. We see there that the effective potential shows a strong minimum that, in fact, persists even when J is zero. Two possible values for the total relative energy E are indicated in the figure.

(a) $E = U_{\min}$. When the total relative energy E equals the effective potential energy at its minimum point, the requirement that $\tfrac{1}{2}\mu\dot{r}^2$ be nonnegative together with Equation 4.65 dictate that the radial momentum vanish identically. Therefore, the distance of separation of the two particles will remain fixed forever at the value of r that makes $U(r)$ a minimum. According to Equation 4.66, this value of r is $(J/\mu\omega_0)^{1/2}$. The relative motion is along a *circle* of radius $(J/\mu\omega_0)^{1/2}$, which is exactly similar to the result we obtained in Equation 4.42. We note that a fixed value of the radial coordinate does *not* mean there is no motion, as it would in a truly one-dimensional problem. Instead, a constant interparticle separation

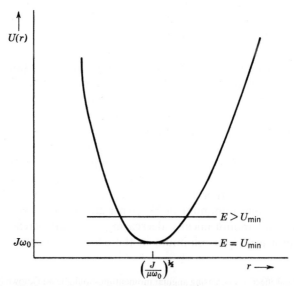

FIGURE 4.8. The effective potential of the isotropic oscillator.

means only that $\dot\phi$ has a fixed value as well, since $J = \mu r^2 \dot\phi$ is a constant of the motion. The orbit is swept out at a uniform rate.

(b) $E > U_{min}$. When the total relative energy is greater than its least value, the motion takes place between the two turning points specified by $E = U(r_m)$, that is, by $\dot r = 0$. This result is exactly similar to the necessary one-dimensional condition for a turning point. By Equation 4.66, we see that r_m is the solution of

$$E = \frac{J^2}{2\mu r_m^2} + \tfrac{1}{2}\mu\omega_0^2 r_m^2$$

or

$$r_m^2 = \frac{E}{\mu\omega_0^2} \pm \left[\left(\frac{E}{\mu\omega_0^2}\right)^2 - \left(\frac{J}{\mu\omega_0}\right)^2\right]^{1/2} \tag{4.67}$$

Equation 4.67 is entirely equivalent to Equations 4.44 if the condition (4.39) is imposed. The relative motion accordingly is seen to trace out an *ellipse* whose semimajor axis and eccentricity are

$$a = \left[\frac{E}{2\mu\omega_0^2} + \frac{J}{2\mu\omega_0}\right]^{1/2} \qquad e = \left[\frac{(E/J\omega_0) - 1}{(E/J\omega_0) + 1}\right]^{1/2} \tag{4.68}$$

respectively.

It should be clear that we have obtained the essential features of the motion of an isotropic oscillator by a means much simpler than what was employed in Section 4.3. Moreover, unlike in the discussion there, the results here are completely independent of the initial conditions. These points should illustrate the advantage of using the concept of the effective potential when describing the motion in a central field.

4.6 THE KEPLER PROBLEM

In its original form the Kepler problem[4] has to do with the relative motion of a single planet and the sun. The two bodies are considered to interact according to a special case of the universal gravitational potential function,

$$V(r) = -G\frac{M_s M_p}{r} \tag{4.69}$$

where r is the distance of separation between the centers of gravity of the sun and planet, G is the gravitational constant:

$$G = 6.670 \pm 0.004 \times 10^{-8} \ \text{cm}^3/\text{g-s}^2$$

---

[4] The name is in honor of Johannes Kepler, the mystic and astronomer who, during the years 1609–1619, worked out the basic phenomenology of planetary motion. His well-known results are called Kepler's Laws. For an interesting historical account of Kepler's work, see C. J. Schneer, *The Evolution of Physical Science*, Grove Press, New York, 1960, Chapter 5.

M_s is the gravitational mass of the sun, and M_p is that of the planet. The potential function $V(r)$ is derived through Equation 4.14 by integrating the celebrated expression for the gravitational force proposed by Isaac Newton in 1666 and finally published by him in detail in 1687. For a perfectly spherical planet and sun, each of uniform density, Equation 4.69 can be shown to be exact[5], aside from corrections required by the theory of relativity. If the bodies in question do not happen to have such regular properties (as, in fact, most do not), then Equation 4.69 is only asymptotically accurate in the limit that the equatorial radii are very much smaller than the distance of separation.[6] We shall assume that this limiting condition is valid in all further discussion.

There are two interesting things about the gravitational potential function worth mentioning at the outset. The first is that $V(r)$, being an asymptotic result, is a potential energy for two *particles* rather than for two *bodies* of finite spacial extension. Nowhere in Equation 4.69 is there an explicit reference to the size, shape, or density of the planet or sun (although a specification of the distance r does require a prior knowledge of their centers of mass). It is for this reason that we can incorporate the Kepler problem at once into the mathematical structure of classical dynamics as we have developed it thus far, despite the obvious fact that neither the sun nor any planet has the infinitesimal point character inherent in the concept of particle. The second interesting aspect of Equation 4.69 is that the masses M_s and M_p are *not* by necessity the inertial masses that appear in the Second Law because the form of $V(r)$ derives from a postulate on gravitation, not a law in dynamics. The gravitational masses M_s and M_p must be related *by experiment* to their inertial counterparts. As is well-known, very carefully done measurements have settled this question and have shown that the relation between gravitational and inertial mass is one of equality.[7]

Because the gravitational potential function depends only on the distance of separation and not on the relative orientation of the sun and planet, we expect that the motion it brings about will exhibit spherical symmetry. It follows immediately from the discussions in previous sections of this chapter that the relative angular momentum of the sun-planet system will be a constant of the motion and that we may describe the planetary orbits with the effective potential

$$U(r) = \frac{J^2}{2\mu r^2} - \frac{K}{r} \tag{4.70}$$

where

$$\mu = \frac{M_s M_p}{M_s + M_p} \qquad K = G M_s M_p \tag{4.71}$$

---

[5] A proof is given in P. M. Fitzpatrick, *Principles of Celestial Mechanics*, Academic Press, New York, 1970, §1.1. This statement was demonstrated first by Newton in 1685 as the last step in his 20-year-old theory of gravitation.

[6] For a good discussion of the gravitational potential near a body of nonspherical shape, see Chapter 12 of the textbook by Fitzpatrick cited in footnote 5.

[7] See R. H. Dicke, *Gravitation and the Universe*, American Philosophical Society, Philadelphia, PA, 1970, for an excellent discussion of the experiments on the relation of the two kinds of mass.

A graph of the function $U(r)$ is shown in Figure 4.9. We can distinguish two different general cases where the relative motion will be bounded.

(a) $E = U_{min}$. The minimum in the effective potential occurs when the distance of separation has the value

$$r_0 = \frac{J^2}{K\mu}$$

as can be seen easily after differentiating $U(r)$ and setting the result equal to zero. When

$$E = U(r_0) = -\frac{K^2\mu}{2J^2} \tag{4.72}$$

it appears that we have a *circular orbit* for the sun-planet system, according to our earlier discussion of this condition as applied to the effective potential. Equation 4.72 represents the least relative total energy the system can have with a real-valued kinetic energy and the corresponding motion is in a circle of radius r_0.

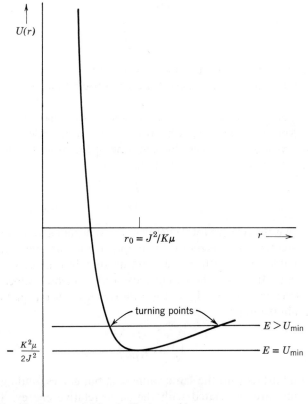

$U(r)$

$r_0 = J^2/K\mu$

$r \longrightarrow$

turning points

$E > U_{min}$

$-\frac{K^2\mu}{2J^2}$

$E = U_{min}$

FIGURE 4.9. The effective potential for the Kepler problem.

(b) $0 > E > U_{min}$. When the relative total energy is greater than its least value, but less than zero, the orbital motion exhibits two turning points known as the *perihelion* and the *aphelion*. They are defined by the equation

$$E = \frac{J^2}{2\mu r_m^{\,2}} - \frac{K}{r_m} \tag{4.73}$$

This condition leads to

$$r_{max} = -\frac{K}{2E}\left[1 + \left(1 + \frac{2J^2E}{K^2\mu}\right)^{1/2}\right] \qquad \text{(aphelion)} \tag{4.74a}$$

$$r_{min} = -\frac{K}{2E}\left[1 - \left(1 + \frac{2J^2E}{K^2\mu}\right)^{1/2}\right] \qquad \text{(perihelion)} \tag{4.74b}$$

and the suggestion that *elliptical orbits* are possible for the system. (Remember that $E < 0$.) Any one of the possible ellipses is described completely by its semimajor axis and eccentricity, which here are given by

$$a = -\frac{K}{2E} \tag{4.75}$$

$$e = \left(1 + \frac{2J^2E}{K^2\mu}\right)^{1/2} \tag{4.76}$$

respectively, according to the fundamental definitions given in Equations 4.48 and 4.49. We note that the orbit gets more eccentric (elongated) as E increases, in agreement with what is expected physically. The smallest value for e (zero) occurs when Equation 4.72 is satisfied, as should be the case for a circular orbit.

Equation 4.75 brings up a very special property of motion under the gravitational potential which makes the Kepler Problem one of the more fascinating ones in classical dynamics. Notice that (4.75) implies that

$$E = -\frac{K}{2a} \tag{4.77}$$

which is to say that the relative total energy of the sun-planet system depends only on *one* of the essential parameters for the elliptic orbit: the semimajor axis. This result is remarkable, since both a and e are absolutely necessary to specify the orbit geometrically. It is as if the eccentricity were somehow superfluous to the energetics of planetary motion. To put it another way, since Equations 4.75 and 4.76 can be combined to give

$$e = \left(1 - \frac{J^2}{K\mu a}\right)^{1/2}$$

it appears that all orbits with the same value of a, but corresponding to arbitrary angular momenta, are associated with the *same* relative energy. This unusual character (technically called an "accidental degeneracy" of E) cannot be attributed solely to the spherical symmetry of the problem because Equation 4.77 was derived ultimately from $E = U(r_m)$, which obviously depends for its concrete expression on what mathematical form $V(r)$ happens to have. Therefore, we must infer that the absence of one of the two orbit-determining parameters from the expression for E is in some way a special attribute of the dependence on the radial coordinate shown by the gravitational potential energy. But how do we determine this special attribute? Fortunately, the rather formidable task of answering this question from first principles has been accomplished. In advanced dynamics two pertinent theorems have been proved. The first of them states that (a) if E depends on only one of the two necessary orbital parameters, an additional constant of the motion can be found that depends on the remaining orbital parameter and (b) this situation occurs whenever the potential function is such as to make a circular orbit always stable against small perturbations (deviations from circularity) of any kind. The second result, known as Bertrand's Theorem, states the remarkable fact that, of all the central-field potential functions one might imagine, only *two* have the property of producing truly stable circular orbits. These two potential functions are

$$V_H(r) = \tfrac{1}{2}kr^2 \qquad \text{(isotropic oscillator)}$$

$$V(r) = -\frac{K}{r} \qquad \text{(Kepler Problem)}$$

Thus only the r^2 and r^{-1} potentials can be expected to cause E to depend only on the semimajor axis and to provide an additional constant of the motion.

What additional constant of the motion can be constructed in the Kepler Problem? Presumably this new quantity will involve the orbital eccentricity in some important way, but otherwise it is difficult to speculate on what its mathematical form ought to be. One candidate which has been found is the Runge-Lenz vector[8], **A**. This vector lies entirely in the plane of motion for the sun-planet system. It has the cartesian coordinates

$$A_1 = x_1(x_1^2 + x_2^2)^{-1/2} - \frac{Jp_2}{K\mu}$$

$$A_2 = x_2(x_1^2 + x_2^2)^{-1/2} + \frac{Jp_1}{K\mu}$$

$$(4.78)$$

provided that the orbit lies in the r-φ plane, making $J = J_3$. By direct computation we find

$$\frac{dA_1}{dt} = v_1(x_1^2 + x_2^2)^{-1/2} - x_1(x_1v_1 + x_2v_2)(x_1^2 + x_2^2)^{-3/2}$$

---

[8] The vector A appeared first in C. Runge, *Vector Analysis*, Methuen, London, 1923, and in a paper published by W. Lenz the following year.

$$-\frac{J}{K\mu}\frac{dp_2}{dt} = x_2(x_2v_1 - x_1v_2)(x_1^2 + x_2^2)^{-3/2} - \frac{1}{K}(x_1v_2 - x_2v_1)\frac{dp_2}{dt} \equiv 0$$

$$\frac{dA_2}{dt} = x_1(x_1v_2 - x_2v_1)(x_1^2 + x_2^2)^{-3/2} + \frac{1}{K}(x_1v_2 - x_2v_1)\frac{dp_1}{dt} \equiv 0$$

We have used the facts that $\mathbf{J}$ is time independent, that $J = \mu(x_1v_2 - x_2v_1)$, and that

$$\frac{dp_1}{dt} = -\frac{Kx_1}{(x_1^2 + x_2^2)^{3/2}}$$

$$\frac{dp_2}{dt} = -\frac{Kx_2}{(x_1^2 + x_2^2)^{3/2}}$$

(4.79)

for the gravitational force $\mathbf{F} = -K\hat{\mathbf{r}}/r^2$. Therefore, both the magnitude and direction of $\mathbf{A}$ must be fixed as time passes. The essential dependence of this conclusion on the precise form of $V(r)$ is seen clearly from the use of the Second Law (4.79) in its proof. The magnitude of $\mathbf{A}$ can be given a dynamical meaning by the following calculation:

$$\|\mathbf{A}\| = (A_1^2 + A_2^2)^{1/2} = \left[\left(\frac{x_1}{r} - \frac{Jp_2}{K\mu}\right)^2 + \left(\frac{x_2}{r} + \frac{Jp_1}{K\mu}\right)^2\right]^{1/2}$$

$$= \left(1 + \frac{2J^2E}{K^2\mu}\right)^{1/2} = e$$

(4.80)

We have used here Equation 4.76, the relation $r \equiv (x_1^2 + x_2^2)^{1/2}$, and the expression

$$E = \frac{(p_1^2 + p_2^2)}{2\mu} - \frac{K}{r}$$

The length of the Runge-Lenz vector turns out to be equal to the eccentricity of the orbit.

The direction of $\mathbf{A}$, along with a plane polar representation of the orbital equation itself, can be established by computing the scalar product $\mathbf{A} \cdot \mathbf{r}$. We find

$$\mathbf{A} \cdot \mathbf{r} = A_1x_1 + A_2x_2 = \frac{x_1^2}{r} - \frac{Jp_2x_1}{K\mu} + \frac{x_2^2}{r} + \frac{Jp_1x_2}{K\mu}$$

$$= r - \frac{J^2}{K\mu}$$

$$= Ar\cos\varphi = er\cos\varphi$$

where the fact that $J = J_3$ has been used to get the second step. We can epitomize this calculation by rewriting it in the form

$$r = \frac{J^2/K\mu}{1 - e\cos\varphi}$$

(4.81)

Equation 4.81 is an expression, written in the plane polar coordinates (r, φ), for an ellipse of eccentricity e and latus rectum $2J^2/K\mu$, whose major axis lies along the x_1 direction (see Figure 4.10). When the polar angle φ is zero, $\mathbf{A}$ is parallel with the vector $\mathbf{r}$. Since r takes on its maximum value when $\varphi = 0$, we conclude that $\mathbf{A}$ always points toward the *aphelion* of the orbit.

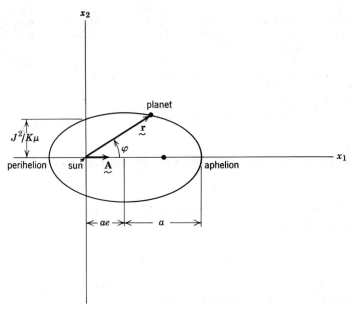

FIGURE 4.10. The general elliptic orbit in the Kepler problem.

With the information now in hand we can derive all of the basic empirical characteristics of planetary motion as discovered by Kepler. Kepler's First Law states that *the orbits are ellipses with the sun at one focus.* This result is derivable immediately from Figure 4.10 if we note that, for most practical purposes, the reduced mass of the sun-planet assembly is that of the planet alone (see Equation 4.71). It then makes sense to say that the planet orbits about the sun. In Table 4.1 we have listed the measured values of the orbital parameters a and e for each of the planets. There we note that the semimajor axis of each orbit is very much larger than the equatorial radius of the sun (always the larger of the two bodies in question) in agreement with our assumption to support the use of Equation 4.69. Moreover, the eccentricity of any planetary orbit is generally quite small, so small, in fact, that the two foci of the elliptical orbit lie inside the perimeter of the sun in several cases.

Kepler's Second Law states that *a line drawn from the sun to a planet will sweep out equal areas in equal times.* To see this, we recall from analytic geometry that

TABLE 4.1   Properties of the Planetary Orbits[a]

Planet	a (A.U.)	e	T (yr)	M_p (sm × 10^{-6})	$R_\odot/a$
Mercury	0.387099	0.205628	0.24085	0.1665	0.01202
Venus	0.723332	0.006787	0.61521	2.449	0.00643
Earth	1.000000	0.016722	1.00004	3.005	0.00465
Mars	1.523691	0.093377	1.88089	0.3230	0.00305
Jupiter	5.202803	0.048452	11.86223	954.9	0.00089
Saturn	9.53884	0.05565	29.4577	285.9	0.00049
Uranus	19.1819	0.04724	84.0139	43.69	0.00024
Neptune	30.0578	0.00858	164.793	51.77	0.00015
Pluto	39.44	0.250	247.7	0.511	0.00011

[a] The data are taken from C. W. Allen, *Astrophysical Quantities*, The Athlone Press, London, 1973. The values of e are slightly time dependent.

Note. 1 A.U. = 1.495979 ± 0.000001 × 10^{13} cm
    1 sm = 1.989 ± 0.002 × 10^{33} g (mass of the sun)
    1 $R_\odot$ = 6.9599 ± 0.0007 × 10^{10} cm (radius of the sun)

the area of an infinitesimal triangle swept out by the motion of a radius vector is

$$d\mathscr{A} = \tfrac{1}{2}r^2\,d\varphi$$

in terms of the diagram shown in Figure 4.11. It follows that the rate at which area is swept out should be

$$\frac{d\mathscr{A}}{dt} = \tfrac{1}{2}r^2\dot\varphi = \frac{J}{2\mu} \tag{4.82}$$

which is constant. Note that Equation 4.82 requires only the constancy of J and, therefore, is valid for *any* central-field potential.

Kepler's Third Law has it that *the period of revolution for an orbit is proportional to the three-halves power of its semimajor axis.* This statement is simply an integral

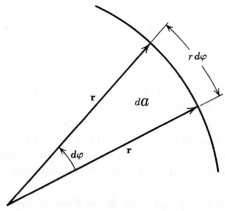

FIGURE 4.11. The element of area swept out by a radius vector moving through $d\varphi$.

form of Equation 4.82. Indeed,

$$\int_0^{\pi ab} d\mathscr{A} = \frac{J}{2\mu} \int_0^T dt$$

by definition, where πab is the area of an ellipse, $b = a(1 - e^2)^{1/2}$ being the length of its semiminor axis, and T is the period of revolution. If we carry out the integrals, we get

$$\pi a^2 (1 - e^2)^{1/2} = \pi a^2 \left(\frac{J^2}{K\mu a} \right)^{1/2}$$

$$= \frac{J}{2\mu} T$$

or

$$T = \left(\frac{4\pi^2 \mu}{K} \right)^{1/2} a^{3/2} \tag{4.83}$$

where we have used Equations 4.75 and 4.76 to reduce the left-hand side of the first equality. Equation 4.83 shows that Kepler's Third Law is not quite exact, since the ratio

$$\frac{\mu}{K} = \frac{1}{G(M_s + M_p)} \tag{4.84}$$

depends very slightly upon the mass of the planet. (See the fifth column of Table 4.1.)

4.7 ARTIFICIAL EARTH SATELLITES

In Table 4.2 are listed the orbital parameters for several artificial Earth satellites. The placing into orbit of these devices has by now become a familiar, almost

TABLE 4.2 Properties of the Orbits of Certain Artificial Earth Satellites[a]

Satellite	a (km)	e	T (min)	m (kg)	Purpose
Sputnik 1	6958	0.052	96.2	83	Exploratory
Explorer 1	7826	0.140	114.8	14	Exploratory
Vostok 1	6625	0.011	89.4	4725	Manned flight
Friendship 7	6583	0.008	88.5	1355	Manned flight
Apollo 7	6591	0.004	88.8	14515	Manned flight
Tiros 1	7094	0.004	99.0	122	Weather observation
OSO 1	6946	0.003	96.0	208	Solar observatory
Alouette 1	7384	0.002	105	145	Ionsphere research
Syncom	42430	0.013	1449	39	Communications
Molniya 1	26310	0.739	707	—	Communications
Vela 1	114260	0.051	6402	135	Nuclear test detection
OGO 1	81200	0.918	3836	487	Geophysical observatory

[a] The data are taken from J. R. Merrill, "Information on Some Interesting Satellites," *American Journal of Physics* 37:826–831 (1969).

routine, event and their scientific utility is well-established in space physics. It is obvious that the study of an orbiting artificial satellite will draw heavily on the theories and techniques of classical dynamics. Here we shall very briefly consider three typical applications.

1. *Calibration of the astronomical unit.* The Gaussian gravitation constant k is defined in terms of the physical parameters for an Earth satellite orbit by the equation

$$k = \frac{2\pi a^{3/2}}{T(m + M_E)^{1/2}} \tag{4.85}$$

where m is the mass of the satellite and M_E is that of the Earth measured in units of the solar mass. The unit of a is, by convention, the astronomical unit (A.U.) and that of T is the day. The value of k has been fixed at that first obtained by Gauss through an analysis of the Earth-moon orbit about the sun:

$$k = 0.01720209895 \ (A.U.)^3/day - sm^{1/2}$$

where *sm* means 1 solar mass. This fixed numerical value then can be used along with orbital data to calibrate the astronomical unit in terms of meters or centimeters.

If the period T is measured in days and the relevant masses are expressed as fractions of the solar mass, the length of the semimajor axis of the orbit of an Earth satellite will be given in A.U. by the equation

$$a = \left[\frac{kT(m + M_E)^{1/2}}{2\pi}\right]^{2/3} \tag{4.86}$$

If this parameter is also known in centimeters, for example, Equation 4.86 provides a means of calibrating the astronomical unit. As an example, consider the data in Table 4.2 for the satellite Tiros 1. We have $T = 99/1440$ day, $m + M_E = 3.005 \times 10^{-6}$ *sm* and, therefore, $a = 0.4739483 \times 10^{-4}$ A.U. The table shows that $a = 7.094 \times 10^8$ cm also. We deduce 1 A.U. $= 1.4968 \times 10^{13}$ cm, in good agreement with other calibrations.

2. *Orbital Stability.* The values of the eccentricity e listed in Table 4.2 are typically very small; the satellite orbits are very nearly perfect circles. An important question is how stable these orbits will be under perturbation by a weak conservative force. Such a force will combine with the potential given by Equation 4.69 (with the masses redesignated appropriately) to produce a general potential $V(r)$.

We begin by considering the effect of the perturbing force on the relative angular momentum $\mathbf{J}_0$ for a circular orbit. The force may be resolved into a component directed along $\mathbf{J}_0$ and one lying in a plane perpendicular to $\mathbf{J}_0$. The part along $\mathbf{J}_0$ will only alter $\mathbf{J}_0$ to a new constant vector and, therefore, will not affect stability. The part in the plane of the orbit will affect $\|\mathbf{J}_0\|$ if it is applied tangentially. This condition we assume to exist.

Now, the quantity J_0^2/μ will be altered to

$$\frac{J^2}{\mu} = \frac{J_0^2}{\mu} + \delta = r_0^3 V'(r_0) + \delta \tag{4.87}$$

by the perturbing force, where $V'(r_0) \equiv (dV/dr)_{r=r_0}$ and δ is the change in $J_0{}^2/\mu$. The radius of the unperturbed orbit is r_0 and $r_0{}^3 V'(r_0)$ is the value of $J_0{}^2/\mu$, as may be seen by taking the derivative of Equation 4.64 with respect to r and setting the result equal to zero. The result (4.87) we shall put into Equation 4.63:

$$E \simeq \tfrac{1}{2}\mu \dot{r}^2 + \frac{1}{2}\frac{(r_0{}^3 V'(r_0) + \delta)}{r^2} + V(r) \tag{4.88}$$

Since the perturbing force is very weak we may set $r = r_0 + \xi$ and expand r^{-2} and $V(r)$ in Taylor series about r_0 to second order in ξ. The expression for the total relative energy then becomes

$$E' \simeq \tfrac{1}{2}\mu \left(\frac{d\xi}{dt}\right)^2 + \frac{1}{2}\left[V''(r_0) + 3\frac{(r_0{}^3 V'(r_0) + \delta)}{r_0{}^4}\right]\xi^2 - \frac{\delta}{r_0{}^3}\xi \tag{4.89}$$

where $E' \equiv E - \tfrac{1}{2}[2V(r_0) + r_0 V'(r_0) + (\delta/r_0{}^2)]$. Equation 4.89 is just an expression for the total energy of a linear harmonic oscillator in a uniform field *provided that*

$$V''(r_0) + 3\frac{V'(r_0)}{r_0} > 0 \tag{4.90}$$

(We assume $r_0{}^3 V'(r_0) \gg \delta$.) If the condition (4.90) is met, the perturbing force will cause the radial coordinate only to oscillate about a value close to r_0 and stability will be ensured. If (4.90) is not satisfied, ξ will grow with time and the orbit will be unstable.

3. *Atmospheric Drag.* An artificial Earth satellite orbiting in the altitude range between 150 to 600 km is subject to an important perturbing force due to atmospheric resistance. The principal effect of this dissipative force is to bring on a monotonic decrease in the length of the semimajor axis that causes the orbit to contract and, eventually, the satellite to plunge to the ground. To calculate this effect we postulate the Newtonian frictional force

$$F = -\alpha v^2$$

where $\alpha \simeq A\rho$, A being an average cross-sectional area of the satellite and ρ, the density of the atmosphere. Now let us write

$$E = -\frac{K}{2a} \qquad \frac{dE}{dt} = Fv$$

and derive the expression

$$\frac{dE}{dt} = \frac{K}{2a^2}\frac{da}{dt} = -\alpha v^3$$

Thus

$$\frac{da}{dt} = -\frac{2\alpha a^2}{K}v^3 \tag{4.91}$$

which shows that the orbit continually shrinks in the presence of friction.

Now let us specialize Equation 4.91 to the case of a circular orbit. Then $a = R$, the radius of the orbit, and $v = 2\pi R/T$, where T is the orbital period. Equation 4.91 becomes

$$\frac{dR}{dt} = -\frac{16\alpha\pi^3 R^5}{KT^3} = -2\alpha\left(\frac{KR}{\mu^3}\right)^{1/2} \qquad (4.92)$$

upon introducing Kepler's Third Law, Equation 4.83. The left-hand side of Equation 4.92 is the satellite decay rate. We see that the decay rate is proportional to the atmospheric density (in the parameter α) and to the square root of the orbital radius. At low altitudes the decay rate increases rapidly because the atmospheric density increases approximately in an exponential fashion with decreasing height above the Earth's surface. When the altitude is less than about 160 km, the orbit will be completely unstable. The lifetime of the orbit, therefore, can be found by integrating Equation 4.92 between the initial altitude and this lower limit, given some reasonable expression for the atmospheric density as a function of height.

FOR FURTHER READING

H. M. Schey — *Div, Grad, Curl, and All That*, W. W. Norton, New York, 1973. This book presents vectors and vector calculus in an informal and applications-oriented manner. Chapter III explains the curl.

J. Marion — *Principles of Vector Analysis*, Academic Press, New York, 1965. This is a more "serious" book than that by Schey. The discussion is introductory, but with many important details included. Chapter 3 gives a good discussion of the line integral.

K. Symon — *Mechanics*, Addison-Wesley, Reading, Mass., 1971. Chapter 3 presents a comprehensive description of vector mechanics and orbital motion. A lengthy physical discussion of the conic sections is given in Section 3.14.

J. M. A. Danby — *Fundamentals of Celestial Mechanics*, Macmillan, New York, 1962. This is a very good introduction to orbital motion in its astronomical context. Chapters 4 and 6, on central-field potentials and the two-body problem, respectively, contain interesting worked examples and problems.

W. M. Kaula — *An Introduction to Planetary Physics*, Wiley, New York, 1968. For any reader wishing to see astrophysical applications of classical dynamics included with much experimental data, this book is a very good source.

K. J. Ball and G. F. Osborne — *Space Vehicle Dynamics*, Oxford University Press, London, 1967. This book presents a fine introduction to the mechanics of artificial satellites. Chapter 10 contains a discussion of atmospheric drag and other orbit-perturbing forces.

PROBLEMS

1. A group of experiments that are not completely unambiguous has led to the following three possibilities for the conservative force acting on a particle under study. Show which of the three is the only physically acceptable one and calculate its associated potential function. (You will have to choose the point where the potential is to be equal to zero, as usual.)

(a) $F = \dfrac{F_0 x_2 \hat{x}_1}{(x_1{}^2 + x_2{}^2)^{1/2}} + \dfrac{F_0 x_1 \hat{x}_2}{(x_1{}^2 + x_2{}^2)^{1/2}}$

(b) $F = \dfrac{F_0 x_1 \hat{x}_1}{(x_1{}^2 + x_2{}^2)^{1/2}} + \dfrac{F_0 x_2 \hat{x}_2}{(x_1{}^2 + x_2{}^2)^{1/2}}$

(c) $F = \dfrac{-F_0 x_2 \hat{x}_1}{(x_1{}^2 + x_2{}^2)^{1/2}} + \dfrac{F_0 x_1 \hat{x}_2}{(x_1{}^2 + x_2{}^2)^{1/2}}$

2. The potential function for an ion of charge q in the vicinity of a water molecule of dipole moment μ is given by

$$V(\mathbf{r}) = \frac{-\mu q \cos \theta}{r^2}$$

where θ is the angle between the dipolar axis of the molecule and the line of centers for the ion-molecule pair. (The molecule may be considered to be fixed at the origin of a spherical polar coordinate system. Then its dipolar axis and the x_3 axis coincide.) Calculate the force exerted by the water molecule on the ion.

3. An object slides without friction up and down a track whose path in space is the parabola expressed by $x_3 = kx_1{}^2$, where k is a positive constant. Calculate the force on the particle at any point on the track and show that this force is conservative.

4. Given the general symmetry property for an exact differential

$$\frac{\partial^2 V}{\partial x_i\, \partial x_j} = \frac{\partial^2 V}{\partial x_j\, \partial x_i} \qquad (i, j = 1, 2, 3)$$

show that Equation 4.16 implies Equations 4.17.

5. Derive Equation 4.23 from Equations 4.20, 4.21, and 4.22.

6. Let the force acting on a particle be divided into a component parallel to the position vector $\mathbf{r}$ and one perpendicular to that vector: $\mathbf{F} = \mathbf{F}_\parallel + \mathbf{F}_\perp$. Show that $d\mathbf{J}/dt = \mathbf{r} \times \mathbf{F}_\perp$, where $\mathbf{J}$ is the angular momentum. The axial vector $\mathbf{r} \times \mathbf{F}_\perp$ is called the *torque* of the force $\mathbf{F}$.

7. Prove the following statement: *if the angular momentum of a particle is a constant of the motion and if the force acting on the particle does not depend on* $\mathbf{v}$ *or* t, *the force is conservative.*

8. Use Equation 4.30, with $\theta = \pi/2$, to calculate the angular motion $\varphi(t)$ of the istropic oscillator. You may assume the same initial condition as in Equation 4.39 to simplify the calculation.

9. Show, for the isotropic oscillator, that

(a) $e = [1 - (r_0/a)^2]^{1/2}$ where $r_0 = (J/m\omega_0)^{1/2}$, and

(b) the speed v_{min}, when $r = r_{max}$, and the speed v_{max}, when $r = r_{min}$, are related by

$$v_{max}{}^2 - v_{min}{}^2 = \frac{4v_0{}^2 e}{1 - e^2}$$

where $v_0 = (J\omega_0/m)^{1/2} = r_0 \omega_0$.

10. (a) Write down the Second Law, in cartesian coordinates, for the isotropic oscillator subject to Stokes damping. You may assume the motion takes place in the x_1-x_2 plane and that the time constant τ is the same for both directions.

(b) Solve the equations of motion for the initial conditions that produced Equation 4.42 and show that $r(t) = (E_0/m\omega_0{}^2)^{1/2} \exp(-t/2\tau)$ in that case, where E_0 is the initial energy of the oscillator. What will the trajectory of the oscillator be in the x_1-x_2 plane?

11. Consider two particles moving under the potential function

$$V(\mathbf{r_A}, \mathbf{r_B}) = V(\mathbf{r_A} - \mathbf{r_B}) + m_A g x_{A3} + m_B g x_{B3}$$

where g is the acceleration due to gravity. Show that the transformation to the center-of-mass coordinate system still separates E_T into E_{CM} and E in this case, but results in a new expression for E_{CM}.

12. Given Equation 4.63 and $J = \mu r^2 \dot{\varphi}$ find the potential function that leads to the orbit $r(\varphi) = r_0 (1 - \cos \varphi)$, where r_0 is a positive constant. The curve described by $r(\varphi)$ in the x_1-x_2 plane is called, for an obvious reason, a cardioid.

13. Repeat Problem 12 for the orbit specified by $r(\varphi) = A/(1 + B \cos \varphi)$ where A and B are positive constants.

14. Use the method of the effective potential to describe the relative motion under the potential function

$$V(r) = -\frac{K}{r} - \frac{B}{r^3}$$

where K is given in Equation 4.71 and $B = J^2 K/(\mu c)^2$, c being the speed of light in vacuum. You may assume $E < 0$ in the discussion. The term in B in $V(r)$ is a correction, due to general relativity theory, to the Kepler potential, $-K/r$.

15. Calculate the values of E and J that correspond to circular orbits under the Yukawa potential

$$V(r) = -\left(\frac{g}{r}\right) e^{-\alpha r}$$

given the condition $\alpha r \ll 1$. The constants g and α are positive.

16. Describe the relative motion under the potential function $V(r) = -A/r^2$, where A is a positive constant, given the conditions $E < 0$, $J < (2\mu A)^{1/2}$, $\dot{r}(0) < 0$, and $r(0)$ equal to its largest possible value.

17. Calculate $r(t)$ for the potential function and conditions given in Problem 16. How long does it take for the particles to collapse together?

18. Consider relative motion under the Yukawa potential when $\alpha r \ll 1$ by expanding the potential function in Problem 15 to first order in αr and calculating $r(t)$. What is the period of an elliptic orbit in this case?

19. Calculate the particle trajectory in the r-φ plane for the situation considered in Problem 18 by applying the method of the Runge-Lenz vector.

20. Show that the Runge-Lenz vector may be written

$$A = \hat{\mathbf{r}} + \frac{(\mathbf{J} \times \mathbf{p})}{K\mu}$$

where $\hat{\mathbf{r}} = \mathbf{r}/\|\mathbf{r}\|$.

21. It is stated in Section 4.6 that at least one additional constant of the motion exists for the isotropic oscillator, besides E and $\mathbf{J}$. Show that

$$T_{ij} = \frac{p_i p_j}{2\mu} + \tfrac{1}{2}kx_i x_j \qquad (i, j = 1, 2, 3)$$

are set of such quantities. The nine functions T_{ij} are, technically, the elements of a tensor of rank two.

22. Derive Equations 4.79.

23. Show, for the Kepler Problem, that
   (a) $e = [1 - (r_0/a)]^{1/2}$, where $r_0 = J^2/K\mu$, and
   (b) the speed v_{max}, when $r = r_{min}$, and the speed v_{min}, when $r = r_{max}$, are related by

$$v_{max}^2 - v_{min}^2 = \frac{4v_0^2 e}{1 - e^2}$$

where $v_0^2 \equiv 2|E|/\mu$.

24. The Earth satellites Tiros 1 and OSO 1, described in Table 4.2, swept out area in their respective orbits at the rate of 26,600 km²/s and 26,300 km²/s. Calculate the orbital angular momentum of Tiros relative to OSO.

25. Using the data in the table below, calculate the speed at apogee (greatest distance from Earth) for each of the Earth satellites. The mass of Explorer 33 is 93.6 kg; that of Earth is 5.977×10^{24} kg.

Satellite	OGO 1	Explorer 33
v (perigee) (km/hr)	38,600	21,030

26. Choose two columns of appropriate data in Table 4.2 and plot their logarithms against one another to determine a value for the mass of the Earth.

27. A pulsar is a rotating star that emits electromagnetic radiation at precise intervals. In a simple model, this radiation originates from an outer layer of the star whose period of rotation equals that of the pulses and is computable from Kepler's Third Law.
   (a) Show that $T^2 = 3\pi/G\rho$ gives the period of rotation of this layer, where ρ is the mean density of the (spherical) star.
   (b) Calculate the density of the Crab pulsar, for which $T = 0.033$ s.

28. Investigate the stability of circular orbits for the following potential functions:
   (a) $V(r) = -K/r$
   (b) $V(r) = -(g/r)e^{-\alpha r}$ $(\alpha r \ll 1)$
   (c) $V(r) = -K/r - B/r^2$

29. A satellite is put into orbit about the Earth at perigee by giving it the speed 29,590 km/hr there at a distance of 360 km from the Earth's surface. Assume the Earth to have the radius 6,378 km and compute the eccentricity, semimajor axis, and period of revolution of the orbit.

30. Calculate the percentage decrease, due to frictional drag, in the semimajor axis of the orbit of Vostok 1 during one period of revolution. The mean orbital speed was 2.79×10^4 km/hr. Take $\alpha \simeq 3 \times 10^{-12}$ kg/m in making your estimate. How long will it take the satellite to fall out of orbit, given this value of α? (Assume a circular orbit to estimate the lifetime.)

5

Unbounded Motion in
Three Dimensions

5.1 ELASTIC COLLISIONS

The most important example of unbounded motion involving two particles occurs when the pair engages in a *scattering process*. This class of phenomena we met only briefly in Chapter 2, where we considered a single particle, incident on a potential barrier, with total energy less than the barrier height. Under this condition we encounter the single turning point so essential to scattering: the particle penetrates the repulsive force-field as deeply as its total energy allows, then is turned about, never to appear again. When we enlarge our perspective to include two particles repelling one another, the situation remains much the same. In fact, the only significant difference comes about through the complicating fact that the scattering is brought on by a mutual interaction of two moving centers of force rather than by a single fixed one. This two-particle aspect of the problem we shall take up now in detail.

From the point of view of the experimental physicist, the simplest description of a two-particle scattering process is carried through in a frame of reference in which one of the particles, called "the target," is initially at rest. In that case, as may be seen in Figure 5.1, the scattering involves a change from the two-particle state $\{\mathbf{v}_A, \mathbf{v}_B = \mathbf{0}\}$ to the state $\{\mathbf{v}'_A, \mathbf{v}'_B\}$, where each velocity is measured relative to the position of the target taken as origin and, most importantly, *is specified only when the particles are far enough apart to be regarded essentially as noninteracting.* Thus the velocities discussed in this scattering process are supposed always to be constant vectors in the frame of reference of the target. This picture of the behavior accordingly is said to occur in the *laboratory coordinate system.* To simplify our investigation of the dynamics in this coordinate system we shall make two broad assumptions that in practice turn out to be valid for many cases of importance. We shall assume first that the two-particle interaction is perfectly well described by a central-field potential and, therefore, that the angles φ_D and φ_R in Figure 5.1 are not necessary to our discussion (i.e., the scattering takes place in a plane). Second, we shall consider only *elastic collisions,* that is, the Energy Theorem for the scattering process will take the form

$$T_{\text{LAB}} \equiv \tfrac{1}{2}m_A v_A{}^2 = \tfrac{1}{2}m_A v'^2_A + \tfrac{1}{2}m_B v_B{}^2 \tag{5.1}$$

where T_{LAB} is, of course, a constant of the motion.[1] Given these two assumptions we may portray the general scattering process as in Figure 5.2. The usual procedure is to settle on which of the dynamical parameters shown there ought to be measured and which should be calculated through some kind of analysis of the problem. Two illustrative choices might be the following. We could measure the masses m_A and m_B, the incident velocity $\mathbf{v}_A$, and the recoil angle θ_R. Then we hope to compute the *recoil energy*

$$\Delta E_R \equiv \tfrac{1}{2}m_B v'^2_B = \tfrac{1}{2}m_A(v_A{}^2 - v'^2_A) \tag{5.2}$$

Otherwise, we could measure just the mass m_A, the velocities $\mathbf{v}_A$ and $\mathbf{v}'_A$, and the deflection angle θ_D with the idea of computing the mass of the target m_B. Let us see how these two possibilities are worked out.

The Momentum Theorem, applied to the situation in Figure 5.2, produces the equations

$$m_A v_A = m_A v'_A \cos \theta_D + m_B v'_B \cos \theta_R \tag{5.3}$$

$$0 = m_A v'_A \sin \theta_D - m_B v'_B \sin \theta_R \tag{5.4}$$

which refer to the x_3 and x_2 directions, respectively. Equations 5.1, 5.3, and 5.4 enable us to determine any three dynamical parameters in terms of the remaining

---

[1] Because $\mathbf{v}_A$ is evaluated before the incident particle interacts with the target, T_{LAB} is numerically equal to the total energy of the two particles. The right-hand side of Equation 5.1 would be smaller than T_{LAB} if the collision were *inelastic*, the difference in energy being either dissipated as heat or used to perturb the internal constituents of the target or the scattered particle.

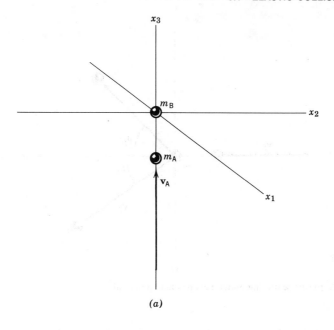

(a)

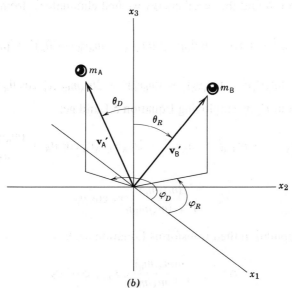

(b)

FIGURE 5.1. The scattering of two particles in the LAB frame of reference. (a) Before the collision. (b) After the collision.

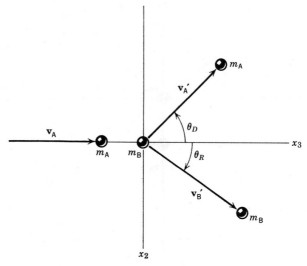

FIGURE 5.2. Elastic scattering under a central-field potential.

ones. If we wish to find the recoil energy we first eliminate θ_D from Equations 5.3 and 5.4:

$$(m_A v'_A \cos \theta_D)^2 + (m_A v'_A \sin \theta_D)^2 = (m_A v_A - m_B v'_B \cos \theta_R)^2 + (m_B v'_B \sin \theta_R)^2$$

or

$$(m_A v'_A)^2 = (m_A v_A)^2 + (m_B v'_B)^2 - 2m_A m_B v_A v'_B \cos \theta_R \tag{5.5}$$

Next we take out v'_A by employing Equation 5.1 and get

$$m_A v_A{}^2 - m_B v'_B{}^2 = m_A v_A{}^2 - 2m_A m_B v_A v'_B \cos \theta_R + \frac{(m_B v'_B)^2}{m_A}$$

or

$$v'_B = \frac{2(m_A/m_B)}{1 + (m_A/m_B)} v_A \cos \theta_R \tag{5.6}$$

A simple manipulation then transforms Equation 5.6 into

$$\Delta E_R = \frac{4(m_A/m_B)}{(1 + m_A/m_B)^2} T_{\text{LAB}} \cos^2 \theta_R \tag{5.7}$$

The recoil energy is completely determined if the mass ratio, the total energy, and the recoil angle are known. We note in passing that the last parameter is restricted to $0 \leqslant \theta_R \leqslant (\pi/2)$ since v'_B in Equation 5.6 must remain nonnegative. Physically this means that the target always recoils in the "forward" direction.

Often θ_R itself is not readily measured, but rather θ_D is known. These two angles are related through the expression

$$\tan \theta_D = \frac{\sin 2\theta_R}{(m_A/m_B) - \cos 2\theta_R} \tag{5.8}$$

which follows from combining Equations 5.3 to 5.5 and using the trigonometric identities

$$\sin 2\theta \equiv 2 \sin \theta \cos \theta \qquad \cos 2\theta \equiv \cos^2 \theta - \sin^2 \theta$$

Equation 5.8 is represented in Figure 5.3 in the form of a plot of θ_D against θ_R for constant mass ratio (m_A/m_B). The interesting aspects of the graph are the discontinuous jump from π to $\pi/2$ in θ_D when $\theta_R = 0$ as the mass ratio approaches the value 1, and the maximum exhibited by θ_D when $(m_A/m_B) > 1$. The former of these characteristics is the result of the velocities $\mathbf{v}_A$, $\mathbf{v}_A'$, and $\mathbf{v}_B'$ being suddenly related to one another through the pythagorean theorem when $m_A = m_B$, according to Equation 5.1. Since v_A is the length of the hypotenuse of the "velocity

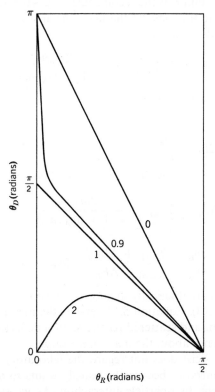

FIGURE 5.3. A graph of Equation 5.8. The mass ratio m_A/m_B is a parameter for each curve.

triangle," we must have

$$\theta_D + \theta_R = \frac{\pi}{2} \tag{5.9}$$

for any scattering event. The case $\theta_R = 0$ gives $\theta_D = \pi/2$, and $v'_A = 0$ according to Equation 5.4 (head-on collision). Otherwise, when $\theta_R = 0$, Equations 5.3 and 5.4 tell us that the three velocities are then collinear and, in fact,

$$v'_A = \frac{(m_A/m_B) - 1}{(m_A/m_B) + 1} v_A$$

$$v'_B = \frac{2(m_A/m_B)}{(m_A/m_B) + 1} v_A \tag{5.10}$$

It follows from the first of these expressions that $\theta_D = \pi$ when $\theta_R = 0$ and $m_A < m_B$. The maximum in θ_D for large mass ratios can also be deduced with the help of Equations 5.10, since they lead to $\theta_D = 0$ when $\theta_R = 0$ and $m_A > m_B$. The fact that $\theta_D = 0$ also when $\theta_R = \pi/2$ then forces us to conclude that the deflection angle must show a maximum as the recoil angle ranges through its possible values.

In order to calculate the mass of the target from scattering data we combine Equations 5.1, 5.2, and 5.7 to get

$$\Delta E_D \equiv \tfrac{1}{2} m_A v'^2_A = T_{LAB} \left\{ 1 - \frac{4(m_A/m_B)}{[1 + (m_A/m_B)]^2} \cos^2 \theta_R \right\} \tag{5.11}$$

where ΔE_D is the *scattering energy*. Now we shall fix the deflection angle at 90° in Equation 5.8 so that

$$\cos 2\theta_R = \frac{m_A}{m_B}$$

$$\equiv \cos^2 \theta_R - \sin^2 \theta_R \equiv 2 \cos^2 \theta_R - 1$$

$$\Delta E_D = T_{LAB} \left\{ 1 - \frac{2(m_A/m_B)}{1 + (m_A/m_B)} \right\} \tag{5.12}$$

and, finally,

$$\frac{m_B}{m_A} = \frac{T_{LAB} + \Delta E_D}{T_{LAB} - \Delta E_D} \quad (m_B > m_A) \tag{5.13}$$

where we have noted parenthetically that $\theta_D = 90°$ is physically impossible unless $m_B > m_A$ (see Figure 5.3). Equation 5.13 permits the measurement of the kinetic energies of the incident and scattered particles, for a deflection angle of 90°, to be converted to an inference about the mass of the target.

The theoretical physicist does not regard the laboratory frame of reference as the simplest for his purposes because he usually wants to analyze the scattering process in terms of the two-particle interaction. As we saw in Chapter 5, the convenient way to look at two-particle motion is in the *center-of-mass coordinate system*. If we adopt that point of view here we must write the equations of transformation

$$\mathbf{v}_A = \mathbf{V} + \frac{m_B}{M} \mathbf{v}_\infty \tag{5.14a}$$

$$\mathbf{v}_B = 0 = \mathbf{V} - \frac{m_A}{M} \mathbf{v}_\infty \tag{5.14b}$$

$$\mathbf{v}'_A = \mathbf{V}' + \frac{m_B}{M} \mathbf{v}'_\infty \tag{5.15a}$$

$$\mathbf{v}'_B = \mathbf{V}' - \frac{m_A}{M} \mathbf{v}'_\infty \tag{5.15b}$$

where

$$\mathbf{V} = \frac{m_A}{M} \mathbf{v}_A$$

$$\mathbf{v}_\infty = \mathbf{v}_A$$

because $v_B = 0$. Now, we know from our previous experience that, in the presence of a two-particle potential alone, the total linear momentum of the system $M\mathbf{V}$ is a constant of the motion. It follows that $\mathbf{V} = \mathbf{V}'$ in the transform equations. Moreover, since both

$$T_{CM} = \tfrac{1}{2} M V^2 \tag{5.16}$$

and

$$T_\infty = \tfrac{1}{2} \mu v_\infty{}^2 \tag{5.17}$$

are constants of the motion as well, we must have $v_\infty = v'_\infty$. Therefore, *the only effect of a collision in the CM frame of reference is to change the direction of* $\mathbf{v}_\infty$, *that is, to rotate the line of centers of the two particles involved* (see Figure 5.4). In terms of the relative motion this important conclusion means that the hypothetical particle of mass μ, which appears in the two-particle problem, undergoes a scattering process very similar to that a single particle would undergo from a *fixed* target. This behavior is shown in Figure 5.5; it is evident there that the *analogs*

$$\mu \longleftrightarrow m_A \qquad v_\infty \longleftrightarrow v_A \qquad \theta \longleftrightarrow \theta_D$$

are valid.

The parameters μ and v_∞ can be calculated directly from their definitions in terms of m_A, m_B, $\mathbf{v}_A$, $\mathbf{v}'_A$, and $\mathbf{v}'_B$. A complete contact between the CM and LAB frames of reference is made by noting that

$$v_\infty \cos \theta = v'_A \cos \theta_D - v'_B \cos \theta_R$$

$$= \left(1 + \frac{m_A}{m_B}\right) v'_A \cos \theta_D - \frac{MV}{m_B} \tag{5.18}$$

FIGURE 5.4. The scattering of two particles in the CM frame of reference.

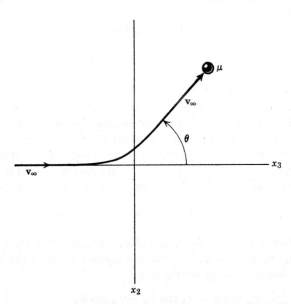

FIGURE 5.5. Scattering in the CM frame in terms of relative coordinates. The origin is a fixed "scattering center."

$$v_\infty \sin \theta = v'_A \sin \theta_D + v'_B \sin \theta_R$$
$$= \left(1 + \frac{m_A}{m_B}\right) v'_A \sin \theta_D \qquad (5.19)$$

where in each case the first equality is derived from breaking the relation

$$\mathbf{v}'_\infty = \mathbf{v}'_A - \mathbf{v}'_B$$

into its components and the second one comes from the conditions for momentum conservation. The two expressions (5.18) and (5.19) combine to form the equation

$$\tan \theta_D = \frac{\sin \theta}{(MV/m_B v_\infty) + \cos \theta}$$

If we add to this expression the fact that $MV = m_A v_\infty$, we get, finally, the rather simple relation

$$\tan \theta_D = \frac{\sin \theta}{(m_A/m_B) + \cos \theta} \qquad (5.20)$$

Two special cases of Equation 5.20 are worth mentioning here. When the mass ratio is very small we may regard the target as effectively immovable and (5.20) gives us

$$\theta_D \simeq \theta \qquad (m_A \ll m_B) \qquad (5.21)$$

This result is entirely consistent with the analogs pointed out above and with the fact that $\mu \simeq m_A$ under the same circumstances. When $m_A = m_B$, we have

$$\tan \theta_D = \frac{\sin \theta}{1 + \cos \theta} = \tan\left(\frac{\theta}{2}\right) \qquad (m_A = m_B) \qquad (5.22)$$

and the LAB deflection angle is always one-half that in the CM frame of reference. Since the maximum value of θ possible is 180°, we conclude that the greatest value of θ_D is 90°—in perfect accord with the graph in Figure 5.3 for the equal-mass case.

5.2 THE SCATTERING CROSS SECTION

The account of collision phenomena given in the previous section has the significant property of being independent of any knowledge of the precise nature of the interaction between the two particles involved. This independence has permitted us to draw some very general conclusions about the relations among the important dynamical parameters in scattering; assuming, of course, that we have evidence that such a phenomenon has actually taken place! However, should we ever choose to ask the question "*Will* two particles of known energy and mutual interaction in fact scatter?" we would be at a loss thus far to give more than a perfunctory answer. Conservation theorems and geometry alone cannot enlighten us on this point because it entails at heart a question of causality. What we need instead (or rather, in addition) is an analysis that is more deeply involved with the exact form of the two-particle potential and which is more directly related to the experimental problem of determining the likelihood of a collision under prescribed conditions. The experimental problem usually presents itself in the following way. The material containing the target particles is arranged to be thin enough to preclude more than one scattering event per incident particle. Then, if the density of the incident beam is low enough to all but prevent interactions between the oncoming particles, the scattering process may be described by

$$N = \sigma_t J \mathcal{N} \tag{5.23}$$

where N is the number of scattered particles produced per unit area of target in unit time, J is the number of incident particles crossing unit area of the target in unit time, and $\mathcal{N}$ is the number of target particles per unit area confronting the incident beam (see Figure 5.6). The coefficient of proportionality between N and

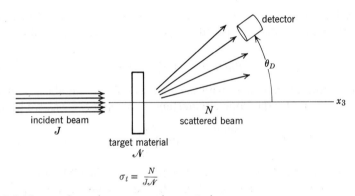

$$\sigma_t = \frac{N}{J\mathcal{N}}$$

FIGURE 5.6. The experimental arrangement in a scattering process.

the product $J\mathcal{N}$, conventionally given the symbol σ_t, has the dimensions of area and is called the total *scattering cross section*. This parameter, despite its name, is not necessarily equal to the cross sectional area of a target particle, but rather is an energy-dependent quantity characteristic of how well the scattering process goes on under given conditions. In agreement with its definition, σ_t is usually inferred from measurements of N, J, and $\mathcal{N}$. For microscopic targets it is recorded in the often convenient units of *barns*, where

$$1 \text{ barn} = 10^{-24} \text{ cm}^2$$

The scattering cross sections of atoms typically are the order of megabarns, while those of atomic nuclei are in the neighborhood of a few barns, and those of fundamental particles lie in the millibarn range.

Quite often the total number of scattered particles N is not measured in a scattering experiment, but instead the number $dN(\Omega_D)$ scattered at an angle θ_D within the element of solid angle[2] $d\Omega_D \equiv 2\pi \sin \theta_D \, d\theta_D$ is the object of the experimentalist's effort. In that case we rewrite Equation 5.23 in the more appropriate form

$$dN(\Omega_D) = \sigma(\theta_D) \, d\Omega_D J \mathcal{N} \tag{5.24}$$

where $\sigma(\theta_D)$ is called the *differential scattering cross section*. This cross section is related to σ_t through

$$\sigma_t = 2\pi \int_0^\pi \sigma(\theta_D) \sin \theta_D \, d\theta_D \tag{5.25}$$

Our remarks should make it clear that we are going to regard the scattering of N particles as N two-particle collisions and that the scattering cross sections are the parameters we expect to calculate in a dynamical analysis. In particular, if we can derive an expression for $\sigma(\theta_D)$ from information about the energetics of the two-particle interaction, we will have at hand a rather complete picture of any scattering process. The differential scattering cross section we get from theory, as suggested earlier, is naturally discussed in the CM frame of reference. Thus we shall consider first the problem of calculating $\sigma(\theta)$, then turn to the purely geometric problem of relating it to $\sigma(\theta_D)$.

In classical dynamics $\sigma(\theta)$ is related to the two-particle interaction through the *impact parameter*.[3] This quantity is shown in Figure 5.7 and is defined[4] to be the perpendicular distance between the origin in the CM frame and the incident path of the hypothetical particle of mass μ. For a given total energy of the incident particle and a given value of b the angle θ is determined precisely. Thus, as shown in Figure 5.8, an incident particle entering the annulus of area $2\pi b \, db$ will be scattered into the element of solid angle $2\pi \sin \theta(-d\theta)$. The minus sign here reminds us that an increase in b implies a less close approach of the particle to the scattering center and, therefore, a *decrease* in θ, because of a less effective repulsive interaction. For a single target, the causal relationship between b and θ may be expressed '

$$J \, d(\pi b^2) = J\sigma(\theta)2\pi \sin \theta(-d\theta)$$

that is, the number of particles entering the annulus per unit time is that scattered

---

[2] A unit solid angle subtends a unit area on the surface of a sphere of unit radius. Usually it is written $d\Omega = \sin \theta \, d\theta \, d\varphi$ and is measured in *steradians*. Here, because of the spherical symmetry assumed in the scattering process, we have integrated $d\Omega$ over φ. For an introductory discussion of $d\Omega$, see A. F. Kip, *Fundamentals of Electricity and Magnetism*, McGraw-Hill Book Co., New York, 1969, p. 44f.

[3] In quantum dynamics the impact parameter has no physical meaning because of the Heisenberg uncertainty principle. The velocity of the incident particle is known precisely here, which prevents any simultaneous knowledge of the trajectory. The quantum-mechanical differential scattering cross section is accordingly related directly to the probability that a collision will occur.

[4] A little thought shows that b also may be defined as the perpendicular distance between the path of the incident particle and the target in the LAB frame.

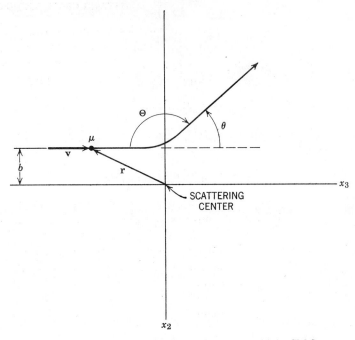

FIGURE 5.7. Geometric quantities in the scattering process as seen in the CM frame.

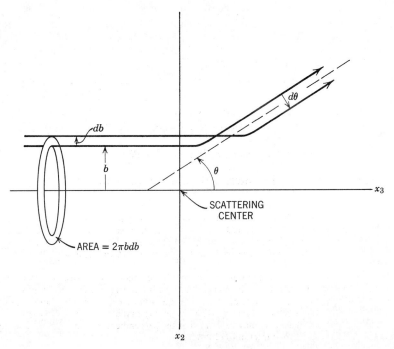

FIGURE 5.8. The relation between the impact parameter and the CM deflection angle. A repulsive particle interaction has been assumed.

into $2\pi \sin \theta(-d\theta)$ per unit time. It follows that

$$\sigma(\theta) = \frac{b(\theta)}{\sin \theta} \left| \frac{db}{d\theta} \right| \qquad (5.26)$$

Through Equation 5.26 we have reduced our problem to that of determining the impact parameter as a function of the deflection angle in the CM frame of reference. This new task we carry out as follows.

Because we have assumed that the two-particle potential function represents a central field we know that the relative angular momentum is a constant of the motion and that the relative energy may be expressed

$$E = \tfrac{1}{2}\mu\dot{r}^2 + \frac{J^2}{2\mu r^2} + V(r) = T_\infty \qquad (5.27)$$

where we have noted the numerical equality between T_∞ and E. (The relative speed $\dot{r}$ that appears in Equation 5.27 must not be confused with the asymptotic relative speed v_∞ that determines T_∞. The former is a variable quantity whose value when r is very large approaches v_∞, but that otherwise is determined by the instantaneous magnitude of r through Equation 5.27). Now, in general,

$$J = \|\mathbf{r} \times \mu\mathbf{v}\| = \mu r v \sin \alpha$$

where α is the angle between r and v in the r-φ plane. From the geometric relations portrayed in Figure 5.9 we see that, when $\dot{r} \sim v_\infty$,

$$\frac{b}{r} = \sin \alpha$$

and

$$J = \mu v_\infty b = (2\mu T_\infty)^{1/2} b \qquad (5.28)$$

On the other hand, it is also quite generally true that

$$J = \mu r^2 \dot{\varphi} = \mu r^2 \dot{r} \frac{d\varphi}{dr}$$

where the chain rule has been employed to write

$$\dot{\varphi} = \dot{r} \frac{d\varphi}{dr}$$

It follows that

$$\frac{d\varphi}{dr} = \frac{J}{\mu r^2} \frac{1}{\dot{r}} = \frac{J}{\mu r^2} \left[\frac{2T_\infty}{\mu} - \frac{J^2}{\mu^2 r^2} - \frac{2V(r)}{\mu} \right]^{-1/2}$$

$$= \frac{b}{r^2} \left[1 - \left(\frac{b}{r} \right)^2 - \frac{V(r)}{T_\infty} \right]^{-1/2} \qquad (5.29)$$

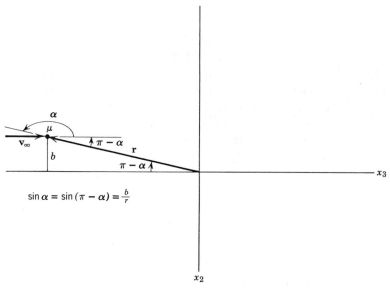

FIGURE 5.9. Geometry of the relation between b and J.

To get the second step above we have solved Equation 5.27 for $\dot{r}$; the third step follows from Equation 5.28. Equation 5.29 is a general parametric equation for the trajectory of the relative motion in the r-φ plane. If we integrate both sides with respect to r between the limits $r_{\min}$ and ∞ we get

$$\frac{\Theta}{2} = \int_{r_{\min}}^{\infty} \frac{(b/r^2)\, dr}{\{1 - (b/r)^2 - [V(r)/T_\infty]\}^{1/2}} \tag{5.30}$$

where Θ is the angle swept out by r in traversing the entire trajectory, as shown in Figure 5.7. It is also evident from that figure that

$$\Theta = \pi - \theta \tag{5.31}$$

Thus Equation 5.30 is really an implicit relation for $b(\theta)$ established—as we wished—in terms of the dynamics of the scattering process. Given a mathematical form for $V(r)$, one can hope to evaluate the integral in (5.30) and thereby obtain the information necessary to compute $\sigma(\theta)$ in Equation 5.26.

Once the differential scattering cross section is calculated in the CM coordinate system, there remains only the problem of relating it to the coordinates which are natural to the experimentalist. This can be done in a straightforward way once it is realized that the quantities dN, J, and $\mathcal{N}$ in Equation 5.24 must be invariant in value under a coordinate transformation. Therefore

$$|2\pi\sigma(\theta) \sin\theta\, d\theta| = |2\pi\sigma(\theta_D) \sin\theta_D\, d\theta_D|$$

or

$$\sigma(\theta_D) = \sigma(\theta) \left| \frac{\sin \theta}{\sin \theta_D} \frac{d\theta}{d\theta_D} \right| = \sigma(\theta) \left| \frac{d(\cos \theta)}{d(\cos \theta_D)} \right|$$

Now, by Equation 5.20,

$$\tan \theta_D = \frac{\sin \theta}{(m_A/m_B) + \cos \theta} \tag{5.20}$$

which, along with the identity

$$\sin^2 \theta_D + \cos^2 \theta_D \equiv 1$$

implies

$$\cos \theta_D = \frac{(m_A/m_B) + \cos \theta}{[1 + (m_A/m_B)^2 + (2m_A/m_B) \cos \theta]^{1/2}} \tag{5.32}$$

Then

$$\frac{d(\cos \theta_D)}{d(\cos \theta)} = \frac{1 + (m_A/m_B) \cos \theta}{[1 + (m_A/m_B)^2 + (2m_A/m_B) \cos \theta]^{3/2}}$$

and we have

$$\sigma(\theta_D) = \frac{[1 + (m_A/m_B)^2 + (2m_A/m_B) \cos \theta]^{3/2}}{|1 + (m_A/m_B) \cos \theta|} \sigma(\theta) \tag{5.33}$$

where the absolute-value sign is to insure the positiveness essential to the definition of the differential scattering cross section. In the special case of equal masses, Equation 5.33 becomes

$$\sigma(\theta_D) = 4 \cos \left(\frac{\theta}{2} \right) \sigma(\theta) \tag{5.34}$$

upon using the relation

$$\cos \left(\frac{\theta}{2} \right) \equiv \left(\frac{1 + \cos \theta}{2} \right)^{1/2}$$

It is interesting that the angular dependence of $\sigma(\theta_D)$ is, in general, rather complicated, even if $\sigma(\theta)$ is a constant.

5.3 HARD-SPHERE SCATTERING

In the simplest approximation one may picture a gas as an assembly of particles whose motions are incessant and chaotic, with interactions limited to elastic collisions induced by a repelling force of zero range. This hypothesis forms the basis for the kinetic theory of gases and places the gas particle in the role of a smooth, hard sphere of given diameter. The properties of the gas then depend only upon the density, average speed, and collision frequency of its constituents.[5] The

[5] Those wishing to learn more about a gas of hard spheres might consult R. D. Present, *Kinetic Theory of Gases*, McGraw-Hill, New York, 1958, Chapter 3.

last-named parameter, of course, must be directly related to the scattering cross section, which itself ought to have the form

$$\sigma_t = \pi d^2 \qquad (5.35)$$

where d is the fixed diameter of the hard-sphere gas particle (see Figure 5.10). Equation 5.35 is the result to be expected on geometric grounds alone for the contact collision of rigid spheres, since it expresses just the cross sectional area of the "sphere of influence" for such a body. On the other hand, the expression should be derivable from Equation 5.30 after the appropriate form of $V(r)$ has been introduced, if the dynamical ideas we have been developing are accurate. The two-particle potential function for the hard-sphere interaction is

$$V(r) = \begin{cases} +\infty & r < d \\ 0 & r > d \end{cases} \qquad (5.36)$$

Equation 5.36 states that an infinite total energy is required for two particles moving under $V(r)$ to approach one another more closely than the distance d. Thus the particles may be regarded as impenetrable spheres of that diameter. Moreover, the very sudden change in $V(r)$ at the distance d (and only there) shows that the repelling force between the particles has no range. The hard-sphere

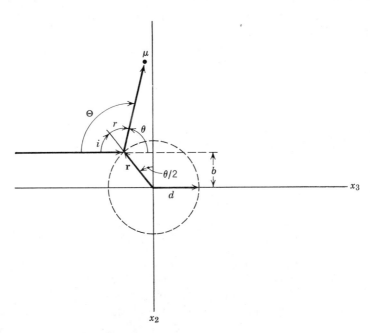

FIGURE 5.10. The scattering of hard spheres.

interaction has a particularly simple effect on the integral in Equation 5.30, for it reduces the quadrature to

$$\frac{\Theta}{2} = \int_d^\infty \frac{(b/r^2)\, dr}{[1 - (b/r)^2]^{1/2}}$$

$$= b \int_d^\infty \frac{dr}{r(r^2 - b^2)^{1/2}}$$

With a table of standard integrals we find easily

$$\frac{\Theta}{2} = \cos^{-1}\left(\frac{b}{r}\right)\Big|_d^\infty = \frac{\pi}{2} - \cos^{-1}\left(\frac{b}{d}\right)$$

and

$$b(\theta) = d \cos\left(\frac{\theta}{2}\right) \tag{5.37}$$

according to Equation 5.31. Equation 5.37 states that, in the CM frame of reference, the impact parameter and the sphere diameter are part of a right triangle with the included angle equal to half the CM angle of deflection (see Figure 5.10). The differential scattering cross section corresponding to (5.37) is

$$\sigma(\theta) = \frac{b(\theta)}{\sin \theta} \left|\frac{db}{d\theta}\right| = \frac{d^2 \cos \theta/2 \sin \theta/2}{2} \frac{1}{\sin \theta} = \frac{d^2}{4} \tag{5.38}$$

The scattering in the CM frame of reference is therefore isotropic, given, of course, that $b \leqslant d$ according to the initial conditions. The scattering cross section follows from (5.38) as

$$\sigma_t = 2\pi \int_0^\pi \sigma(\theta) \sin \theta \, d\theta = \pi d^2 \tag{5.39}$$

in agreement with Equation 5.35. Classical dynamics is seen to confirm fully our expectations based purely on geometric ideas.

There are three points worth mentioning in connection with the calculation we have just gone through. The first is that Equation 5.37 leads to the conclusion that hard-sphere scattering satisfies the "law of reflection" in the CM frame of reference. To see this fact, we refer to Figure 5.10 and note that the angle of incidence i is equal to the complement of the angle $\theta/2$ because it is vertical to that complement in the b-d triangle. Therefore

$$i + \frac{\theta}{2} = \frac{\pi}{2}$$

Moreover, we also can write

$$i + r + \theta = \pi$$

according to Equation 5.31, because $\Theta = i + r$ by definition, where r is the angle of reflection. These two expressions immediately reduce to

$$i = r$$

which is the "law of reflection."

A second aspect of our calculation involves the differential scattering cross section for hard spheres of equal mass in the LAB frame of reference. By Equation 5.34 this quantity is

$$\sigma(\theta_D) = d^2 \cos\left(\frac{\theta}{2}\right) \qquad (m_A = m_B)$$

With Equation 5.32 (for the case $m_A = m_B$) we can transform this expression to the more useful form:

$$\sigma(\theta_D) = d^2 \cos \theta_D \tag{5.40}$$

Hard-sphere scattering in the LAB frame is *not* isotropic and, in fact, is limited to the forward direction, since $0 \leqslant \theta_D \leqslant \pi/2$ in order that $\sigma(\theta_D) \geqslant 0$. However,

$$\sigma_t = 2\pi \int_0^{\pi/2} \sigma(\theta_D) \sin \theta_D \, d\theta_D = \pi d^2$$

as should be, owing to the coordinate-invariance of the differential scattering cross section.

Finally, let us mention that Equation 5.39 is not valid in quantum dynamics. Instead, one finds the scattering cross section for a hard sphere, in the appropriate limit of large T_∞, consists of a geometric part πd^2 and a part arising because of diffraction of the matter wave associated with the incident particle. This latter part is also equal to πd^2 and provides a value for σ_t which is twice that given by classical dynamics.[6]

5.4 COULOMB SCATTERING

One of the most important phenomena studied in the experimental investigation of fundamental particles is the scattering brought on by the Coulomb interaction. As is well known, this interaction exists between any pair of electrically charged particles and is described by the potential function

$$V_C(r) = \frac{q_A q_B}{r} \tag{5.41}$$

where q is the charge on a particle, measured in the CGS unit of statcoulomb, and r is the distance between the centers of the two particles. If T_{LAB} is not too large,

---

[6] For a discussion of this interesting phenomenon, see R. H. Dicke and J. P. Wittke, *Introduction to Quantum Mechanics*, Addison-Wesley, Reading, Mass., 1960, p. 289 ff.

Equation 5.41 provides an excellent accounting of the interactions between protons, electrons, and atomic nuclei in all possible combinations.[7] For these kinds of particles, q is expressed as a multiple of the elementary quantum of charge

$$e = -4.80324 \pm 0.00001 \times 10^{-10} \text{ statcoulomb}$$

which is the charge on the electron. Thus in general we shall write

$$q = Z \, |e|$$

where Z is a whole number of the appropriate sign.

According to Equations 5.30 and 5.41 the impact parameter for Coulomb scattering is determined by the integral

$$\frac{\Theta}{2} = \int_{r_{min}}^{\infty} \frac{(b/r^2)\, dr}{[1 - (b/r)^2 - (r_0/r)]^{1/2}} \tag{5.42}$$

where

$$r_0 \equiv \frac{Z_A Z_B e^2}{T_\infty} \tag{5.43}$$

Because $T_\infty > 0$ the parameter r_0 will be negative for two particles of opposite charge (attractive interaction) and positive for two particles of like charge (repulsive interaction). The integral for the impact parameter may be written

$$b \int_{r_{min}}^{\infty} \frac{dr}{r(r^2 - r_0 r - b^2)^{1/2}}$$

$$= \sin^{-1}\left[\frac{-r_0 r - 2b^2}{r(r_0{}^2 + 4b^2)^{1/2}}\right]_{r_{min}}^{\infty}$$

$$= \sin^{-1}\left[\frac{-r_0}{(r_0{}^2 + 4b^2)^{1/2}}\right] - \sin^{-1}\left[\frac{-r_0 r_{min} - 2b^2}{r_{min}(r_0{}^2 + 4b^2)^{1/2}}\right]$$

The argument of the second arcsine is seen to be equal to -1 after it is recalled that, by definition,

$$E = \frac{J^2}{2\mu r_{min}{}^2} + \frac{r_0}{r_{min}} T_\infty = T_\infty$$

$$= \left(\frac{b^2}{r_{min}{}^2} + \frac{r_0}{r_{min}}\right) T_\infty$$

since r_{min} is a turning point. If we solve this expression for b we get

$$b = r_{min}\left(1 - \frac{r_0}{r_{min}}\right)^{1/2}$$

[7] When T_{LAB} is sufficiently large (about 1.6×10^{-7} ergs for proton-proton scattering) the Coulomb interaction must be augmented by the short-ranged, but very strong, attractive nuclear interaction.

which reduces the argument of the arcsine as suggested. Therefore

$$\frac{\Theta}{2} = \sin^{-1}\left[\frac{-r_0}{(r_0^2 + 4b^2)^{1/2}}\right] + \frac{\pi}{2} = \frac{\pi}{2} - \frac{\theta}{2}$$

or

$$b(\theta) = \frac{r_0}{2}\cot\left(\frac{\theta}{2}\right) \tag{5.44}$$

upon unraveling everything in terms of $b(\theta)$. With the help of Equations 5.26 and 5.44 we find directly

$$\sigma(\theta) = \left(\frac{r_0}{4}\right)^2 \csc^4\left(\frac{\theta}{2}\right) \tag{5.45}$$

Equation 5.45 gives the differential scattering cross section for the Coulomb interaction. This expression was derived first by Ernest Rutherford in 1911 as a part of his celebrated investigation of the scattering of α particles ($_2^4$He nuclei) by the atomic nucleus. For this reason $\sigma(\theta)$ is often designated as $\sigma_R(\theta)$—the Rutherford scattering cross section. Its importance in the theory of low-energy scattering by fundamental particles is only enhanced by the coincidental fact that it remains valid even in quantum dynamics.

There are two interesting aspects of Equation 5.45 worth discussing here. The first is that $\sigma_R(\theta)$ does not depend on the sign of the parameter r_0 and, therefore, is insensitive to whether the Coulomb interaction is attractive or repulsive. It follows that measurements of $\sigma_R(\theta)$ cannot be used to distinguish the sign of the electric charge on the target. A second point is that $\sigma_R(\theta)$ becomes infinite for a zero deflection angle in the CM frame of reference. This unusual occurrence is possible only because of the very long range of $V_C(r)$ that permits zero deflection only when the impact parameter is infinitely large (cf. Equation 5.44). In reality, this situation never exists on the microscopic level because of the propensity for electrical neutrality in matter. Thus a proton interacting with a hydrogen atom is subjected to the full nuclear charge only when it approaches closely enough to overcome the neutralizing effect of the electron in the atom; for sufficiently large, finite proton-atom separations the Coulomb interaction disappears and Equation 5.42 becomes invalid. Consequently $\sigma(\theta)$ will no longer diverge and we need not take too seriously the fact that

$$2\pi \int_0^\pi \sigma_R(\theta)\sin\theta\,d\theta = \frac{\pi r_0^2}{4}\int_0^\pi \cos\left(\frac{\theta}{2}\right)\sin^{-3}\left(\frac{\theta}{2}\right)d\theta \to \infty$$

The small-angle divergence of $\sigma_R(\theta)$ appears in an interesting light when the energy-loss cross section

$$\sigma(\Delta E_D) \equiv \left|\frac{d\sigma_t}{d\,\Delta E_D}\right|$$

is calculated, where ΔE_D is the scattering energy,

$$\Delta E_D = T_{LAB} \left\{ 1 - \frac{4(m_A/m_B)}{[1 + (m_A/m_B)]^2} \cos^2 \theta_R \right\} \tag{5.11}$$

derived in Section 5.1. The computation of $\sigma(\Delta E_D)$ may be carried out by making use of the chain-rule expression

$$\frac{d\sigma_t}{d\,\Delta E_D} = \frac{d\sigma_t}{d\theta} \frac{d\theta}{d\,\Delta E_D} = 2\pi\sigma[\theta(\Delta E_D)] \sin \theta(\Delta E_D) \frac{d\theta}{d\,\Delta E_D} \tag{5.46}$$

once the scattering energy has been given in terms of CM parameters. To accomplish this, we note that

$$T_{LAB} = \tfrac{1}{2}m_A v_A{}^2 = \tfrac{1}{2}\frac{M\mu}{m_B} v_\infty{}^2 = \frac{M}{m_B} T_\infty \tag{5.47}$$

$$\frac{m_A/m_B}{[1 + (m_A/m_B)]^2} = \frac{\mu}{M} \tag{5.48}$$

$$\theta_R = \tfrac{1}{2}(\pi - \theta) \tag{5.49}$$

Equation 5.49 follows from combining Equations 5.8 and 5.20. Thus

$$\Delta E_D = \frac{M}{m_B} T_\infty \left[1 - \frac{4\mu}{M} \sin^2 \left(\frac{\theta}{2}\right) \right] \tag{5.50}$$

and

$$\begin{aligned} \frac{d\,\Delta E_D}{d\theta} &= -\frac{4\mu}{m_B} T_\infty \sin \left(\frac{\theta}{2}\right) \cos \left(\frac{\theta}{2}\right) \\ &= -\frac{2\mu}{m_B} T_\infty \sin \theta \end{aligned} \tag{5.51}$$

If we now use Equation 5.50 to express $\sigma_R(\theta)$ in terms of the scattering energy we get

$$\sigma_R[\theta(\Delta E_D)] = \left(\frac{\mu r_0 T_\infty/m_B}{T_{LAB} - \Delta E_D} \right)^2 \tag{5.52}$$

The results we have obtained are finally combined to produce

$$\begin{aligned} \sigma_R(\Delta E_D) &= 2\pi \left(\frac{\mu r_0 T_\infty/m_B}{T_{LAB} - \Delta E_D} \right)^2 \left(\frac{2\mu T_\infty}{m_B} \right)^{-1} \\ &= \frac{m_A}{m_B} \frac{\pi}{T_{LAB}} \left(\frac{Z_A Z_B e^2}{T_{LAB} - \Delta E_D} \right)^2 \end{aligned} \tag{5.53}$$

This expression shows that collisions wherein there is very little energy loss, that is, where $E_D \simeq T_{LAB}$, are highly favored. The preponderance of these glancing collisions is entirely consonant with Equations 5.45 and 5.50 that tell us that the small-angle divergence of $\sigma_R(\theta)$ is closely related to the small-angle maximization of ΔE_D.

The Coulomb potential has the same dependence on the radial coordinate as does the gravitational potential. Thus it is worthwhile to pursue how the relative trajectory under $V_C(r)$ compares with what we encountered in the Kepler Problem. We can discover that by returning to Equation 5.29 and integrating it to get a parametric equation for the trajectory in the r-φ plane. If the limits $r(\varphi)$ and $r(0)$ are placed upon the integral of both sides of (5.29), we have

$$\varphi = b \int_{r(0)}^{r(\varphi)} \frac{dr}{r(r^2 - r_0 r - b^2)^{1/2}}$$

$$= \sin^{-1}\left[\frac{-r_0 r(\varphi) - 2b^2}{r(\varphi)(r_0^2 + 4b^2)^{1/2}}\right] - \sin^{-1}\left[\frac{-r_0 r(0) - 2b^2}{r(0)(r_0^2 + 4b^2)^{1/2}}\right]$$

We can simplify matters by orienting the CM frame of reference so that

$$b = r(0)\left(1 - \frac{r_0}{r(0)}\right)^{1/2}$$

Then

$$\frac{r_0 r(\varphi) + 2b^2}{r(\varphi)(r_0^2 + 4b^2)^{1/2}} = -\sin\left(\varphi - \frac{\pi}{2}\right) = \cos\varphi$$

and

$$r(\varphi) = \frac{-J^2/Z_A Z_B e^2 \mu}{1 - e\cos\varphi} \tag{5.54}$$

where

$$e = \left(1 + \frac{4b^2}{r_0^2}\right)^{1/2} = \left[1 + \frac{2J^2 T_\infty}{\mu(Z_A Z_B e^2)^2}\right]^{1/2} \tag{5.55}$$

and we have employed Equations 5.28 and 5.43 to express everything in terms of dynamical parameters. Equation 5.55 is in precise analogy with Equation 4.76, just as Equation 5.54 corresponds to Equation 4.81. Thus e is the eccentricity of a conic section in the r-φ plane. Because we have insisted that $E = T_\infty > 0$ we must have $e > 1$. Therefore, we associate (5.54) with the equation for an *hyperbola*.[8] Each branch of the hyperbola corresponds with one of the two possibilities for the Coulomb interaction. If the interaction is attractive, $\cos\varphi < e^{-1}$ and the trajectory for the relative motion is that shown in Figure 5.11a. If the interaction is repulsive, $\cos\varphi > e^{-1}$ and the trajectory is as shown in Figure 5.11b.

---

[8] In the critical case $T_\infty = 0$ we cannot have a repulsive interaction. The trajectory accordingly degenerates to a single curve for the attractive interaction—a parabola.

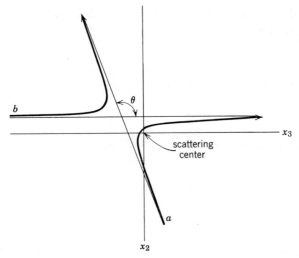

FIGURE 5.11. The trajectories in Coulomb scattering. (*a*) Attractive interaction. (*b*) Repulsive interaction.

FOR FURTHER READING

A. P. French	*Newtonian Mechanics*, W. W. Norton, New York, 1971. Chapter 9 of this book presents a very good introduction to collision phenomena.
T. W. B. Kibble	*Classical Mechanics*, Halsted Press, New York, 1973. A brief, but very useful, discussion of the basic physics of elastic scattering appears in Chapter 7.
J. B. Marion	*Classical Dynamics*, Academic Press, New York, 1970. Chapter 9 gives a description of two-particle scattering at the same level as the present chapter.
E. Rutherford, J. Chadwick, and C. D. Ellis	*Radiations from Radioactive Substances*, Cambridge University Press, London, 1951. Chapter VIII of this classic book gives a detailed description of the original theory and experiments on Rutherford scattering.

PROBLEMS

1. Make a vector diagram that expresses the conservation of momentum for a collision in the LAB frame and derive Equation 5.5 using only trigonometry.

2. It is an experimental fact that hydrogenous materials (such as water or paraffin) exhibit a great stopping power for neutrons. Explain how this can be by considering the quantity $\Delta E_R / T_{LAB}$ as a function of the mass ratio.

3. (a) Show that the maximum loss of energy to a resting helium nucleus ($^4_2$He) by a proton colliding with it elastically is 64 percent of the incident proton's energy.

(b) Show that the maximum loss of energy to a resting proton by an electron colliding with it elastically is 0.218 percent of the incident electron's energy.

4. A helium nucleus ($^4_2$He) scatters elastically from a proton in a cloud chamber. Prove that the deflection angle must be less than 14.5°.

5. The physicist James Chadwick in 1932 determined a value for the mass of the neutron by observing its elastic collisions with protons and nitrogen nuclei. He observed that the protons recoiled with a maximum speed of $3.3 \pm .3 \times 10^7$ m/s and that the $^{14}_7$N nuclei recoiled with a maximum speed of $4.7 \pm .5 \times 10^6$ m/s after scattering the neutron beam. Calculate the mass of the neutron and its initial speed.

6. A proton scatters from an atomic nucleus with a LAB deflection angle equal to 155°. The recoil angle of the unknown nucleus is observed to be 30°. Calculate the mass of this target nucleus.

7. Prove that the maximum LAB deflection angle is the solution of

$$\theta_D = \sin^{-1}\left(\frac{m_B}{m_A}\right)$$

provided that $m_B/m_A \leqslant 1$. What is the maximum angle of elastic scattering of a proton by a resting electron?

8. Show that $\tan \theta_R = (\Delta E_D/\Delta E_R)^{1/2}$ for elastic scattering involving particles of equal mass.

9. Helium ($^4_2$He) nuclei are scattered elastically with a LAB deflection angle of 90° from more massive target nuclei. The ratio $\Delta E_D/\Delta E_R$ is observed to be equal to 11. Calculate the mass of the target nucleus.

10. One means of detecting the quantity of water in soil is to lower into a test hole a probe that produces fast neutrons and detects very slow ones. The fast neutrons are made into slow neutrons by elastic collisions with the hydrogen nuclei in water. Therefore, the rate of detection of slow neutrons should be proportional to the amount of water present. How many such collisions would be required to reduce the fast neutron energy from 1 MeV to 0.025 eV assuming $\theta_D = 75°$ on the average?

11. A theoretical physicist needs data on elastic proton-proton scattering for $\theta = 40°$ and $T_\infty = 100$ MeV (1 MeV $= 10^{-12}$ erg). What values of θ_D and T_{LAB} should he give to an experimental physicist in order that the latter may set up his equipment appropriately?

12. Derive expressions in the LAB frame for:

(a) the probability that a particle will be scattered through any angle larger than a certain angle β, and

(b) the probability, per unit area of detector, that a particle will be scattered into a detector located at the distance R along a radial line making an angle θ_D with the incident beam direction.

13. Calculate the ratio of the differential scattering cross section in the LAB frame to that in the CM frame for the scattering process described in Problem 11.

14. Calculate the probability that a helium atom, regarded as a hard sphere of radius 1.09 Å, will be scattered through an angle greater than 60° by another helium atom (assumed to be at rest).

15. The fractional energy loss in an elastic collision is $\Delta E_D/T_{LAB}$, a quantity that depends on θ_D. The average fractional energy loss for collisions at all possible angles is then

$$\left\langle \frac{\Delta E_D}{T_{LAB}} \right\rangle = \int p(\theta_D) \frac{\Delta E_D(\theta_D)}{T_{LAB}} d\theta_D$$

where $p(\theta_D)\, d\theta_D = 2\pi\sigma(\theta_D) \sin\theta_D\, d\theta_D/\sigma_t$ is the probability of scattering between θ_D and $\theta_D + d\theta_D$. Calculate the average fractional energy loss for hard-sphere scattering.

16. Determine the cross section in barns for the elastic scattering of helium atoms, taken as hard spheres, through angles less than $45°$ in the CM frame. The atomic radius is given in Problem 14.

17. Derive an expression for $\sigma(\theta)$ in the case of an interaction potential given by

$$V(r) = \begin{cases} -V_0 & 0 \leqslant r < d \\ 0 & r > d \end{cases}$$

where V_0 is a positive constant.

18. In the list below are some of the measurements of Hans Geiger and Edward Marsden on the elastic scattering of $^4_2$He nuclei by a thin gold foil. The data were reported in 1913 as the first systematically determined evidence for the Rutherford model of the atom. Devise a graphical method to demonstrate that the Rutherford cross section will describe these data.

θ (deg.):	45	60	75	105	120	135	150
$\dfrac{dN}{d\Omega}$:	1435	477	211	69.5	51.9	43	33.1

19. Calculate the rate of detection of $^4_2$He nuclei scattered at an angle of $30°$ in the LAB frame by a copper target of thickness 10^{-6} m, given that the incident flux is 10^5 particles/s-cm^2, each with an energy of 0.1 MeV. The density of copper is 8.8 g/cm^3. The detector is circular, with a radius of 2 cm, and is placed 1 m away from the target along the radial line making an angle of $30°$ with the direction of the incident beam.

20. How closely can a $^4_2$He nucleus with an initial energy of 0.1 MeV approach a copper nucleus ($^{63}_{29}$Cu), assuming a head-on collision?

21. Derive an expression for the differential scattering cross section, in the LAB frame, which describes the elastic scattering of two protons.

22. A beam of protons, flux 3×10^6 particles/s-cm^2, strikes a target containing 5×10^{22} protons. Given an incident particle energy of 1 MeV, calculate the rate of proton detection at a LAB deflection angle of $5°$. The detector has the same geometry as described in Problem 19.

23. Compute the differential scattering cross section in the CM frame for two particles interacting according to

$$V(r) = \frac{k}{r^2}$$

where k is a positive constant.

24. For sufficiently low energies of the incident particles, the elastic scattering of protons by neutrons is hard-sphere scattering. For an incident flux of 10^6 neutrons/s-cm^2 on a target containing 10^{20} protons/cm^2, a total scattering rate of neutrons equal to 8×10^3 particles/s-cm^2 is observed. Calculate the radius of the neutron according to these data.

25. Prove that the ratio $\sigma(\Delta E_D)/\sigma(\theta)$ is a constant, for a given value of T_{LAB}.

6

Assemblies of Particles

6.1 COUPLED HARMONIC OSCILLATORS

We shall begin our consideration of the dynamics of assemblies of particles by discussing a very simple model of a solid: a one-dimensional chain of atoms bound together through forces acting only between nearest neighbors. This model will be even further simplified by the assumptions that (1) the linear array comprises pairs of identical, light atoms separated from one another by very heavy atoms, as shown in Figure 6.1, and (2) the forces of interaction are proportional to the spatial separations of the atoms. Given assumption 1, we can neglect in a first approximation the motion of the very heavy atoms, regarding them as fixed in their equilibrium positions. Thus we need consider only the motion of a single pair of the light atoms in order to describe the assembly. With assumption 2 we understand the light atoms to be coupled harmonic oscillators. Although these restrictions may seem severe, it will turn out, for reasons to be made clear in the next section, that our analysis of the model leads to a very general and useful method for describing the oscillatory motions of an arbitrary number of

165

FIGURE 6.1. A one-dimensional solid comprising two kinds of atom.

particles whose mutual interactions can be represented by potential wells. Now let us look at the problem.

Given that the positive x direction is to the right, the potential function for the two oscillating atoms is

$$V(\xi_A, \xi_B) = \tfrac{1}{2}k\xi_A{}^2 + \tfrac{1}{2}k\xi_B{}^2 + \tfrac{1}{2}k_{AB}(\xi_A - \xi_B)^2 \tag{6.1}$$

where we have chosen the zero of potential energy to occur where the particles are at their equilibrium positions. The physical significance of $V(\xi_A, \xi_B)$ can be appreciated by computing from it the forces on oscillators A and B. They are:

$$F_A = -k\xi_A - k_{AB}\xi_A + k_{AB}\xi_B \tag{6.2a}$$

$$F_B = -k\xi_B - k_{AB}\xi_B + k_{AB}\xi_A \tag{6.2b}$$

The first term in each of these expressions is the restoring force brought about through the interaction of the oscillator with the fixed atom nearest it; the appropriate force constant is k in either case. The second term represents the restoring force arising because of the interaction (or "coupling") of one oscillator, through the force constant k_{AB}, to the other. This term exists simply because there are interactions on both sides of each oscillator. The third term in the force expressions is more subtle. It represents the effect on one oscillator of a displacement of the other oscillator. The relevant force constant is k_{AB} since it is only through the coupling that the displacement of one atom could influence that of another. The sign of the third term is positive because it is not a restoring force but rather is a force that tends to make one of the atoms follow the other. In this sense we should regard the third term as the genuine representative of the coupling. Were it to disappear, our problem would reduce mathematically to what was considered in Section 2.5.

The total energy of the coupled oscillators follows from Equation 6.1 as

$$E = \frac{p_A{}^2}{2m} + \frac{p_B{}^2}{2m} + \tfrac{1}{2}m(\omega_0{}^2 + \omega_{AB}{}^2)(\xi_A{}^2 + \xi_B{}^2) - m\omega_{AB}{}^2\xi_A\xi_B \tag{6.3}$$

where

$$\omega_0{}^2 = \frac{k}{m} \qquad \omega_{AB}{}^2 = \frac{k_{AB}}{m}$$

The equations of motion that derive from Equations 6.2 are

$$\frac{d^2\xi_A}{dt^2} + (\omega_0{}^2 + \omega_{AB}{}^2)\xi_A(t) - \omega_{AB}{}^2\xi_B(t) = 0 \tag{6.4a}$$

$$\frac{d^2\xi_B}{dt^2} + (\omega_0{}^2 + \omega_{AB}{}^2)\xi_B(t) - \omega_{AB}{}^2\xi_A(t) = 0 \qquad (6.4b)$$

according to the usual prescription of the Second Law. Equations 6.4 expose clearly the nature of our problem. In the absence of coupling $\omega_{AB} = 0$ and the equations of motion describe two independent linear oscillators with the same natural frequency ω_0. But, when the coupling is "switched on," the equations of motion are no longer independent: the time variation of $\xi_A(t)$ depends on what value $\xi_B(t)$ has at the same instant, and vice versa. This mathematical coupling of the equations of motion is a direct reflection of the physical coupling of the oscillators and is what makes our problem a nontrivial one.

It is apparent that we could attempt to solve Equations 6.4 by some kind of technique involving successive approximations. We could, for example, assume at the outset that $k_{AB} \ll k$, thereby justifying the use of the solution for non-interacting oscillators as a zeroth-order approximation. This approach, despite its physical simplicity, is quite likely to be misleading in general because it does not really come to grips with the fundamentally *cooperative* character of the motion produced by the coupling. In other words, an approximation based on the description of a noninteracting assembly of particles often fails inherently to account for any new behavior that the assembly as a whole may exhibit because of the interactions. What is truly needed, therefore, is a method that allows for a cooperative, "total" motion in some way.

Since the particles we wish to describe are identical oscillators and couple through an oscillatory force as well, we might consider the possibility that their cooperative motion will also be periodic in nature.[1] In particular, suppose that both particles were capable of undergoing concurrent displacements corresponding to one and the same vibrational frequency. Evidently there would be a different frequency for each genuinely different set of displacements and these sets would be at least in part dictated by the physical arrangement of the oscillators themselves. A look at Figure 6.2 suggests that, for the two-particle, one-dimensional assembly we are considering, there should be just two distinct kinds of *cooperative* displacement. One occurs when the particles move in phase and thereby reduce their coupling potential energy to a minimum. (See Equation 6.1 again.) The other occurs at the opposite extreme, when the particles move π radians out of phase and maximize their coupling potential energy. Thus we are led to postulate two different general displacements and to write

$$\eta_1(t) = a_{11}\xi_A(t) + a_{21}\xi_B(t) \qquad (6.5)$$

$$\eta_2(t) = a_{12}\xi_A(t) + a_{22}\xi_B(t) \qquad (6.6)$$

---

[1] This possibility is reasonable, if not compelling, when the oscillators are identical. If the force constants and masses were different for each oscillator, then there would be no reason *in general* to suppose that the total motion would be periodic. In that case periodic motion would be expected to occur only when certain initial conditions were satisfied by the coordinates and velocities.

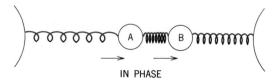

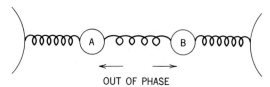

FIGURE 6.2. The two cooperative modes of vibration of a pair of coupled oscillators.

where the a_{ij} are constants yet to be determined. The "generalized displacement coordinates" $\eta_i(t)$ $(i = 1, 2)$ are linear combinations of concurrent displacements of the two oscillators and, therefore, refer to the assembly as a whole. These two quantities are called *normal coordinates*. The cooperative oscillations they describe are referred to as *normal modes of vibration*.

The normal coordinates are to replace the individual particle coordinates as being the more appropriate to the interacting assembly. By hypothesis they satisfy the equations of motion

$$\ddot{\eta}_i + \omega_i^2 \eta_i(t) = 0 \qquad (i = 1, 2) \tag{6.7}$$

where $\dot{\eta} \equiv d\eta/dt$ and ω_i is the ith of the two normal mode vibrational frequencies. The problem we must turn to now is the determination of the ω_i and the a_{ij} in order that the physical meaning of the normal coordinates be precise. At the same time we expect that a result of our effort will be a complete solution of Equations 6.4. Now, if the displacements $\xi_A(t)$ and $\xi_B(t)$ contribute to a normal mode of frequency ω, it follows that Equations 6.4 can be written

$$-\omega^2 \xi_A(t) + (\omega_0^2 + \omega_{AB}^2)\xi_A(t) - \omega_{AB}^2 \xi_B(t) = 0$$

$$-\omega^2 \xi_B(t) + (\omega_0^2 + \omega_{AB}^2)\xi_B(t) - \omega_{AB}^2 \xi_A(t) = 0$$

or more compactly,

$$(\omega_0^2 + \omega_{AB}^2 - \omega^2)\xi_A(t) - \omega_{AB}^2 \xi_B(t) = 0 \tag{6.8a}$$

$$-\omega_{AB}^2 \xi_A(t) + (\omega_0^2 + \omega_{AB}^2 - \omega^2)\xi_B(t) = 0 \tag{6.8b}$$

Equations 6.8 are a set of linear, homogeneous equations for the displacements. According to the theory of algebraic equations,[2] they can have no solution except $\xi_A(t) = \xi_B(t) \equiv 0$ unless the determinant formed from the coefficients of the unknowns is identically zero. Thus we must have

$$\begin{vmatrix} (\omega_0{}^2 + \omega_{AB}{}^2) - \omega^2 & -\omega_{AB}{}^2 \\ -\omega_{AB}{}^2 & (\omega_0{}^2 + \omega_{AB}{}^2) - \omega^2 \end{vmatrix} \equiv 0 \tag{6.9}$$

that is,

$$(\omega^2 - \omega_0{}^2 - \omega_{AB}{}^2)^2 - \omega_{AB}{}^4 = 0$$

and

$$\omega^2 = \begin{cases} \omega_0{}^2 + 2\omega_{AB}{}^2 \equiv \omega_1{}^2 \\ \omega_0{}^2 \qquad\qquad \equiv \omega_2{}^2 \end{cases} \tag{6.10}$$

The determinant in Equation 6.9 is known as the *secular determinant* for the normal mode frequencies.[3] We see that one of the ω is greater than the natural frequency while the other is exactly the same as ω_0. This fact tells us that an iterative solution of Equations 6.4 would have been very misleading, since the iterations would have stopped at the zeroth approximation and led to the conclusion that the coupling between the oscillators made absolutely no difference! The physical reason for the coincidental equality between ω_2 and ω_0 will become clear shortly.

To calculate the coefficients in Equations 6.5 and 6.6 we note that Equations 6.7 imply that the total energy expressed in normal coordinates must be

$$E = \frac{p_1{}^2}{2m} + \frac{p_2{}^2}{2m} + \tfrac{1}{2}m\omega_1{}^2\eta_1{}^2 + \tfrac{1}{2}m\omega_2{}^2\eta_2{}^2 \tag{6.11}$$

where $p_i = m\dot{\eta}_i$ $(i = 1, 2)$. Since E in this equation expresses a constant of the motion for the assembly, it has to be numerically equal to the total energy given by Equation 6.3. If we put (6.5) and (6.6) into (6.11) we get

$$\begin{aligned} E = {}& (a_{11}{}^2 + a_{12}{}^2)\frac{p_A{}^2}{2m} + (a_{21}{}^2 + a_{22}{}^2)\frac{p_B{}^2}{2m} + (a_{21}a_{11} + a_{12}a_{22})\frac{p_A p_B}{m} \\ & + \left[\omega_0{}^2(a_{11}{}^2 + a_{12}{}^2) + 2\omega_{AB}{}^2 a_{11}{}^2\right]\tfrac{1}{2}m\xi_A{}^2 \\ & + \left[\omega_0{}^2(a_{21}{}^2 + a_{22}{}^2) + 2\omega_{AB}{}^2 a_{21}{}^2\right]\tfrac{1}{2}m\xi_B{}^2 \\ & + \left[\omega_0{}^2(a_{21}a_{11} + a_{12}a_{22}) + 2\omega_{AB}{}^2 a_{11}a_{21}\right]m\xi_A\xi_B \end{aligned}$$

[2] For a mathematical discussion of Equations 6.8, see, for example, H. Margenau and G. M. Murphy, *The Mathematics of Physics and Chemistry*, D. Van Nostrand Co., Princeton, 1956, Vol. I, p. 314. The mathematically minded reader may wonder how Equations 6.8 can determine $\xi_A(t)$ and $\xi_B(t)$, since no initial conditions have been given. The answer lies with the homogeneity of the equations, which fact makes them reducible to *one* equation for the two unknowns and thus able to determine only the *ratio* of $\xi_A(t)$ and $\xi_B(t)$.

[3] The name is a holdover from celestial mechanics, where an equation like (6.9) makes an appearance in the theory of successive approximations.

This expression reduces to Equation 6.3 only if we require the coefficients a_{ij} to satisfy generally

$$a_{11}{}^2 + a_{12}{}^2 = 1 = a_{21}{}^2 + a_{22}{}^2 \tag{6.12}$$

$$a_{21}a_{11} + a_{12}a_{22} = 0 \tag{6.13}$$

and, specifically,

$$a_{11}{}^2 = a_{21}{}^2 = \tfrac{1}{2} \qquad a_{11}a_{21} = -\tfrac{1}{2}$$

It follows directly that:

$$a_{11} = (2)^{-1/2} \qquad a_{12} = (2)^{-1/2} \qquad a_{21} = -(2)^{-1/2} \qquad a_{22} = (2)^{-1/2} \tag{6.14}$$

and

$$\eta_1(t) = (2)^{-1/2}[\xi_A(t) - \xi_B(t)] \qquad (\omega = \omega_1) \tag{6.15a}$$

$$\eta_2(t) = (2)^{-1/2}[\xi_A(t) + \xi_B(t)] \qquad (\omega = \omega_2) \tag{6.15b}$$

The normal coordinate $\eta_1(t)$ changes sign when the labels A and B are interchanged on the individual oscillators and for this reason is called the *antisymmetric* linear combination of displacements. Similarly $\eta_2(t)$ is the *symmetric* linear combination. If we impose the initial condition that $\xi_A(t)$ and $\xi_B(t)$ have the same amplitude and phase, then $\eta_1(t) \equiv 0$ and only $\eta_2(t)$ survives. Therefore, the symmetric normal mode corresponds to the oscillators moving in phase at all times. That the frequency of this mode, ω_2, is equal to the natural frequency should come as no surprise, since the coupling potential energy is maintained at zero. The oscillators *are* moving naturally with no apparent effect of the coupling other than the mainte- nance of their phase relationship. (See Figure 6.2.) On the other hand, if we set $\xi_A(t) \equiv -\xi_B(t)$, so that the oscillators are always π radians out of phase, we get $\eta_2(t) \equiv 0$ with $\eta_1(t)$ remaining. The antisymmetric normal mode corresponds to the greatest coupling energy, when the oscillators use one another as "spring- boards" during their motion.

If we invert Equations 6.15 we find

$$\xi_A(t) = (2)^{-1/2}[\eta_1(t) + \eta_2(t)] \tag{6.16a}$$

$$\xi_B(t) = (2)^{-1/2}[\eta_1(t) - \eta_2(t)] \tag{6.16b}$$

These equations are readily shown to satisfy Equations 6.4 identically once Equa- tions 6.7 and 6.10 are used. Since the $\eta_i(t)$ must be of the form

$$\eta_i(t) = C_i \cos(\omega_i t - \delta_i) \qquad (i = 1, 2) \tag{6.17}$$

we can also write

$$\xi_A(t) = (2)^{-1/2}[C_1 \cos(\omega_1 t - \delta_1) + C_2 \cos(\omega_2 t - \delta_2)] \tag{6.18a}$$

$$\xi_B(t) = (2)^{-1/2}[C_1 \cos(\omega_1 t - \delta_1) - C_2 \cos(\omega_2 t - \delta_2)] \tag{6.18b}$$

as the complete solution of our problem, where the arbitrary constants are determined by

$$C_1 \cos \delta_1 = (2)^{-1/2}(\xi_A(0) + \xi_B(0))$$

$$C_2 \cos \delta_2 = (2)^{-1/2}(\xi_A(0) - \xi_B(0))$$

$$C_1\omega_1 \sin \delta_1 = (2)^{-1/2}(v_A(0) + v_B(0))$$

$$C_2\omega_2 \sin \delta_2 = (2)^{-1/2}(v_A(0) - v_B(0))$$

An interesting special case of these equations comes about when the condition $k_{AB} \ll k$ is valid. When this is true we may write

$$\omega_1 \simeq \omega_0 + \left(\frac{\omega_{AB}}{\omega_0}\right)\omega_{AB}$$

approximately. Now let us consider the initial conditions

$$\xi_A(0) = A_0 \qquad \xi_B(0) = 0 \qquad v_A(0) = 0 \qquad v_B(0) = 0$$

which correspond to an initial displacement of one oscillator alone and a releasing of it from rest. Then

$$C_1 = C_2 = (2)^{-1/2} A_0 \qquad \delta_1 = \delta_2 = 0$$

and Equations 6.18 become

$$
\begin{aligned}
\xi_A(t) &= \frac{A_0}{2} \left[\cos(\omega_1 t) + \cos(\omega_2 t)\right] \\
&= A_0 \cos\left[\tfrac{1}{2}(\omega_1 - \omega_2)t\right] \cos\left[\tfrac{1}{2}(\omega_1 + \omega_2)t\right] \\
&\simeq A_0 \cos\left[\frac{1}{2}\left(\frac{\omega_{AB}}{\omega_0}\right)\omega_{AB}t\right] \cos(\omega_0 t)
\end{aligned}
\tag{6.19a}
$$

$$
\begin{aligned}
\xi_B(t) &= \frac{A_0}{2} \left[\cos(\omega_1 t) - \cos(\omega_2 t)\right] \\
&= -A_0 \sin\left[\tfrac{1}{2}(\omega_1 - \omega_2)t\right] \sin\left[\tfrac{1}{2}(\omega_1 + \omega_2)t\right] \\
&\simeq -A_0 \sin\left[\frac{1}{2}\left(\frac{\omega_{AB}}{\omega_0}\right)\omega_{AB}t\right] \sin(\omega_0 t)
\end{aligned}
\tag{6.19b}
$$

respectively, where we have approximated ω_1 and written $\omega_1 + \omega_2 \simeq 2\omega_0$. Equations 6.19 describe displacements that are periodic functions of the time with a frequency equal to the natural frequency ω_0, but whose amplitude is a slowly varying function of the time as well. In Figure 6.3 these equations are plotted. We see that, when the coupling between the oscillators is weak, the vibratory motion "resonates" from one particle to the other. This phenomenon may be considered as the result of a gradual and repetitive transfer of energy between the oscillators. In the theory of sound, the behavior shown in Figure 6.3 is called *beats*, the beat frequency being $\omega_1 - \omega_2 \simeq (\omega_{AB}/\omega_0)\omega_{AB}$ here.

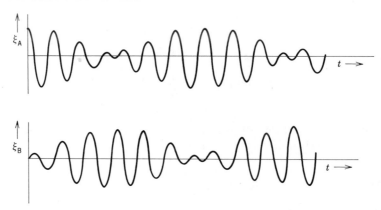

FIGURE 6.3. A sketch of Equations 6.19, illustrating the resonant transfer of vibrational energy and the beat phenomenon.

Our analysis has shown that it is possible to describe the behavior of coupled oscillators by replacing the individual particle displacements with a set of normal coordinates

$$\eta_j(t) = \sum_{i=1}^{2} a_{ij}\xi^i(t) \qquad (j = 1, 2) \tag{6.20}$$

where we have written $\xi^1 \equiv \xi_A$ and $\xi^2 \equiv \xi_B$ in order to use the summation sign. The coefficients in this transformation must satisfy Equations 6.12 and 6.13, and otherwise must be such as to leave the total energy invariant in numerical magnitude. Because Equations 6.20 are linear, it is known that they can always be represented as the *matrix expression*

$$(\boldsymbol{\eta}) = (a) \times (\boldsymbol{\xi}) \tag{6.21}$$

where

$$(\boldsymbol{\eta}) = \begin{pmatrix} \eta_1 \\ \eta_2 \end{pmatrix} \qquad (\boldsymbol{\xi}) = \begin{pmatrix} \xi^1 \\ \xi^2 \end{pmatrix} = \begin{pmatrix} \xi_A \\ \xi_B \end{pmatrix}$$

and

$$(a) = \begin{pmatrix} a_{11} & a_{12} \\ a_{21} & a_{22} \end{pmatrix} \tag{6.22}$$

Moreover, the matrix (a) possesses the special property

$$(a) \times (\tilde{a}) = \begin{pmatrix} a_{11} & a_{12} \\ a_{21} & a_{22} \end{pmatrix} \times \begin{pmatrix} a_{11} & a_{21} \\ a_{12} & a_{22} \end{pmatrix} = \begin{pmatrix} a_{11}^2 + a_{12}^2 & a_{21}a_{11} + a_{12}a_{22} \\ a_{21}a_{11} + a_{12}a_{22} & a_{21}^2 + a_{22}^2 \end{pmatrix}$$

$$= \begin{pmatrix} 1 & 0 \\ 0 & 1 \end{pmatrix} = (I) \equiv (a) \times (a)^{-1}$$

according to Equations 6.12 and 6.13. This tells us that $(\tilde{a}) = (a)^{-1}$ and, therefore, that (a) is an *orthogonal matrix*.[4] Equation 6.22, then, represents a kind of rotation wherein the coordinates ξ^1 and ξ^2 of the "vector" $\boldsymbol{\xi}$ are transformed to those of the "vector" $\boldsymbol{\eta}$. Of course, $\boldsymbol{\xi}$ is not a vector in ordinary space since ξ^1 and ξ^2 refer to *different* particles. Instead, $\boldsymbol{\xi}$ and $\boldsymbol{\eta}$ are vectors in a general mathematical sense because they transform into one another as vectors do.

With this geometric picture of the normal coordinates we can state a method of solving Equations 6.4. We begin by writing down the equations for the normal coordinates (6.20) with the specification that the matrix of the transformation coefficients be orthogonal:

$$\sum_{j=1}^{2} a_{ij}a_{kj} = \begin{cases} 1 & i = k \\ 0 & i \neq k \end{cases} \tag{6.23}$$

Then we note that *the primary purpose of the transformation is to permit the potential energy terms in Equation 6.3 to be written as a simple sum of squared coordinates.* If we write this condition,

$$\sum_{i=1}^{2} \tfrac{1}{2}m\omega_i^2\eta_i^2 = \sum_{i=1}^{2}\sum_{j=1}^{2} \tfrac{1}{2}m\hat{\omega}_{ij}^2\xi^i\xi^j \tag{6.24}$$

in matrix form, where

$$\hat{\omega}_{11}^2 \equiv \omega_0^2 + \omega_{AB}^2 \qquad \hat{\omega}_{12}^2 \equiv -\omega_{AB}^2 \equiv \hat{\omega}_{21}^2 \qquad \hat{\omega}_{22}^2 \equiv \omega_0^2 + \omega_{AB}^2$$

we have the expression

$$(\eta_1 \quad \eta_2) \times \begin{pmatrix} \omega_1^2 & 0 \\ 0 & \omega_2^2 \end{pmatrix} \times \begin{pmatrix} \eta_1 \\ \eta_2 \end{pmatrix} = (\xi^1\xi^2) \times \begin{pmatrix} \hat{\omega}_{11}^2 & \hat{\omega}_{12}^2 \\ \hat{\omega}_{21}^2 & \hat{\omega}_{22}^2 \end{pmatrix} \times \begin{pmatrix} \xi^1 \\ \xi^2 \end{pmatrix}$$

or, simply,

$$(\tilde{\boldsymbol{\eta}}) \times (\omega^2) \times (\boldsymbol{\eta}) = (\tilde{\boldsymbol{\xi}}) \times (\hat{\omega}^2) \times (\boldsymbol{\xi}) \tag{6.25}$$

Equation 6.25 expresses the basic property of the transformation. If we put into it Equation 6.21, we have the result

$$(\tilde{\boldsymbol{\xi}}) \times (\tilde{a}) \times (\omega^2) \times (a) \times (\boldsymbol{\xi}) = (\tilde{\boldsymbol{\xi}}) \times (\hat{\omega}^2) \times (\boldsymbol{\xi})$$

or

$$(\tilde{a}) \times (\omega^2) \times (a) = (\hat{\omega}^2)$$

which also may be written

$$(\omega^2) \times (a) = (a) \times (\hat{\omega}^2) \tag{6.26}$$

if we multiply from the left by (a) on both sides of the equation. Written out, Equation 6.26 is the set of equations

$$\omega_1^2 a_{11} = a_{11}\hat{\omega}_{11}^2 + a_{12}\hat{\omega}_{21}^2$$

$$\omega_1^2 a_{12} = a_{11}\hat{\omega}_{12}^2 + a_{12}\hat{\omega}_{22}^2$$

[4] Orthogonal matrices and rotations are discussed in Section A.2 of the Appendix.

$$\omega_2^2 a_{21} = a_{21}\hat{\omega}_{11}^2 + a_{22}\hat{\omega}_{21}^2$$

$$\omega_2^2 a_{22} = a_{21}\hat{\omega}_{12}^2 + a_{22}\hat{\omega}_{22}^2$$

or

$$(\hat{\omega}_{11}^2 - \omega_1^2)a_{11} + \hat{\omega}_{21}^2 a_{12} = 0$$

$$(\hat{\omega}_{22}^2 - \omega_1^2)a_{12} + \hat{\omega}_{12}^2 a_{11} = 0$$

(6.27)

$$(\hat{\omega}_{11}^2 - \omega_2^2)a_{21} + \hat{\omega}_{21}^2 a_{22} = 0$$

$$(\hat{\omega}_{22}^2 - \omega_2^2)a_{22} + \hat{\omega}_{12}^2 a_{21} = 0$$

(6.28)

Equations 6.27 and 6.28 are independent sets of linear, homogeneous equations for the coefficients a_{ij}. For either normal mode frequency, they have no useful solutions unless

$$\begin{vmatrix} \hat{\omega}_{11}^2 - \omega^2 & \hat{\omega}_{21}^2 \\ \hat{\omega}_{12}^2 & \hat{\omega}_{22}^2 - \omega^2 \end{vmatrix} \equiv 0$$

(6.29)

Equation 6.29 is precisely the same as Equation 6.9, which led to the calculation of ω^2. Once ω^2 has been found in this way, we can return it to Equations 6.27 and 6.28 in order to solve these for the a_{ij} [subject to the condition (6.23)]. Thus Equations 6.20, 6.23, and 6.27 to 6.29 specify a complete solution of Equations 6.4. The method we have used is quite general and is based only on the desired mathematical property (6.25) and the orthogonality of the matrix (a). For this reason it ought to be useful for the problem of *many* coupled oscillators even in three-dimensional space, provided we can find the appropriate generalizations of the equations given above. This idea we shall substantiate in the next section.

6.2 NORMAL COORDINATE ANALYSIS

Now let us consider an assembly of N particles in three-dimensional space. We shall make the basic assumptions that (1) the potential function for the interaction of any two of the particles displays a single minimum and (2) there are no external forces such as those created by applied fields or enclosing walls. The total energy of the assembly will be

$$E = \tfrac{1}{2} \sum_{\alpha=1}^{N} \sum_{i=1}^{3} m_\alpha (v_i^\alpha)^2 + V(x_1^1, x_2^1, x_3^1, \ldots, x_1^N, x_2^N, x_3^N)$$

(6.30)

where x_i^α is the ith cartesian coordinate of the αth particle and the velocity coordinate v_i^α is labeled similarly. The equations of motion associated with this expression will be extraordinarily difficult—in fact, impossible—to solve without making further approximations. The simplest ones we could make would be either to set the potential energy equal to a constant or to expand it about its minima in a Taylor series to terms of second order. The first possibility is relevant to the situation wherein particle interactions are very weak, such as what occurs in a gas of low density. The second possibility describes the opposite extreme, such as occurs in a cold solid, wherein particle interactions are so strong that large deviations from the equilibrium positions are unlikely. In this latter case we can replace the potential function for the assembly with

$$V_H(x_1^1, \ldots, x_3^N) = \frac{1}{2} \sum_{\alpha=1}^{N} \sum_{\beta=1}^{N} \sum_{i=1}^{3} \sum_{j=1}^{3} \phi_{ij}^{\alpha\beta}(x_i^\alpha - x_{i0}^\alpha)(x_j^\beta - x_{j0}^\beta) \quad (6.31)$$

where

$$\phi_{ij}^{\alpha\beta} = \left(\frac{\partial^2 V}{\partial x_i^\alpha \partial x_j^\beta} \right)_{x_{i0}^\alpha x_{j0}^\beta} \quad (6.32)$$

and x_{i0}^α refers to the ith equilibrium coordinate of the αth particle. Equation 6.31 is the result of a Taylor expansion of the total potential energy about its minima and the renormalization of the potential relative to $V(x_{10}^1, \ldots, x_{30}^N)$. We note that linear terms in $(x_i^\alpha - x_{i0}^\alpha)$ do not appear because of the equilibrium condition

$$\left(\frac{\partial V}{\partial x_i^\alpha} \right)_{x_{i0}^\alpha} = 0 \quad (\alpha = 1, \ldots, N; i = 1, 2, 3) \quad (6.33)$$

To simplify the notation we shall set $\xi_i^\alpha \equiv x_i^\alpha - x_{i0}^\alpha$ ($\alpha = 1, \ldots, N; i = 1, 2, 3$). Then the total energy of the assembly may be written

$$E = \frac{1}{2} \sum_{\alpha=1}^{N} \sum_{i=1}^{3} \left[m_\alpha(v_i^\alpha)^2 + \sum_{\beta=1}^{N} \sum_{j=1}^{3} \phi_{ij}^{\alpha\beta} \xi_i^\alpha \xi_j^\beta \right] \quad (6.34)$$

in the approximation we have made. It is evident from the mathematical form of this equation that the coefficients $\phi_{ij}^{\alpha\beta}$ play the role of force constants. Thus $-\phi_{ij}^{\alpha\beta}$ is the force on particle α in the i direction if just particle β is displaced by unit distance in the j direction while all the other particles remain at their equilibrium positions.

At this point we have defined the physics of our problem. We wish to describe an assembly of particles, each of which may possess a different mass, wherein interactions are exclusively through harmonic forces whose force constants may be different for each pair of particles. The total energy for the assembly bears a strong formal resemblance to the total energy in Equation 6.3. This fact suggests both that equations of motion coming from (6.34) will show the same coupling property as did Equations 6.4 and that a set of uncoupled equations of motion might be found if a transformation analogous to (6.20) can be developed. Indeed, if the "cross terms" in Equation 6.34 can be eliminated, the total energy of the assembly will be expressible simply as a sum of the total energies of $3N$ independent harmonic oscillators. The procedure that leads to such an expression is called *normal coordinate analysis*. Let us see now how it may be carried through.

First we note that matters are made more elegant if we arrange the $3N$ displacement coordinates ξ_i^α in a definite numerical order such that

$$\{\xi_1^1, \xi_2^1, \xi_3^1, \ldots, \xi_1^N, \xi_2^N, \xi_3^N\} \equiv \{q_1, q_2, q_3, \ldots, q_{3N-2}, q_{3N-1}, q_{3N}\}$$

Then Equation 6.34 may be written

$$E = \frac{1}{2} \sum_{i=1}^{3N} \sum_{j=1}^{3N} (M_{ij}\dot{q}_i\dot{q}_j + \phi_{ij}q_iq_j) \tag{6.35}$$

in terms of the new cartesian coordinates q_i, and M_{ij} is equal to zero unless $i = j$, whereupon it becomes equal to the mass of the particle whose coordinate is q_i. We wish to have things end up with the total energy expressed in the form

$$E = \frac{1}{2} \sum_{i=1}^{3N} (m_{ii}\dot{\eta}_i\dot{\eta}_i + k_{ii}\eta_i\eta_i) \tag{6.36}$$

where $\eta_i(t)$ is the ith normal coordinate for the assembly and m_{ii} and k_{ii} are constants. The $\eta_i(t)$ are related to the $q_i(t)$ through

$$\eta_j(t) = \sum_{i=1}^{3N} a_{ij}q_i(t) \qquad (j = 1, \ldots, 3N) \tag{6.37}$$

or, in matrix notation,

$$(\mathbf{\eta}) = (a) \times (\mathbf{q}) \tag{6.38}$$

Here $\mathbf{\eta} = \{\eta_1, \ldots, \eta_{3N}\}$ is a "vector" with $3N$ coordinates, just as is $\mathbf{q}$, and (a) is a $3N$ by $3N$ orthogonal matrix whose elements are subject to[5]

$$\sum_{j=1}^{3N} a_{ij}a_{kj} = \delta_{ik} \equiv \begin{cases} 1 & i = k \\ 0 & i \neq k \end{cases} \tag{6.39}$$

where δ_{ik} is the Kronecker delta,[6] a very useful quantity for expressing orthogonality relations. The inverses of Equations 6.37 are found by multiplying both sides with a_{kj} and summing over j with an eye on Equation 6.39:

$$q_k(t) = \sum_{j=1}^{3N} a_{kj}\eta_j(t) \qquad (k = 1, \ldots, 3N) \tag{6.40}$$

Because the masses of the particles are not necessarily the same for each particle in the assembly and because the $3N$ force constants for their interactions need not be identical, the time variation of the $q_k(t)$ does not have to be periodic. If each of the $\eta_j(t)$ is "going its own way" with a different frequency, phase, and amplitude, there is no particular reason why a linear combination of them should produce a periodic function. On the other hand, if the initial conditions are chosen carefully, so that, for example, all of the a_{kj} except a_{km} ($1 \leqslant m \leqslant 3N$) are left out of the sum, then it is clear from Equation 6.40 that the assembly will carry out

---

[5] Actually, if there turn out to be several $\eta_j(t)$ corresponding to the same normal mode vibrational frequency, the associated a_{ij} need not satisfy Equation 6.39. However, even in that case it is always possible to create by an "orthogonalization process" a set of a_{ij} that do satisfy (6.39). Therefore, we can assume that (a) is an orthogonal matrix with no difficulty. For more on this technical point, see, for example, H. Goldstein, *Classical Mechanics*, Addison-Wesley, Reading, Mass., 1959, pp. 327–329.

[6] The name is in honor of the mathematician Leopold Kronecker, who introduced the symbol.

a normal mode of vibration whose displacement coordinate is $\eta_m(t)$ and each particle will vibrate with the normal mode frequency ω_m. Therefore, we may conclude quite generally that the assembly will show periodic behavior only under special circumstances, when the particles have been put into motion with a set of phase and amplitude relationships that are in some sense cooperative. Otherwise the motion is construed simply as a linear combination or superposition of normal modes according to Equations 6.40.

The equations of motion that follow from the potential in (6.36) are

$$\ddot{\eta}_i + \frac{k_{ii}}{m_{ii}} \eta_i(t) = 0 \qquad (i = 1, \ldots, 3N) \tag{6.41}$$

These have the solutions (for k_{ii} not equal to zero):

$$\eta_i(t) = C_i \cos(\omega_i t - \delta_i) \qquad (i = 1, \ldots, 3N) \tag{6.42}$$

where

$$\omega_i^2 \equiv \omega_{ii}^2 \equiv \frac{k_{ii}}{m_{ii}} \qquad (i = 1, \ldots, 3N) \tag{6.43}$$

defines the normal mode vibrational frequency. We see that *there are 3N normal mode frequencies for an assembly of N particles.* However, not all of these frequencies need be distinct, and inevitably some of them will be equal to zero. The reason for the latter situation will be discussed shortly.

The quantities m_{ii} and k_{ii}, which are central to our discussion, are the elements along the principal diagonals of the matrices

$$(m) = \begin{bmatrix} m_{11} & 0 & 0 & \cdots \\ 0 & m_{22} & 0 & \cdots \\ 0 & 0 & m_{33} & \cdots \\ \vdots & \vdots & \vdots & \end{bmatrix} \qquad (k) = \begin{bmatrix} k_{11} & 0 & 0 & \cdots \\ 0 & k_{22} & 0 & \cdots \\ 0 & 0 & k_{33} & \cdots \\ \vdots & \vdots & \vdots & \end{bmatrix}$$

respectively. According to Equations 6.35 to 6.38, these matrices are related to the matrices comprising the coefficients M_{ij} and ϕ_{ij} through

$$(m) = (a) \times (M) \times (a)^{-1} = (M) \qquad (k) = (a) \times (\phi) \times (a)^{-1} \tag{6.44}$$

where the elements of (M) are the M_{ij} and those of (ϕ) are ϕ_{ij}. The equation for (m) contains no real information because $m_{ij} = M_{ij}$. However, the equation for (k) states that the matrix (ϕ) is cast into *diagonal form* (i.e., that only elements along the principal diagonal are different from zero) by the orthogonal matrix (a). The elements of (a) are calculated by writing

$$(k) \times (a) = (a) \times (\phi)$$

which is the set of linear equations

$$\sum_{j=1}^{3N} k_{ij} a_{jm} = \sum_{j=1}^{3N} a_{ij} \phi_{jm} \qquad \text{or} \qquad \sum_{j=1}^{3N} a_{ij}(\phi_{jm} - k_{ii}\delta_{jm}) = 0 \tag{6.45}$$

where i and m separately take on all values from 1 to $3N$. We know from previous experience that this set of equations for the a_{ij} has only trivial solution unless the determinant $|\phi_{jm} - k_{ii}\,\delta_{jm}|$ vanishes identically for each element of (k):

$$\begin{vmatrix} \phi_{11} - k & \cdots & \phi_{13N} \\ \vdots & & \\ \phi_{3N1} & \cdots & \phi_{3N3N} - k \end{vmatrix} \equiv 0 \qquad (k = \text{any one of the } k_{ii}) \qquad (6.46)$$

This equation serves to determine the k_{ii} since the ϕ_{ij} are presumed known already. Then the elements of (a) are found by solving Equations 6.45 subject to Equations 6.39.

An alternative method for calculating the normal mode frequencies, which has the advantage of permitting a direct calculation of the $\omega_{ii}{}^2$, makes use of the fact that

$$(k) = (\omega^2 m) = (\omega^2) \times (m) \qquad (6.47)$$

where (ω^2) is the matrix

$$(\omega^2) \;=\; \begin{bmatrix} \omega_{11}{}^2 & 0 & 0 & \cdots \\ 0 & \omega_{22}{}^2 & 0 & \cdots \\ 0 & 0 & \omega_{33}{}^2 & \cdots \\ \vdots & \vdots & & \end{bmatrix}$$

Equation 6.47 is simply a matrix version of Equations 6.43. It follows from (6.44) and (6.47) that

$$(a) \times (\phi) \times (a)^{-1} = (\omega^2) \times (m)$$
$$= (\omega^2) \times (M) = (\omega^2) \times (a) \times (M) \times (a)^{-1}$$

and, therefore, that

$$(a) \times (\phi) = (\omega^2) \times (a) \times (M) \qquad (6.48)$$

Equation 6.48 is the set of homogeneous, linear equations

$$\sum_{j=1}^{3N} a_{ij}\phi_{jm} - \sum_{n=1}^{3N}\sum_{j=1}^{3N} \omega_{in}{}^2 a_{nj} M_{jm} = 0$$

or

$$\sum_{j=1}^{3N} a_{ij}(\phi_{jm} - \omega_{ii}{}^2 M_{jm}) = 0 \qquad (i, m = 1, \ldots, 3N) \qquad (6.49)$$

since $\omega_{in}{}^2 = \omega_{ii}{}^2\,\delta_{in}$. The secular determinant must also vanish for these equations if a nontrivial set of a_{ij} is to be found. Therefore, with $M_{jm} = m_{jj}\,\delta_{jm}$,

$$\begin{vmatrix} \phi_{11} - \omega^2 m_{11} & \cdots & \phi_{13N} \\ \vdots & & \\ \phi_{3N1} & \cdots & \phi_{3N3N} - \omega^2 m_{3N3N} \end{vmatrix} \equiv 0 \qquad (6.50)$$

determines the $3N$ possible values of ω^2. Once these are known, the solution for the a_{ij} proceeds as before.

Whenever Equations 6.45 or 6.50 are solved, it is always found that certain of the normal mode frequencies are equal to zero. A look at Equations 6.41 tells us that such frequencies do not correspond to oscillations at all, but rather to cooperative motions of the assembly described by

$$\eta(t) = \eta(0) + \dot{\eta}(0)t \qquad (6.51)$$

where $\eta(0)$ and $\dot{\eta}(0)$ refer to the initial conditions. Equation 6.51 refers to a translation or a rotation displacement. Therefore, it accounts for the situation when the entire assembly moves as a *rigid unit*. Evidently the motions corresponding to $\omega_{ii} = 0$ involve no change in the relative separation of the particles, as would be observed in a true oscillation, but instead have to do either with the linear motion of the center of mass of the assembly or with the rotation of the assembly about a set of fixed axes. Conversely, a normal mode of nonvanishing frequency will not involve a change in the position of the center of mass or a rotation of the assembly as a whole. Both of these conclusions are consistent with the idea that a wholesale translation or rotation of the assembly provokes no restoring forces and thus *should* correspond to zero frequency. The number of zero-frequency modes to be expected can be computed once it is realized that a translation involves a displacement vector with components along three axes and that a rotation involves one with the same number of components, unless the assembly is constrained to lie along a line. In the latter case the number of components reduces to two, since the orientation of the line is completely specifiable by two parameters. Therefore, the number of zero values for ω^2 is usually six, but only five if the assembly lies along a line (as does, for example, a diatomic molecule). The ω^2 with zero values, of course, will appear as factors in Equation 6.50 and are of no direct interest in normal coordinate analysis. For this reason it is customary to define a quantity, called the number of *internal degrees of freedom* of an assembly, by

$$f \equiv 3N - 6 \qquad \text{or} \qquad f_L \equiv 3N - 5 \qquad \text{(linear assembly)}$$

The quantity f expresses the number of genuine normal modes of vibration for N particles interacting through harmonic forces. We note that the modifier "internal" used above is appropriate, since the *external* degrees of freedom would be those associated with the rigid-body motion of an assembly.

When the number of particles in an assembly is very large (e.g., 10^{23}), the difference between f and $3N$ is ignorable. But, for a several-particle assembly, such as a single molecule, f may be comparable with $3N$ and it is desirable to have a method for eliminating the zero-valued ω^2 from the normal coordinate analysis at the outset. Although no universal technique exists for accomplishing this task, it is possible often, through a careful choice of coordinates that reflects the spatial symmetry of the assembly, to come up with a secular equation that has no vanishing roots. This possibility will be illustrated by the example in the next section.

6.3 THE VIBRATING WATER MOLECULE

The water molecule is an assembly of three interacting particles that are constrained by chemical bonding to take up the planar triangle configuration shown
in Figure 6.4. When the two hydrogen and single oxygen atoms are at their
equilibrium positions the OH distance l is 0.958×10^{-8} cm and the HOH angle
2α is 104.5°. Otherwise, so long as their chemical bonds are not subject to a severe
perturbation, the absorption of electromagnetic radiation can cause the atoms to
carry out vibrations of small amplitude that should be susceptible to a normal
coordinate analysis.

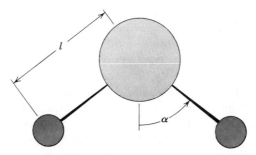

FIGURE 6.4. The atomic configuration of the water molecule. (Atomic sizes not to scale.)

The degree of labor involved with the procedure outlined in Section 6.2 depends
sensitively on how well we are able to choose the coordinates with which to express
the total energy (6.35). It is clear that the most useful set of these coordinates will
reflect to the greatest extent possible the inherent symmetry of the molecule under
consideration. If we accept the arrangement shown in Figure 6.4 as a starting point,
then it follows that the cartesian displacement coordinates $\{\xi_1{}^{\alpha}, \xi_2{}^{\alpha}; \alpha = 1, 2, 3\}$
with their natural rectangular symmetry are not really appropriate. Instead, it
appears sensible to postulate that a set of coordinates that describe small displacements from an equilibrium triangular arrangement ought to be devised. One
possibility is the set $\{Q_1, Q_2, \delta\}$, where

$$Q_1 = (\xi_1{}^2 - \xi_1{}^3) \cos \beta + (\xi_2{}^3 - \xi_2{}^2) \sin \beta \qquad (6.52)$$

$$Q_2 = (\xi_1{}^3 - \xi_1{}^1) \cos \beta + (\xi_2{}^3 - \xi_2{}^1) \sin \beta \qquad (6.53)$$

in terms of the quantities portrayed in Figure 6.5. The coordinate Q_1 expresses the
small displacement of the oxygen atom and one of the hydrogen atoms (that on
the right-hand side in the figure) along the direction of the chemical bond holding
them together. The coordinate Q_2 does the same for the remaining OH pair, while
δ refers to a displacement of the HOH angle. Assuming harmonic forces, we may

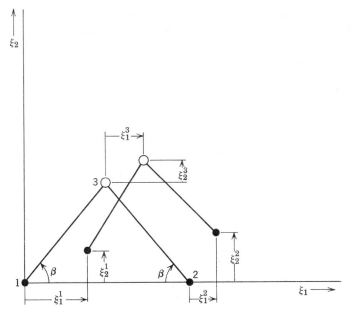

FIGURE 6.5. A general displacement of the atoms in a water molecule.

write the potential energy of the molecule in terms of these coordinates as

$$V(Q_1, Q_2, \delta) = \tfrac{1}{2}k(Q_1{}^2 + Q_2{}^2) + \tfrac{1}{2}\kappa\,\delta^2 \tag{6.54}$$

where k and κ are force constants. Equation 6.54 is the direct result of taking advantage of the geometric configuration of the molecule brought on by chemical bonding. We note that there are no cross terms of the form Q_iQ_j ($i \neq j$) or $Q_i\delta$ in the expression and that the number of these new coordinates is *three* rather than nine, the number of cartesian displacement coordinates for three particles. This latter result derives from our use of Figure 6.5. There we have implicitly set all of the $\xi_3{}^\alpha$ ($\alpha = 1, 2, 3$) equal to zero and have removed from consideration any displacement of the atoms that corresponds to a translation or a rotation of the molecule as a whole. (Actually, setting $\xi_3{}^\alpha \equiv 0$ already prohibits any translation along the x_3 direction and rotation about the x_1 and x_2 axes.) Therefore, we have automatically subtracted the six external degrees of freedom from the $3N = 9$ total degrees of freedom by casting everything in terms of displacements along the chemical bonds. This is one obvious advantage of the procedure.

If we were to compute the cartesian velocity coordinates in terms of Q_1, Q_2, and δ, and their derivatives with respect to time, we would get a very complicated expression for the kinetic energy that would offset completely the simplicity of Equation 6.54. For this reason we shall instead exploit even further the symmetry of the water molecule in order to come up with a second set of coordinates whose mathematical handling will be even more facile than that of the set $\{Q_1, Q_2, \delta\}$. Ultimately, however, we shall want to know the normal mode frequencies in terms of k and κ in (6.54), since these quantities directly are a measure of the strength of the chemical bonds in the molecule.

The water molecule is invariant in appearance with respect to three related *symmetry operations*.[7] The first of these is a rotation by multiples of 180° about an axis that bisects the molecule through the oxygen atom. (See Figure 6.6.) These rotations produce arrangements of the atoms that are identical to that observed when no rotation at all has been carried out. A second operation that leaves the molecule looking the same is reflection in its own plane, an action that merely exchanges one hydrogen atom for the other. The third symmetry operation is reflection in the plane perpendicular to the plane of the molecule. It is apparent that these three operations are not independent; indeed, the line of intersection of the two planes of symmetry is just the axis of symmetry around which the rotations occur. Therefore, we need consider (for example) only the rotations by 180° and reflection in the plane of the molecule in subsequent discussions of the symmetry operations.

The displacements that the atoms of the water molecule undergo in an arbitrary motion can be classified as to whether they are invariant (appear the same) after

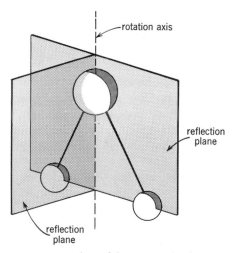

FIGURE 6.6. The three symmetry operations of the water molecule.

---

[7] The determination of the symmetry operations for a molecule is one of the most valuable tools in normal coordinate analysis. Quite general methods have been developed for relating symmetries to vibrational modes, of which the description here is a special case. For an authoritative, technical discussion of the procedures and their applications, see G. Herzberg, *Infrared and Raman Spectra of Polyatomic Molecules*, D. Van Nostrand, Princeton, 1945.

the symmetry operations. Moreover, those motions that are equivalent to a translation or a rotation of the molecule as a whole must be eliminated in a determination of the normal modes. These two general principles provide enough constraints for us to introduce a set of *symmetry coordinates* $\{S_1, S_2, S_3\}$ to replace the coordinates $\{Q_1, Q_2, \delta\}$. To see how this comes about, consider a displacement of the atoms that is *not* invariant under either rotation by 180° or a reflection in the plane of the molecule, and does not correspond to a rotation or a translation. Such a displacement is shown in Figure 6.7a. In order that the motions not be invariant under the two symmetry operations we need only have the two hydrogen atoms displaced with horizontal components in the same direction. To avoid translation we then must have the oxygen atom move horizontally in the opposite direction from the hydrogen atoms. (The vertical components of the displacements of the hydrogen atoms are arranged so as to cancel one another and prevent a translation along the x_2 direction.) But this condition requires that the displacements of the hydrogen atoms be along the OH bonds in order that no rotation occurs. Thus we arrive at the overall displacement shown in Figure 6.7a, which we shall say corresponds to the symmetry coordinate S_3.

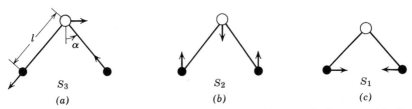

FIGURE 6.7. Symmetry coordinates for the water molecule: (*a*) a completely asymmetric displacement, (*b*) and (*c*) two independent displacements, that are each invariant under two of the symmetry operations.

Now let us consider displacements that *are* invariant under the two symmetry operations. In order to create them we must have the hydrogen atoms move parallel with the axis of symmetry or in opposite horizontal directions as depicted, for example, in Figures 6.7b and c. If the hydrogen atoms move vertically, the oxygen atom must move vertically as well (but in the opposite direction) in order to preclude any motion of the whole molecule. If the hydrogen atoms move horizontally, no further motion of the oxygen atom is necessary. Of course, the two sets of displacements described here are not unique. Any pair of motions where the respective displacements of the hydrogen atoms are perpendicular will do, for these will form a basis for expressing in a linear combination an arbitrary opposing motion of the two atoms. The symmetry coordinates in Figures 6.7b and c, which we shall call S_2 and S_1, respectively, are chosen for convenience in relating them to the cartesian displacement coordinates.

What about displacements of the atoms that do not give rise to translations or rotations, but that are invariant under only one of the two symmetry operations? A look at Figure 6.6 should convince anyone that, for the water molecule, such displacements are impossible. Since reflection in the plane of the molecule is the same as a rotation by 180° around the symmetry axis, what is invariant (or not invariant) under one must be so under the other. Therefore, we may conclude that we have exhausted the possibilities for vibratory motions of the molecule with the symmetry coordinates $\{S_1, S_2, S_3\}$.

Figure 6.7 suggests that the transformation from the cartesian displacement coordinates to the symmetry coordinates is

$$\xi_1{}^1 = S_1 - S_3 \sin \alpha \qquad \xi_1{}^2 = -S_1 - S_3 \sin \alpha \qquad \xi_1{}^3 = \frac{2m_H}{m_O} S_3 \sin \alpha \quad (6.55a)$$

$$\xi_2{}^1 = S_2 - S_3 \cos \alpha \qquad \xi_2{}^2 = S_2 + S_3 \cos \alpha \qquad \xi_2{}^3 = \frac{-2m_H}{m_O} S_2 \quad (6.55b)$$

where S_i means "the displacement of the atom in question along the direction under consideration according to the symmetry coordinate S_i," m_H is the mass of the hydrogen atom, and m_O is that of the oxygen atom. The expressions for $\xi_1{}^3$ and $\xi_2{}^3$ derive from the fact that, in order to prevent a translation of the molecule, the weighted displacement of the oxygen atom always must cancel the sum of the weighted displacements of the hydrogen atoms. Thus

$$m_O\xi_1{}^3 + m_H\xi_1{}^1 \sin \alpha + m_H\xi_1{}^2 \sin \alpha = 0$$
$$m_O\xi_2{}^3 + m_H\xi_2{}^1 + m_H\xi_2{}^2 = 0 \qquad (6.56)$$

and the last of Equations 6.55a and 6.55b follow on solving for $\xi_1{}^3$ and $\xi_2{}^3$, respectively. Equations 6.56 stipulate that the center of mass of the molecule shall not move during a vibration.

With Equations 6.52, 6.53, and 6.55, we can write

$$Q_1 = -\sin \alpha \, S_1 - a \cos \alpha \, S_2 - bS_3 \qquad (6.57)$$
$$Q_2 = -\sin \alpha \, S_1 - a \cos \alpha \, S_2 + bS_3 \qquad (6.58)$$

where

$$a \equiv 1 + \frac{2m_H}{m_O} \qquad b \equiv 1 + \frac{2m_H}{m_O} \sin^2 \alpha$$

and we have noticed that $\beta = (\pi/2) - \alpha$. The relation between δ and the symmetry coordinates is found less easily. We must examine the effect on α of a very small angular displacement. Consider first the change in α corresponding to the symmetry coordinate S_1. Let us write

$$\cot \alpha = \frac{h}{d}$$

where $h = l \cos \alpha$ is the length of the altitude of the triangle formed by HOH, and $d = l \sin \alpha$ is the length of the base of that triangle. Then, for a change in α by $\delta_1/2$, we have

$$\cot\left(\alpha + \frac{\delta_1}{2}\right) - \cot \alpha = \frac{h}{d - S_1} - \frac{h}{d}$$

$$\simeq \frac{hS_1}{d^2} = \frac{S_1 \cos \alpha}{l \sin^2 \alpha}$$

But a Taylor expansion of $\cot \theta$ about $\theta = \alpha$ gives, to first order,

$$\cot \theta \equiv \cot\left(\alpha + \frac{\delta_1}{2}\right) \simeq \cot \alpha - \frac{\delta_1}{2 \sin^2 \alpha}$$

Therefore

$$\delta_1 = -2 \frac{\cos \alpha}{l} S_1$$

In the case of S_2 we have

$$\cot\left(\alpha + \frac{\delta_2}{2}\right) - \cot \alpha = \frac{\left(h - S_2 - 2\frac{m_H}{m_O} S_2\right)}{d} - \frac{h}{d}$$

$$\simeq \frac{-aS_2}{l \sin \alpha}$$

$$\delta_2 = 2a \frac{\sin \alpha}{l} S_2$$

The small displacement δ is then

$$\delta = \delta_1 + \delta_2 = -2 \frac{\cos \alpha}{l} S_1 + 2a \frac{\sin \alpha}{l} S_2 \qquad (6.59)$$

Now we can rewrite Equation 6.54 in terms of the symmetry coordinates. Using Equations 6.57 to 6.59 we get

$$V(S_1, S_2, S_3) = \tfrac{1}{2}(V_{11}S_1{}^2 + V_{12}S_1S_2 + V_{21}S_2S_1 + V_{22}S_2{}^2 + V_{33}S_3{}^2) \qquad (6.60)$$

where

$$V_{11} = 2k \sin^2 \alpha + 4\frac{\kappa}{l^2} \cos^2 \alpha$$

$$V_{12} = 2a \sin \alpha \cos \alpha \left(k - 2\frac{\kappa}{l^2}\right) = V_{21}$$

$$V_{22} = 2a^2 \left(k \cos^2 \alpha + 2\frac{\kappa}{l^2} \sin^2 \alpha\right)$$

$$V_{33} = 2b^2 k$$

The kinetic energy of the molecule may be calculated after taking the first derivative with respect to time of Equations 6.55. Doing so, we can put together the expression

$$E_{int} = \tfrac{1}{2} \sum_{i=1}^{3} \sum_{j=1}^{3} (M_{ij}\dot{S}_i\dot{S}_j + \phi_{ij}S_iS_j) \tag{6.61}$$

where

$$M_{11} = 2m_O \qquad M_{22} = 2am_H \qquad M_{33} = 2bm_H$$

and E_{int} refers to the total energy derived from internal degrees of freedom. All the remaining M_{ij} and the ϕ_{ij} not already given are equal to zero. Equation 6.61 represents a considerable simplification over what we would have obtained with cartesian displacement coordinates (complicated potential energy) or the coordinates $\{Q_1, Q_2, \delta\}$ (complicated kinetic energy). We note in passing that Equation 6.61 involves only the $3N-6$ degrees of freedom relevant to vibration. This reduction in the number of terms is one of the main advantages of the symmetry coordinates. Indeed, if the symmetries of the molecule permit it, the coordinates $\{S_1, S_2, S_3\}$ could very well be just the normal coordinates themselves.

Now that we know we can follow the procedure of Section 6.2 without hesitation. The normal mode frequencies are determined directly from Equation 6.54, which here has the form

$$\begin{vmatrix} \phi_{11} - \omega^2 M_{11} & \phi_{12} & 0 \\ \phi_{21} & \phi_{22} - \omega^2 M_{22} & 0 \\ 0 & 0 & \phi_{33} - \omega^2 M_{33} \end{vmatrix} \equiv 0 \tag{6.62}$$

Equation 6.62 leads to

$$\omega^2 \equiv \omega_3{}^2 = \frac{\phi_{33}}{M_{33}} \tag{6.63}$$

$$\omega^4 - \frac{\phi_{22}M_{11} + \phi_{11}M_{22}}{M_{11}M_{22}}\omega^2 + \frac{\phi_{11}\phi_{22}}{M_{11}M_{22}} \tag{6.64}$$

Usually the two roots of Equation 6.64 are calculated and recombined into the pair of equations

$$\omega_1{}^2 + \omega_2{}^2 = \frac{\phi_{22}M_{11} + \phi_{11}M_{22}}{M_{11}M_{22}} \tag{6.65}$$

$$\omega_1{}^2\omega_2{}^2 = \frac{\phi_{11}\phi_{22}}{M_{11}M_{22}} \tag{6.66}$$

which are convenient for numerical work. Equations 6.63, 6.65, and 6.66 may be rewritten more specifically in terms of the molecular parameters k, (κ/l^2), m_O, m_H, and α as

$$\omega_3{}^2 = \left(1 + 2\frac{m_H}{m_O}\sin^2\alpha\right)\frac{k}{m_H} \tag{6.67}$$

$$\omega_1{}^2 + \omega_2{}^2 = \left(1 + 2\frac{m_H}{m_O}\cos^2\alpha\right)\frac{k}{m_H} + \frac{2}{m_H}\left(1 + 2\frac{m_H}{m_O}\sin^2\alpha\right)\frac{\kappa}{l^2} \quad (6.68)$$

$$\omega_1{}^2\omega_2{}^2 = 2\left(1 + 2\frac{m_H}{m_O}\right)\frac{k}{m_H{}^2}\frac{\kappa}{l^2} \quad (6.69)$$

If the molecular parameters are known, the normal mode frequencies follow directly. Usually, however, it is the frequencies and the angle α that have been measured, while the force constants k and (κ/l^2) remain to be deduced from Equations 6.67 and 6.69, respectively. In that case (6.68) serves as a check on self-consistency since the left-hand side is known from experiment and the right-hand side is known from the calculation of the force constants. For the water molecule, experiment gives

$$\frac{\omega_1}{2\pi c} = 3652 \text{ cm}^{-1} \qquad \frac{\omega_2}{2\pi c} = 1595 \text{ cm}^{-1} \qquad \frac{\omega_3}{2\pi c} = 3756 \text{ cm}^{-1}$$

where c is the speed of light in vacuum. These values, together with the value of α given earlier, lead to

$$k = 7.76 \times 10^5 \text{ dyne/cm} \qquad \kappa = 0.69 \times 10^5 \ l^2 \text{ dyne cm}$$

The consistency check results in the value 9.4 for the left-hand side of (6.68) multiplied by m_H for convenience sake and the value 9.6 for the right-hand side of (6.68) similarly multiplied. Consistency is attained to within 2 percent, which we may attribute to anharmonic forces. We note that the effective force constant (κ/l^2) for the bending motion of the molecule is but a tenth that for the stretching motion. This fact suggests that the chemical bonds are easier to rotate than to stretch.

The normal coordinates for the water molecule are obtained in the usual way by solving Equations 6.49. Actually, the form of Equation 6.62 tells us that in one case no calculation whatsoever will be necessary. S_3 is a normal coordinate already. (This is obvious from Equation 6.61, too, since the terms in S_3 do not couple with the other symmetry coordinates.) In this special instance symmetry arguments alone have produced the form of a normal coordinate. As for the remaining two, only a straightforward calculation is needed to show that S_1 and S_2 are in ratio as 1.2 to 1 for the mode of frequency ω_1, and that they are in ratio as -0.875 to 1 for the mode of frequency ω_2. The normal modes are displayed accordingly in Figure 6.8.

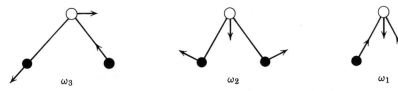

FIGURE 6.8. A sketch of the normal vibrational modes of the water molecule.

6.4 THE CONSERVATION THEOREMS

During the course of our discussion of the N-particle assembly interacting according to purely harmonic forces, we dealt in a tacit way with some general characteristics that, on a closer look, appear to be related to conservation principles. To begin with, we have based the essential properties of normal coordinate analysis on the invariance of the total energy. This would certainly be pointless if the forces among the particles were not conservative. Although a normal coordinate analysis may sometimes be possible in the presence of dissipative or driving forces,[8] the Energy Theorem would seem to be a basic requirement for this many-particle technique. Another, perhaps more significant relationship to a conservation principle developed when we noted that several of the normal mode frequencies ought to be equal to zero; that is, those associated with a translation or a rotation of the assembly as a whole. The possibility of a free bodily translation suggests that the total linear momentum of the assembly is constant in time, if the arguments of Section 4.5 can be broadened to apply to N particles. Similarly, a free rotation of the assembly should mean that the total angular momentum is fixed. Thus, the vanishing of a normal mode frequency can be interpreted as a sign that a conservation theorem is at work.

But these ideas are only suggestive. What must be done now is to establish them in detail under as general conditions as possible. We begin by assuming that the total energy of an N-particle assembly has the form

$$
E = \sum_{\alpha=1}^{N} \sum_{i=1}^{3} \frac{(p_i{}^{\alpha})^2}{2m_{\alpha}} + \sum_{\alpha=1}^{N} U(x_1{}^{\alpha}, x_2{}^{\alpha}, x_3{}^{\alpha})
$$
$$
+ \tfrac{1}{2} \sum_{\substack{\alpha=1 \\ (\alpha \neq \beta)}}^{N} \sum_{\beta=1}^{N} V(x_1{}^{\alpha}, x_2{}^{\alpha}, x_3{}^{\alpha}; x_1{}^{\beta}, x_2{}^{\beta}, x_3{}^{\beta}) \tag{6.70}
$$

when expressed in cartesian coordinates. The first term in (6.70) is the total kinetic energy. The second term is the potential function generated by forces *not* arising from interactions among the particles themselves, but rather from externally applied fields that persist even in the absence of the particles. (These might be electric or gravitational fields.) The last term in Equation 6.70 represents the potential energy of the forces of mutual interaction between the particles. These forces we assume to be conservative and to be decomposable into a sum of two-particle interactions, $V(\mathbf{r}^{\alpha}, \mathbf{r}^{\beta})$. Moreover, $V(\mathbf{r}^{\alpha}, \mathbf{r}^{\beta}) \equiv 0$ when $\alpha = \beta$, because a particle does not directly interact with itself, and

$$
V(\mathbf{r}^{\alpha}, \mathbf{r}^{\beta}) = V(\mathbf{r}) \qquad (\mathbf{r} = \mathbf{r}^{\alpha} - \mathbf{r}^{\beta}) \tag{6.71}
$$

in order that the Third Law be satisfied. To appreciate the property (6.71), we note

---

[8] For a discussion of this advanced topic, see H. Goldstein, *Classical Mechanics*, Addison-Wesley, Reading, Mass., 1959, §10.5.

that the force on particle α due to particle β is, by definition,

$$\mathbf{F}_{\alpha\beta} \equiv -\mathbf{V}_\alpha V(\mathbf{r}^\alpha, \mathbf{r}^\beta)$$

while that on β due to α is

$$\mathbf{F}_{\beta\alpha} \equiv -\mathbf{V}_\beta V(\mathbf{r}^\alpha, \mathbf{r}^\beta)$$

where $\mathbf{V}_\alpha$ is the gradient operator expressed in the coordinates of particle α and $\mathbf{V}_\beta$ is comparably defined. If we put $V(\mathbf{r}^\alpha, \mathbf{r}^\beta) \equiv V(\mathbf{r})$, then

$$\mathbf{F}_{\alpha\beta} = -\mathbf{V}V(\mathbf{r}) \qquad \mathbf{F}_{\beta\alpha} = \mathbf{V}V(\mathbf{r})$$

and the Third Law is supported. Because of the symmetry property (6.71) we have inserted the factor $\frac{1}{2}$ before the last sum in Equation 6.70, thereby avoiding double-counting of the interactions.

In Section 4.4 we found it very useful to consider a transformation from the cartesian coordinates for a two-particle assembly to the center-of-mass coordinates. It is a remarkable fact that the same kind of transformation is possible and almost as valuable (it does not, unfortunately, lead to an exact solution of the equations of motion) for an assembly containing an arbitrary number of particles. By analogy with Equations 4.54 and 4.55, we define the center-of-mass vector $\mathbf{R}$ and the N relative vectors $\bar{\mathbf{r}}^\alpha$ by

$$\mathbf{R} \equiv \frac{1}{M} \sum_{\alpha=1}^{N} m_\alpha \mathbf{r}^\alpha \tag{6.72}$$

$$\bar{\mathbf{r}}^\alpha \equiv \mathbf{r}^\alpha - \mathbf{R} \qquad (\alpha = 1, \ldots, N) \tag{6.73}$$

respectively, where the sum in (6.72) means vector addition and

$$M = \sum_{\alpha=1}^{N} m_\alpha \tag{6.74}$$

is the total mass of the assembly. The vector $\bar{\mathbf{r}}^\alpha$ is not a position vector established relative to the position of another particle, as was $\mathbf{r}$ in Equation 4.55, but instead denotes the position of the αth particle relative to that of the center of mass. In the spirit of Section 4.4, let us now introduce Equations 6.72 and 6.73 into the total kinetic energy:

$$T = \frac{1}{2} \sum_{\alpha=1}^{N} m_\alpha v_\alpha^2 = \frac{1}{2} \sum_{\alpha=1}^{N} m_\alpha \left(\frac{d\mathbf{r}^\alpha}{dt} \cdot \frac{d\mathbf{r}^\alpha}{dt} \right)$$

$$= \frac{1}{2} \sum_{\alpha=1}^{N} m_\alpha (\dot{\bar{\mathbf{r}}}^\alpha + \dot{\mathbf{R}}) \cdot (\dot{\bar{\mathbf{r}}}^\alpha + \dot{\mathbf{R}})$$

$$= \frac{1}{2} \sum_{\alpha=1}^{N} m_\alpha \bar{v}_\alpha^2 + \frac{1}{2} M V^2 \tag{6.75}$$

where

$$\bar{v}_\alpha \equiv \frac{d\bar{\mathbf{r}}^\alpha}{dt} \qquad \mathbf{V} \equiv \frac{d\mathbf{R}}{dt}$$

We have noted that, by Equations 6.73 and 6.72,

$$\sum_{\alpha=1}^{N} m_\alpha \dot{\bar{\mathbf{r}}}^\alpha = \frac{d}{dt} \sum_{\alpha=1}^{N} m_\alpha (\mathbf{r}^\alpha - \mathbf{R})$$

$$= \frac{d}{dt} \left(\sum_{\alpha=1}^{N} m_\alpha \mathbf{r}^\alpha - M\mathbf{R} \right)$$

$$= \frac{d}{dt} (M\mathbf{R} - M\mathbf{R}) = 0 \qquad (6.76)$$

Therefore, the kinetic energy of an N-particle assembly may be expressed as the kinetic energy of a hypothetical particle, whose mass is that of the assembly and whose velocity is that of the center of mass, plus the sum of kinetic energies of the N particles expressed relative to the center of mass.

Expressed in the CM coordinate system, the total energy (6.70) has the form

$$E = \frac{P^2}{2M} + \sum_{\alpha=1}^{N} \frac{\bar{p}_\alpha^{\,2}}{2m} + \sum_{\alpha=1}^{N} \bar{U}(\bar{\mathbf{r}}^\alpha, \mathbf{R}) + \tfrac{1}{2} \sum_{\substack{\alpha,\,\beta \\ (\alpha=\beta)}} \bar{V}(\bar{\mathbf{r}}^\alpha, \bar{\mathbf{r}}^\beta) \qquad (6.77)$$

where $\mathbf{P} \equiv M\mathbf{V}$, $\bar{\mathbf{p}}_\alpha \equiv m_\alpha \bar{\mathbf{v}}_\alpha$, $\bar{U}(\bar{\mathbf{r}}^\alpha, \mathbf{R}) \equiv U(\mathbf{r}^\alpha)$, and $\bar{V}(\bar{\mathbf{r}}^\alpha, \bar{\mathbf{r}}^\beta) \equiv V(\mathbf{r}^\alpha, \mathbf{r}^\beta)$. Since $\bar{V}$ represents the *mutual* interaction of a pair of particles, it is inconceivable that it could depend in any way on their absolute positions in the CM frame of reference. Therefore, we have deleted $\mathbf{R}$ from the argument of $\bar{V}$ in Equation 6.77. On the other hand, we cannot say the same for $\bar{U}(\bar{\mathbf{r}}_\alpha, \mathbf{R})$ because this potential function is generated by a force applied externally to the αth particle. The potential energy $\bar{U}$ thus will depend on the absolute location of the particle in the applied field and, therefore, on the vector $\mathbf{R}$ as well as $\bar{\mathbf{r}}^\alpha$. Bearing these conclusions in mind we now may state:

The Many-Particle Linear Momentum Theorem   If the potential energy generated by externally applied forces is zero, the total linear momentum of an N-particle assembly is a constant of the motion.

Proof   First we note that

$$\mathbf{P} = M\mathbf{V} = M\dot{\mathbf{R}} = \sum_{\alpha=1}^{N} m_\alpha \mathbf{v}_\alpha \qquad (6.78)$$

by Equation 6.72, so that $\mathbf{P}$ is indeed the total linear momentum of the assembly. The time variation of $\mathbf{P}$ is expressed by the equation

$$\dot{P}_i = -\frac{\partial}{\partial x_i} \sum_{\alpha=1}^{N} \bar{U}(\bar{\mathbf{r}}^\alpha, \mathbf{R}) \qquad (i = 1, 2, 3) \qquad (6.79)$$

where $\mathbf{R} = \{X_1, X_2, X_3\}$, because $\bar{V}$ in (6.77) is independent of R. Since we have assumed by hypothesis that U is zero, the time derivative of P_i in (6.79) vanishes and the theorem is proved.

The constancy of the total linear momentum is seen to depend on the non-existence of *externally applied forces*. When these forces are not present, the assembly is free to drift through space as a unit with the constant velocity **V**. In normal coordinate analysis it is always assumed that only forces of mutual interaction are active. Therefore, it is easy to see why the three degrees of translational freedom separate from the secular equation: they reflect the conservation of total linear momentum. Moreover, Equations 6.56, which, according to Equation 6.72 and Figure 6.7, express the components of the displacements of particles from the center of mass, are a direct result of our freedom to choose $\mathbf{V} \equiv 0$, without a loss in generality, whenever the total linear momentum is conserved.

The total angular momentum of an N-particle assembly is evidently

$$\mathbf{J} = \sum_{\alpha=1}^{N} \mathbf{J}_\alpha = \sum_{\alpha=1}^{N} (\mathbf{r}^\alpha \times \mathbf{p}_\alpha) \tag{6.80}$$

by way of generalization of Equation 4.27, the sum again referring to vector addition. With the help of Equations 6.72, 6.73, and 6.76, we can write

$$
\begin{aligned}
\mathbf{J} &= \sum_{\alpha=1}^{N} [(\bar{\mathbf{r}}^\alpha + \mathbf{R}) \times m_\alpha(\bar{\mathbf{v}}_\alpha + \mathbf{V})] \\
&= \sum_{\alpha=1}^{N} m_\alpha[(\bar{\mathbf{r}}^\alpha \times \bar{\mathbf{v}}_\alpha) + (\mathbf{R} \times \mathbf{v}) + (\mathbf{R} \times \bar{\mathbf{v}}_\alpha) + (\bar{\mathbf{r}}^\alpha \times \mathbf{V})] \\
&= \sum_{\alpha=1}^{N} [(\bar{\mathbf{r}}^\alpha \times \bar{\mathbf{p}}_\alpha) + (\mathbf{R} \times \mathbf{P})] \equiv \sum_{\alpha=1}^{N} \bar{\mathbf{J}}_\alpha + \bar{\mathbf{J}}_{CM}
\end{aligned}
\tag{6.81}
$$

Therefore, the total angular momentum of the assembly is the vector sum of the angular momentum of the center of mass, $\mathbf{J}_{CM}$, and that of the assembly relative to the center of mass. It follows directly that we may state:

The Many-Particle Angular Momentum Theorem (a) If the potential function for the externally applied forces does not depend on the orientation of the center-of-mass vector **R**, then the angular momentum $\mathbf{J}_{CM}$ is a constant of the motion.

(b) If the potential functions $\bar{U}(\bar{\mathbf{r}}^\alpha, \mathbf{R})$ and $\bar{V}(\bar{\mathbf{r}}^\alpha, \bar{\mathbf{r}}^\beta)$ do not depend on the orientation of the vectors $\bar{\mathbf{r}}^\alpha$ ($\alpha = 1, \ldots, N$), then the angular momentum relative to the center of mass is a constant of the motion.

Proof To establish statement (a) we need only recall the discussion of Section 4.2 in relation to central-field potentials. If $\bar{U}(\bar{\mathbf{r}}^\alpha, \mathbf{R}) = \bar{U}(\bar{\mathbf{r}}^\alpha, R)$, then we may write $\mathbf{R} = \{R, \theta, \varphi\}$. It follows that

$$\mathbf{J}_{CM}{}^2 = P_\theta{}^2 + \frac{P_\varphi{}^2}{\sin^2 \theta}$$

as well as the direction of $\mathbf{J}_{CM}$ will be a constant of the motion, according to our previous arguments. To prove statement (b) we apply the same reasoning separately to each of the $\bar{\mathbf{J}}_\alpha$.

Before doing a normal coordinate analysis, we assume that no external forces are present and that the $V(\mathbf{r}^\alpha, \mathbf{r}^\beta)$ are functions only of the distance of separation of particles α and β. Therefore, the conditions of the conservation theorem given above are met and the total angular momentum is a constant of the motion. But this result means that the assembly encounters no resistance to a bodily rotation about the axis with respect to which $\mathbf{J}_{CM}$ is determined. It follows that the degrees of freedom corresponding to rotation will separate from the secular equation and that the attendant normal mode frequencies will be zero. We note also that, if the assembly is confined to lie along a line, one of the angles $\{\theta, \varphi\}$ must be fixed and the three orientational degrees of freedom reduce to two in number.

6.5 THE RIGID BODY

A rigid body, as we have already implied in the discussion at the end of Section 6.2, is an assembly of particles *whose distances of mutual separation are fixed*. This condition prevents there being any relative motion of the particles and results in the conclusion that the number of degrees of freedom of a rigid body is six—three for translation and three for rotation. In geometric terms, this means that we specify the position and orientation of the rigid body in space by choosing three noncollinear points in the body relative to a fixed set of axes. The first of these points, established in terms of three coordinates, may be at the center of mass. A second point in the body then may be selected to lie on the surface of a sphere that is centered on the first point. The orientation of a line drawn between the two points will be specified by two coordinates (e.g., two spherical polar angles), bringing the total number of coordinates employed thus far to five. Finally, to prescribe the orientation of the body about the axis made by the line between points "one" and "two," we add a third point that is off that axis. The orientation of a perpendicular drawn between this point and the axis can be specified by a single angular coordinate measured relative to a line that is also perpendicular to the axis and lies in the plane comprising the latter and the x_3 axis of our original fixed set. This brings the total number of coordinates required to six. An alternative to this whole procedure is to form the rigid body by binding particles together, one at a time, and then compute the net number of coordinates added each time a particle is added. (See Problem 15 at the end of this chapter.) The net total number of coordinates for the assembly again comes out to be six.

If the additional constraint is imposed that the assembly of particles form a linear array, the number of degrees of freedom is reduced to five. This corresponds to describing the assembly with two collinear points instead of three and, therefore, five coordinates instead of six. Therefore, we have the general relationship:

$F + f = 3N$, where F is the number of rigid-body degrees of freedom, including any external constraint on the shape of the assembly, and f is the number of internal degrees of freedom as defined in Section 6.2. It is clear from this expression why the adjective "internal" is used in describing f.

The Second Law for the conservative translational motion of a rigid body must have the form

$$\frac{d\mathbf{P}}{dt} = \mathbf{F}(\mathbf{R}) \tag{6.82}$$

according to Equation 6.79, where $\mathbf{F} = -\mathbf{V}_R \sum_{\alpha=1}^{N} \bar{U}(\bar{\mathbf{r}}^\alpha, \mathbf{R})$ is an externally applied force and $\mathbf{V}_R$ is the gradient operator in center-of-mass coordinates. We note that Equation 6.82 makes no reference to the individual particles making up the rigid body, but refers only to the motion of their center of mass. The force $\mathbf{F}$, in particular, acts on that point and may be regarded only as an influence on the body as a whole. For example, if the body is in a uniform gravitational field, $\mathbf{F} = M\mathbf{g}$, where $\mathbf{g}$ points along the negative x_3 direction and M is the mass of the body. The mathematical techniques for solving Equation 6.82 are precisely the same as those outlined in Chapters 1 and 4.

The problem of rotational motion is somewhat more complicated. To consider it in detail we must first prescribe a method for expressing the orientation of the rigid body in a fixed frame of reference. One possibility that would come to mind immediately is to employ three angles of rotation, each related to one axis of a set of three mutually perpendicular axes whose origin is at the center of mass. Rotations about these axes would bring the body into any desired orientation. The difficulty with thise choice is that a set of three such angles of rotation is not unique; the final orientation will depend on the *order* in which the rotations are carried out. For example, a rotation of a body about an x_1 axis followed by a rotation about an x_2 axis produces a different orientation from what results by reversing this order of rotation. It is not trivial to find a good method for representing the orientation of a rigid body relative to its center of mass.

A convenient set of coordinates for prescribing orientation is that comprising the *Euler angles*. Consider a set of coordinate axes $X_1 X_2 X_3$, which is fixed in space, and a set $X'_1 X'_2 X'_3$, which is fixed in the rigid body, with both having their origins at the center of mass. If the body has single axis of symmetry, that will be the X'_3 axis. Assuming that the two sets initially are congruent, a series of three rotations about the body axes (two for orienting the symmetry axis and one for orienting the body relative to this axis), carried out in a defined order, produces the Euler angles (see Figure 6.9). These rotations are as follows:

(a) A positive rotation through the angle ϕ about the X'_3 axis. This results in the transformation of $X'_1 X'_2 X'_3$ into $X''_1 X''_2 X'_3$.

(b) A positive rotation through the angle θ about the X''_1 axis. This results in the transformation of $X''_1 X''_2 X'_3$ into $X''_1 X'''_2 X''_3$.

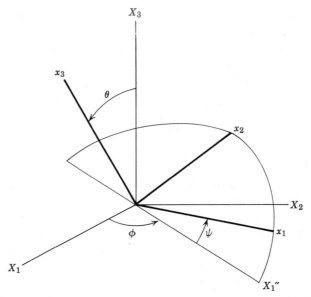

FIGURE 6.9. Rotations that define the Euler angles ϕ, θ, and ψ.

(c) A positive rotation through the angle ψ about the X_3'' axis. This results in the transformation of $X_1''X_2'''X_3''$ into $x_1x_2x_3$.

These three rotations define the Euler angles ϕ, θ ψ. Since the Euler angle ϕ represents a rotation about the X_3' axis, we know from the discussion in Section A.2 of the Appendix that the transformation of $X_1'X_2'X_3'$ into $X_1''X_2''X_3'$ may be written in the matrix form

$$
\begin{pmatrix} X_1'' \\ X_2'' \\ X_3' \end{pmatrix} = \begin{pmatrix} \cos\phi & \sin\phi & 0 \\ -\sin\phi & \cos\phi & 0 \\ 0 & 0 & 1 \end{pmatrix} \times \begin{pmatrix} X_1' \\ X_2' \\ X_3' \end{pmatrix}
\tag{6.83a}
$$

Similarly, the transformations described in (b) and (c) above are expressed as

$$
\begin{pmatrix} X_1'' \\ X_2''' \\ X_3'' \end{pmatrix} = \begin{pmatrix} 1 & 0 & 0 \\ 0 & \cos\theta & \sin\theta \\ 0 & -\sin\theta & \cos\theta \end{pmatrix} \times \begin{pmatrix} X_1'' \\ X_2'' \\ X_3' \end{pmatrix}
\tag{6.83b}
$$

and

$$
\begin{pmatrix} x_1 \\ x_2 \\ x_3 \end{pmatrix} = \begin{pmatrix} \cos\psi & \sin\psi & 0 \\ -\sin\psi & \cos\psi & 0 \\ 0 & 0 & 1 \end{pmatrix} \times \begin{pmatrix} X_1'' \\ X_2''' \\ X_3'' \end{pmatrix}
\tag{6.83c}
$$

respectively. The complete transformation from $X_1'X_2'X_3' = X_1X_2X_3$ into the oriented set $x_1 x_2 x_3$ is then specified by the matrix product $(x) = (CBA)(X)$, where

$$(x) = \begin{pmatrix} x_1 \\ x_2 \\ x_3 \end{pmatrix} \qquad (X) = \begin{pmatrix} X_1 \\ X_2 \\ X_3 \end{pmatrix}$$

and (A), (B), and (C) are the rotation matrices in Equations 6.83a, 6.83b, and 6.83c, respectively. If the elements of the matrix (CBA) are labeled R_{ij} $(i, j = 1, 2, 3)$, then we have for the matrix elements the expressions:

$$R_{11} = \cos\phi\cos\psi - \sin\phi\cos\theta\sin\psi$$
$$R_{12} = \sin\phi\cos\psi + \cos\phi\cos\theta\sin\psi$$
$$R_{13} = \sin\theta\sin\psi$$
$$R_{21} = -\cos\phi\sin\psi - \sin\phi\cos\theta\cos\psi$$
$$R_{22} = -\sin\phi\sin\psi + \cos\phi\cos\theta\cos\psi \qquad (6.84)$$
$$R_{23} = \sin\theta\cos\psi$$
$$R_{31} = \sin\phi\sin\theta$$
$$R_{32} = -\cos\phi\sin\theta$$
$$R_{33} = \cos\theta$$

It is very important to note here that the orientation of a rigid body resulting from rotations through the Euler angles depends on the rotations having been carried out *precisely* as described in the steps (a), (b), and (c). Any change in the order of rotation will change the resultant orientation of the body. In mathematical terms, we are saying that the matrix elements R_{ij} will be entirely different if the product of (A), (B), and (C) is formed in any order different from (CBA).

Now, the total angular momentum of a rigid body about its center of mass is given by

$$\bar{\mathbf{J}} = \sum_{\alpha=1}^{N} m_\alpha (\bar{\mathbf{r}}^\alpha \times \bar{\mathbf{v}}_\alpha) \qquad (6.85)$$

according to Equation 6.81. As will be shown in detail in the next chapter, the velocity $\bar{\mathbf{v}}_\alpha$ may be written as the axial vector $\bar{\mathbf{v}}_\alpha = \bar{\omega} \times \bar{\mathbf{r}}^\alpha$ where $\bar{\omega}$ is a vector pointing along the direction of $\bar{\mathbf{J}}$ and having a magnitude equal to the angular rate of rotation of the body about its center of mass. With this expression for $\bar{\mathbf{v}}_\alpha$ introduced, Equation 6.85 becomes

$$\bar{\mathbf{J}} = \sum_{\alpha=1}^{N} m_\alpha (\bar{\mathbf{r}}^\alpha \times (\bar{\omega} \times \bar{\mathbf{r}}^\alpha))$$
$$= \sum_{\alpha=1}^{N} m_\alpha [\bar{\omega}(\bar{\mathbf{r}}^\alpha)^2 - \bar{\mathbf{r}}^\alpha(\bar{\mathbf{r}}^\alpha \cdot \bar{\omega})] \qquad (6.86)$$

where the second step comes from using the vector identity $\mathbf{A} \times (\mathbf{B} \times \mathbf{C}) \equiv \mathbf{B}(\mathbf{A} \cdot \mathbf{C}) - \mathbf{C}(\mathbf{A} \cdot \mathbf{B})$. If the vector $\bar{\omega}$ is decomposed into its components along the axes x_1, x_2 and x_3 of the rigid body, Equation 6.86 reduces to the set of three equations

$$\bar{J}_1 = I_{11}\bar{\omega}_1 + I_{12}\bar{\omega}_2 + I_{13}\bar{\omega}_3 \tag{6.87a}$$

$$\bar{J}_2 = I_{21}\bar{\omega}_1 + I_{22}\bar{\omega}_2 + I_{23}\bar{\omega}_3 \tag{6.87b}$$

$$\bar{J}_3 = I_{31}\bar{\omega}_1 + I_{32}\bar{\omega}_2 + I_{33}\bar{\omega}_3 \tag{6.87c}$$

where

$$I_{ii} = \sum_{\alpha=1}^{N} m_\alpha [(\bar{r}^\alpha)^2 - (\bar{x}_i^\alpha)^2] \qquad (i = 1, 2, 3) \tag{6.88}$$

is called a *moment of inertia coefficient* and

$$I_{ij} = - \sum_{\alpha=1}^{N} m_\alpha \bar{x}_i^\alpha \bar{x}_j^\alpha \qquad (i, j = 1, 2, 3; i \neq j) \tag{6.89}$$

is called a *product of inertia*. The moments of inertia $I_{11} = \sum_\alpha m_\alpha(\bar{x}_2{}^2 + \bar{x}_3{}^2)$, and the like, should be familiar from introductory mechanics. They are always positive and, in general, depend on the time. The products of inertia may be of either sign and also may be time dependent. In most cases of practical interest, the sum over particles in Equations 6.88 and 6.89 is replaced by an integral over the volume of the rigid body:

$$I_{ii} = \iiint_v \rho(\bar{r})(\bar{r}_2 - \bar{x}_i{}^2) \, dv \qquad (i = 1, 2, 3) \tag{6.90}$$

$$I_{ij} = -\iiint_v \rho(\bar{r})\bar{x}_i\bar{x}_j \, dv \qquad (i, j = 1, 2, 3) \tag{6.91}$$

where $\rho(\bar{r})$ is the density of the body at the point $\bar{r}$ within the volume v.

Equations 6.87 may be summarized in a matrix form by analogy with what was done for Equations 6.20:

$$(\bar{J}) = (I) \times (\bar{\omega}) \tag{6.92}$$

where

$$(\bar{J}) = \begin{pmatrix} \bar{J}_1 \\ \bar{J}_2 \\ \bar{J}_3 \end{pmatrix} \qquad (\bar{\omega}) = \begin{pmatrix} \bar{\omega}_1 \\ \bar{\omega}_2 \\ \bar{\omega}_3 \end{pmatrix}$$

and the matrix elements of (I) are given by Equations 6.88 and 6.89. The moments and products of inertia are parameters whose values depend on the origin and orientation of the body axes $x_1x_2x_3$. This being the case, it is evident that the most desirable choice of body axes would be that which causes all of the products of inertia to vanish and thereby reduces Equations 6.87 to the simple forms

$$\bar{J}_1 = I_A\bar{\omega}_1 \qquad \bar{J}_2 = I_B\bar{\omega}_2 \qquad \bar{J}_3 = I_C\bar{\omega}_3 \tag{6.93}$$

where I_A is the value of I_{11} for this optimal set of axes, and so on. The transformation of Equations 6.87 into Equations 6.93 is, mathematically, just the problem of casting the matrix (I) into diagonal form. *It is precisely the same kind of problem as was considered for the matrix $(\hat{\omega}^2)$ in Section 6.1.* Here, the analogs of the squares of the normal mode frequencies ω_{ii}^2, which lay along the principal diagonal of the diagonalized $(\hat{\omega}^2)$, are the principal moments of inertia I_A, I_B, I_C, which lie along the principal diagonal of the diagonalized (I). The normal coordinate transformation of Section 6.2 is here called a *principal axis transformation*, the "principal axes" being those particular $x_1x_2x_3$ that permit the matrix (I) to appear in diagonal form. The analog of Equation 6.29 is, then, the expression

$$\begin{vmatrix} I_{11} - I & I_{12} & I_{13} \\ I_{21} & I_{22} - I & I_{23} \\ I_{31} & I_{32} & I_{33} - I \end{vmatrix} \equiv 0 \qquad (6.94)$$

where $I = I_A, I_B,$ or I_C. This equation determines the principal moments. From this point on we shall assume that the transformation has actually been carried through and, therefore, that $x_1x_2x_3$ and the Euler angles ϕ, θ, ψ refer to the rigid body *oriented along its principal axes.*

In the case of the rigid body that rotates freely about its center of mass, we know that the angular momentum is a constant of the motion. As will be shown in the next chapter, this condition, when referred to the rotating system of body axes, is

$$\frac{d\bar{\mathbf{J}}}{dt} + \bar{\omega} \times \bar{\mathbf{J}} = \mathbf{0} \qquad (6.95)$$

or, equivalently,

$$I_A \frac{d\bar{\omega}_1}{dt} + \bar{\omega}_2\bar{\omega}_3(I_C - I_B) = 0 \qquad (6.96a)$$

$$I_B \frac{d\bar{\omega}_2}{dt} + \bar{\omega}_3\bar{\omega}_1(I_A - I_C) = 0 \qquad (6.96b)$$

$$I_C \frac{d\bar{\omega}_3}{dt} + \bar{\omega}_1\bar{\omega}_2(I_B - I_A) = 0 \qquad (6.96c)$$

upon introducing Equations 6.93. Equations 6.96 are known as *Euler's equations* for a rigid body that rotates about its own center of mass. These expressions can be related to the three angular speeds $\dot{\phi}, \dot{\theta},$ and $\dot{\psi}$ if the angular speeds $\bar{\omega}_1, \bar{\omega}_2, \bar{\omega}_3$ are known in terms of the Euler angles. The relationship is determined by the inverse of the transformation matrix (R), whose elements are given in Equations 6.84.[9] The result is:

$$\bar{\omega}_1 = \dot{\phi} \sin\theta \sin\psi + \dot{\theta} \cos\psi$$
$$\bar{\omega}_2 = \dot{\phi} \sin\theta \cos\psi - \dot{\theta} \sin\psi \qquad (6.97)$$
$$\bar{\omega}_3 = \dot{\phi} \cos\theta + \dot{\psi}$$

[9] See H. Goldstein, op. cit., p. 134, for the somewhat complicated details of the calculation.

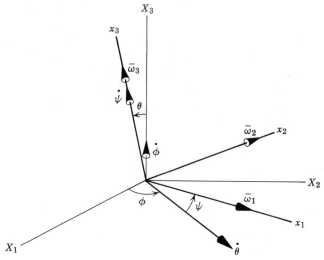

FIGURE 6.10. Angular velocity coordinates in terms of the Euler angles.

The relationships in Equations 6.97 also may be deduced, with a little thought, directly from Figure 6.10.

6.6 DYNAMICS OF A SPACE VEHICLE IN ITS COASTING PHASE

A problem of considerable interest in space flight is the motion of a symmetric space vehicle in its coasting phases. By "symmetric" it is meant that $I_A = I_B$; that is, the vehicle appears as a rigid body of revolution symmetric about its x_3 axis. "Coasting phase" refers to any condition under which all external forces on the vehicle are negligibly different from zero. Then we expect that the center of mass of the vehicle will be moving with constant velocity, as prescribed by Equation 6.82, and that the rotation of the vehicle will be governed by Equations 6.96. (See Figure 6.11.) Therefore, we have to solve Equations 6.96a and 6.96b together with

$$I_C \dot{\bar{\omega}}_3 = 0 \qquad (6.98)$$

in order to determine the rotational motion. Equation 6.98 tells us immediately that the vehicle spins about its x_3 axis at a steady rate: $\bar{\omega}_3$ is a constant. If we then set

$$\omega_0 \equiv \bar{\omega}_3 \frac{(I_C - I_B)}{I_A} = \bar{\omega}_3 \frac{(I_C - I_A)}{I_A}$$

Equations 6.96a and 6.96b may be rewritten as

$$\frac{d\bar{\omega}_1}{dt} + \omega_0 \bar{\omega}_2 = 0 \qquad (6.99a)$$

$$\frac{d\bar{\omega}_2}{dt} - \omega_0\bar{\omega}_1 = 0 \tag{6.99b}$$

respectively. Upon operating on Equation 6.99a with d/dt and introducing Equation 6.99b into the result we find

$$\frac{d^2\bar{\omega}_1}{dt^2} + \omega_0{}^2\bar{\omega}_1 = 0 \tag{6.100a}$$

and, similarly,

$$\frac{d^2\bar{\omega}_2}{dt^2} + \omega_0{}^2\bar{\omega}_2 = 0 \tag{6.100b}$$

It follows that $\bar{\omega}_1$ and $\bar{\omega}_2$ are sinusoidal functions of the time subject to the constraint

$$(\bar{\omega}_1)^2 + (\bar{\omega}_2)^2 \equiv \omega_{12}{}^2 = \text{constant} \tag{6.101}$$

The component of $\bar{\omega}$ in the x_1-x_2 plane rotates at the steady rate ω_0 while the component along the x_3 axis remains fixed.

In order to visualize the rotation of the vehicle we must express the equations of motion in terms of the Euler angles. This is done simply if the X_3 axis is assumed to coincide with the direction of $\mathbf{J}$. Then the angle between the X_3 axis and the

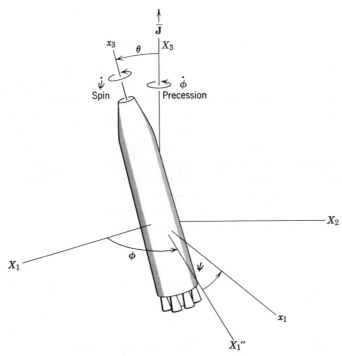

FIGURE 6.11. A space vehicle with cylindrical symmetry.

x_3 axis is fixed (because $\mathbf{J}$ and $\bar{\omega}_3$ are fixed) and, since that angle is now θ, we infer $\dot{\theta} = 0$. Equations 6.97 reduce to

$$\bar{\omega}_1 = \dot{\phi} \sin \theta \sin \psi \qquad \bar{\omega}_2 = \dot{\phi} \sin \theta \cos \psi \qquad \bar{\omega}_3 = \dot{\phi} \cos \theta + \dot{\psi} \qquad (6.102)$$

and it follows that

$$\frac{d\bar{\omega}_1}{dt} = \dot{\phi}\dot{\psi} \sin \theta \cos \psi \qquad \frac{d\bar{\omega}_2}{dt} = -\dot{\phi}\dot{\psi} \sin \theta \sin \psi \qquad \frac{d\bar{\omega}_3}{dt} = 0 \qquad (6.103)$$

(Compare Equation 6.103 to Equations 6.98 and 6.99.) A relation between the rate of spin of the vehicle about its symmetry axis, $\dot{\psi}$, and the rate of precession of the symmetry axis about the X_3 axis, $\dot{\phi}$, may now be found by substituting from Equations 6.102 and 6.103 into Equation 6.96a:

$$I_A \dot{\phi}\dot{\psi} \sin \theta \cos \psi + (\dot{\phi}^2 \sin \theta \cos \theta \cos \psi + \dot{\phi}\dot{\psi} \sin \theta \cos \psi)(I_C - I_B) = 0$$

which is just

$$\dot{\phi} = \frac{I_C}{(I_B - I_C) \cos \theta} \dot{\psi} \qquad (6.104)$$

The rate of precession is proportional to the rate of spin. The direction of precession depends on the sign of $(I_B - I_C)$. If $I_C > I_B$ the precession is opposite the spin and is called *retrograde*. If $I_C < I_B$, which is usual for space vehicles, the precession is in the same rotational sense as the spin and is called *direct*.

FOR FURTHER READING

K. Symon	*Mechanics*, Addison-Wesley, Reading, Mass., 1971. A comprehensive introduction to normal coordinates and rigid bodies is given in Chapters 4, 5, 10, 11, and 12 of this book.
D. T. Greenwood	*Principles of Dynamics*, Prentice-Hall, Englewood Cliffs, N. J., 1965. Chapters 7 and 8 present the dynamics of rigid bodies in an applications-oriented context. Good reading for the engineering minded.
R. E. D. Bishop and D. C. Johnson	*The Mechanics of Vibration*, Cambridge University Press, London, 1960. A detailed presentation of normal coordinates appears in Chapter 3 of this weighty tome.
K. J. Ball and G. F. Osbourne	*Space Vehicle Dynamics*, Oxford University Press, London, 1967. Space vehicles as rigid bodies abound in this fine introduction.

PROBLEMS

1. Apply normal coordinate analysis to a pair of linear oscillators, each of mass m, whose total potential energy is given by

$$V(\xi_A, \xi_B) = \tfrac{1}{2}k_A \xi_A{}^2 + \tfrac{1}{2}k_B \xi_B{}^2 + \tfrac{1}{2}k_{AB}(\xi_A - \xi_B)^2$$

Under what conditions will the motion of the assembly be periodic?

2. Show that the "rotation" that transforms ξ_A and ξ_B in Equations 6.5 and 6.6 into the normal coordinates η_1 and η_2 is a "rotation" by $45°$. Represent this situation in a graph.

3. The "unperturbed state" of two identical coupled oscillators is usually chosen as that wherein each particle is bound to a single fixed point through a restoring force characterized by k. Alternatively we may choose this state to be one where one of the oscillators has infinite mass and the other is bound to two fixed points through restoring forces characterized by k and k_{AB}. The frequencies given in Equation 6.10 are then displaced above and below the frequency of the unperturbed state. Prove this last statement.

4. Two identical pendula hang from a ceiling and are coupled together laterally by a very light spring. Its ends are fastened to the pendulum rods at a distance h below the ceiling. Given that the lengths of the rods are each l and that the pendulum bobs are each of mass m and swing in the same plane, compute the normal mode frequencies of the assembly. (Assume small oscillations.)

5. Imagine that a pendulum of rod length l and bob mass M has been attached to the center of the harmonic oscillator shown in Figure 2.3. Gravity is to affect only the pendulum. The small oscillations of this system are governed by the equations

$$m\frac{d^2\xi}{dt^2} + k\xi(t) - Mg\theta(t) = 0$$

$$\frac{d^2\xi}{dt^2} + l\frac{d^2\theta}{dt^2} + g\theta(t) = 0$$

Calculate the normal mode frequencies for the assembly.

6. Consider a pair of LC circuits, as discussed in Special Topic 3, which have a common side with a capacitor of capacitance C_3 inserted. The Kirchoff voltage law for these "capacitatively coupled" circuits produces the two equations

$$L_m\frac{di_m}{dt} + \frac{q_m(t)}{C_m} + \frac{1}{C_3}(q_m(t) - q_n(t)) = 0 \qquad (m \neq n; m = 1, 2, n = 1 \text{ or } 2)$$

where all quantities are identified in Table 3.2.

(a) Calculate the normal mode frequencies for the system.

(b) Assume $C_3 \gg C_m$ ($m = 1, 2$) and describe the "resonant charge transfer" between the circuits in terms of $q_1(t)$ and $q_2(t)$.

7. The total energy of a linear triatomic molecule is

$$E = \frac{p_A{}^2}{2m_A} + \frac{p_B{}^2}{2m_B} + \frac{p_C{}^2}{2m_A} + \tfrac{1}{2}k[(\xi_B - \xi_A)^2 + (\xi_C - \xi_B)^2]$$

(a) Calculate the three normal mode frequencies for the molecule.

(b) Compute the normal modes themselves and interpret them physically in terms of the displacements ξ_A, ξ_B, and ξ_C.

8. Apply the results of Problem 7a to the molecules CO_2 and CS_2. The observed normal mode frequencies are:

$$\frac{\omega_1}{2\pi c} = 1337 \text{ cm}^{-1} \qquad \frac{\omega_3}{2\pi c} = 2349 \text{ cm}^{-1} \qquad \text{for } CO_2$$

$$\frac{\omega_1}{2\pi c} = 657 \text{ cm}^{-1} \qquad \frac{\omega_3}{2\pi c} = 1523 \text{ cm}^{-1} \qquad \text{for } CS_2$$

Check these results to see if they agree with the theory and comment on the likely causes of any disagreement.

9. Suppose that CO_2 is a nonlinear molecule and apply the normal coordinate analysis of Section 6.3. The observed value for $\omega_2/2\pi c$ is 667 cm^{-1}. Compute the force constants k and κ/l^2 using this frequency and those given in Problem 8 along with $\alpha = 90°$. Then perform the check on self-consistency associated with Equation 6.68. What are the implications of your results?

10. The observed vibrational frequencies of D_2O are:

$$\frac{\omega_1}{2\pi c} = 2666 \text{ cm}^{-1} \qquad \frac{\omega_2}{2\pi c} = 1179 \text{ cm}^{-1} \qquad \frac{\omega_3}{2\pi c} = 2784 \text{ cm}^{-1}$$

Calculate the force constants k and κ/l^2 under the assumption that the angle α is the same as in H_2O.

11. The potential energy of the water molecule may be written

$$V(Q_1, Q_2, Q_3) = \tfrac{1}{2}k(Q_1^2 + Q_2^2) + \tfrac{1}{2}k_1 Q_3^2$$

where $Q_3 = \xi_1^2 - \xi_1^1$ is the displacement along the line of centers of the H atoms, as shown in Figure 6.5.
 (a) Show that $Q_3 = -2S_1$.
 (b) Find the new values of the coefficients in Equation 6.60.
 (c) Find the new versions of Equations 6.67 to 6.69.

12. Apply the results of Problem 11c to H_2O, including the consistency check associated with Equation 6.68. How well does $V(Q_1, Q_2, Q_3)$ describe the water molecule, compared with Equation 6.54?

13. Apply the results of Problem 11c to CO_2, assuming that $\alpha = 90°$. Does $V(Q_1, Q_2, Q_3)$ adequately represent a linear molecule? (The relevant data are given in Problem 8.)

14. Prove the following theorem:
   Let the total force on the αth particle in an N-particle assembly be

$$\mathbf{F}_\alpha = -\nabla_\alpha U - \sum_{\substack{\beta=1 \\ (\beta \neq \alpha)}}^{N} \nabla_\alpha V \equiv \mathbf{F}_\alpha^{(e)} + \sum_{\substack{\beta=1 \\ (\beta \neq \alpha)}}^{N} \mathbf{F}_{\alpha\beta}$$

where $\mathbf{F}_\alpha^{(e)}$ is the force applied externally and the $\mathbf{F}_{\alpha\beta}$ are central forces. Then

$$\frac{d\mathbf{J}}{dt} = \mathbf{N}^{(e)}$$

where $\mathbf{J}$ is the total angular momentum of the assembly and

$$\mathbf{N}^{(e)} \equiv \sum_{\alpha=1}^{N} (\mathbf{r}^\alpha \times \mathbf{F}_\alpha^{(e)})$$

is the *external torque* about some fixed axis.

15. Show that the maximum number of degrees of freedom of a rigid body is six by considering the degrees of freedom added and the conditions of constraint subtracted when three, then four, then five, ..., and finally N particles are attached rigidly together.

16. The inertia matrix for a rigid body has the form

$$(I) = \begin{pmatrix} 225 & -30 & 50 \\ -30 & 250 & 3 \\ 50 & 3 & 275 \end{pmatrix}$$

in units of kg-m². Calculate the principal moments of inertia.

17. Calculate the principal moments of inertia and the magnitude of the angular momentum relative to the center of mass for the rigid body, which is described by

$$(I) = \begin{pmatrix} 75 & 0 & -50 \\ 0 & 125 & 0 \\ -50 & 0 & 150 \end{pmatrix} \text{(units of kg-m²)}$$

and which rotates with a constant angular speed such that $\bar{\omega}_1 = 5 \text{ rad/s}$, $\bar{\omega}_2 = 5 \text{ rad/s}$, $\bar{\omega}_3 = 0$.

18. Prove that the inertia matrix (I) satisfies $I_{ij} = I_{ji}$.

19. Calculate the moments of inertia I_{ii} for an assembly of four particles rigidly bound to the vertices of a rectangle with sides of length $2a$ and $2b$, respectively. The particles each have mass M. The rectangle may be taken to lie in the x_1-x_2 plane with one vertex at the origin and with the two sides emanating from that vertex aligned along the x_1 and x_2 axes, respectively.

20. For a rigid body made up of a continuum of particles, Equations 6.88 and 6.89 become Equations 6.90 and 6.91:

$$I_{ii} = \iiint_v \rho(\bar{r})(\bar{r}^2 - \bar{x}_i{}^2) \, dv$$

$$I_{ij} = -\iiint_v \rho(\bar{r})\bar{x}_i\bar{x}_j \, dv$$

respectively, where $\rho(\bar{r})$ is the density of the body at the point $\bar{r}$ within the volume v. Calculate the moments of inertia for a homogeneous solid sphere of radius R.

21. Repeat Problem 20 for a homogeneous solid cube of side L.

22. The Earth may be regarded as a rigid body that is symmetrical about the polar axis and slightly flattened at the poles. Given that $I_A = I_B = 0.3329MR^2$, $I_C = 0.3340MR^2$, and $\dot{\psi} \simeq 1 \text{ day}^{-1}$, where M is the mass and R is the equatorial radius of the planet, calculate the expected period of precession of the axis of spin about the North Pole. An observed value of this period is 427 days. What assumption did you have to make about the spin axis?

23. A space vehicle with "pitch and yaw" moments of inertia $I_A = I_B = 10^3 \text{ kg-m}^2$ spins at 15 rad/s and precesses directly with a period of 2s about an axis inclined by 10° relative to its own axis of symmetry. Calculate the "roll" moment of inertia, I_C.

24. Derive Equations 6.103.

25. An axially symmetric space vehicle rotates about its center of mass according to

$$\bar{\omega}_1(t) = -10 \sin(4t) \qquad \bar{\omega}_2(t) = 5 \cos(4t) \qquad \bar{\omega}_3(t) = 8 \text{ rad/s}$$

The x_3 axis is the axis of symmetry. Calculate I_C/I_A for this vehicle.

7

Noninertial Frames of Reference

7.1 DYNAMICS IN A ROTATING FRAME OF REFERENCE

The validity of the Laws of Motion depends on our ability to locate a frame of reference in which to observe the motion of a particle such that the particle goes forward with constant velocity when it is free of influence from matter and radiation. In contradistinction to this situation, should we place ourselves in the point of view of a noninertial, or accelerated, frame of reference, we may presume that the Laws of Motion will not describe our observations of particle motion. For example, a particle free of influences will not move uniformly, but rather will accelerate in some fashion. On this ground it would appear that noninertial frames of reference should not be considered, when doing dynamical theory, in order that confusion be avoided. This simple conclusion, however, is unmindful of a most important fact: that many dynamical phenomena, including those that occur on planet Earth, are observed most conveniently in significantly noninertial frames of reference. Thus, in order to apply the predictions of the Laws of Motion effectively to observed phenomena in a real world, we ought to consider at least the relation between those Laws as presented traditionally in an inertial frame and their modification when extended to an arbitrarily accelerated frame. This rather practical suggestion is reason enough to justify this chapter, but there is another motive behind the investigation of motion relative to a noninertial frame. Experience has shown that quite often the description of particle motion actually is *simplified* when it is carried out in an appropriate, noninertial frame instead of the perhaps more obvious inertial frame. Consider, for example, the motion of a particle settling through the liquid in a tube spun about in a centrifuge. It is clearly the simpler alternative to choose the origin of position coordinates for the particle at, for example, the bottom of the spinning tube and analyze the motion in terms of centrifugal, buoyancy, and frictional forces, rather than work things out from the point of view of a coordinate system resting at the center of the centrifuge. Indeed, such a coordinate system as the latter would not only complicate matters but would produce irrelevant information as well, insofar as rates of settling are concerned.

Let us consider now the motion of a particle located in space by a vector **r** relative to the origin of an inertial frame of reference, which is itself initially congruent with a noninertial frame that rotates about the x_3 axis, as shown in Figure 7.1. If the noninertial frame undergoes an infinitesimal rotation, the vector **r**, *as analyzed from the inertial frame*, can generally change in two ways. The first type of change comes about because the particle may be moving relative to the rotating frame. The second type of change is due only to the rotation itself. Suppose that the particle is at rest in the rotating frame and that the change from rotation

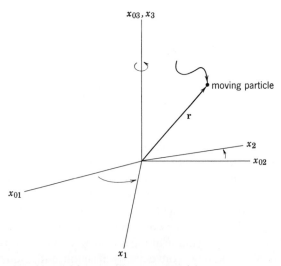

FIGURE 7.1. Rotating and inertial frames with congruent origins.

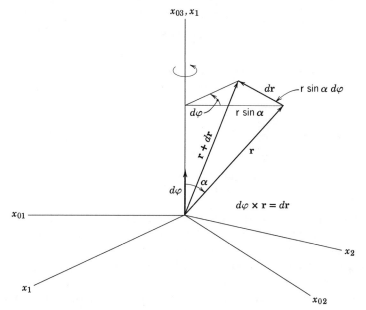

FIGURE 7.2. An infinitesimal rotation that defines the vector $d\varphi$. The vector **r** is assumed to be fixed in the rotating frame for this construction.

is **r** to **r** + $d\mathbf{r}$. The vector $d\mathbf{r}$, shown in Figure 7.2, is perpendicular to the plane containing **r** and the axis of rotation. It has the length

$$||d\mathbf{r}|| \equiv dr = r \sin \alpha \, d\varphi \qquad (7.1)$$

Because of these two properties of $d\mathbf{r}$, we may express it formally as the axial vector

$$d\mathbf{r} = d\boldsymbol{\varphi} \times \mathbf{r} \qquad (7.2)$$

where the direction of $d\boldsymbol{\varphi}$ is specified to be along the axis of rotation and the magnitude $||d\boldsymbol{\varphi}||$ is simply equal to $d\varphi$. Of course, just writing down Equation 7.2 does not automatically ensure that $d\boldsymbol{\varphi}$ is, in fact, a vector. A rigorous proof of this contention requires showing that $d\boldsymbol{\varphi}$ satisfies the axioms for a vector space. This demonstration is carried out in Section A.2 of the Appendix; accordingly, we shall accept the vector quality of $d\boldsymbol{\varphi}$ from now on without concern.

With Equation 7.2 we can write down immediately the time-derivative of **r** due to rotation. This is

$$\left(\frac{d\mathbf{r}}{dt}\right)_{\text{rot}} = \frac{d\boldsymbol{\varphi}}{dt} \times \mathbf{r}$$

or

$$\left(\frac{d\mathbf{r}}{dt}\right)_{\text{rot}} = \boldsymbol{\omega} \times \mathbf{r} \qquad (7.3)$$

where

$$\omega \equiv \frac{d\varphi}{dt} \tag{7.4}$$

is called the *angular velocity* of the point at $\mathbf{r}$, by analogy with the velocity of purely linear motion. (Note, however, that the direction of ω defines the axis of rotation, *not* the direction of motion.) Equation 7.3 expresses the velocity of the particle, due to the rotation, in terms of intrinsically angular quantities. In order to deduce the entire velocity of the particle, we need only add to $(d\mathbf{r}/dt)_{\text{rot}}$ the velocity of the particle relative to the rotating frame, $(d\mathbf{r}/dt)_{\text{rel}}$. Thus:

$$\left(\frac{d\mathbf{r}}{dt}\right)_0 = \frac{d\mathbf{r}}{dt} + \omega \times \mathbf{r} \tag{7.5}$$

where, for simplicity of notation we have dropped immediately the subscript "rel" and have introduced Equation 7.3. The subscript on the left-hand side of (7.5) is to remind us that we actually are calculating the time derivative of $\mathbf{r}$ in the inertial frame of reference, denoted by the subscript "0". This latter time derivative is the true velocity of the particle relative to the origin of the inertial frame, while $d\mathbf{r}/dt$ is the apparent velocity relative to the rotating frame. In a more convenient notation, Equation 7.5 may be written

$$\mathbf{v}_0 = \mathbf{v} + \omega \times \mathbf{r} \tag{7.6}$$

We see that the time derivative of $\mathbf{r}$, as seen in the inertial frame, is just that seen in the rotating frame plus the velocity arising from rotational motion.

The force on a particle, relative to the inertial frame, now may be expressed in terms of Equation 7.6 with no difficulty. Because the Second Law is valid in an inertial frame we have

$$\mathbf{F}_0 = m\left(\frac{d\mathbf{v}_0}{dt}\right)_0$$

Therefore,

$$\mathbf{F}_0 = m\left(\frac{d\mathbf{v}}{dt}\right)_0 + m\omega \times \left(\frac{d\mathbf{r}}{dt}\right)_0 \qquad \left(\frac{d\omega}{dt} = \mathbf{0}\right) \tag{7.7}$$

where, for convenience (and no practical loss in generality), we have assumed that ω does not change with time. The first term on the right-hand side of Equation 7.7 is the change in the velocity of the particle, to be expressed relative to the rotating frame, *as calculated in the inertial frame* and, therefore, is an analog of the left-hand side of Equation 7.5. Indeed, a little thought shows that our derivation of the latter equation, taken as a purely mathematical exercise, could just as well be applied to *any* vector emanating from the origin of a rotating frame of reference. It follows that we are justified in writing

$$\left(\frac{d\mathbf{v}}{dt}\right)_0 = \frac{d\mathbf{v}}{dt} + \omega \times \mathbf{v} \tag{7.8}$$

With Equations 7.6 and 7.8 put in, Equation 7.7 becomes

$$\mathbf{F}_0 = m\frac{d\mathbf{v}}{dt} + m\boldsymbol{\omega} \times \mathbf{v} + m\boldsymbol{\omega} \times \mathbf{v} + m\boldsymbol{\omega} \times (\boldsymbol{\omega} \times \mathbf{r})$$

or

$$\mathbf{F}_a = \mathbf{F}_0 - m\boldsymbol{\omega} \times (\boldsymbol{\omega} \times \mathbf{r}) - 2m(\boldsymbol{\omega} \times \mathbf{v}) \qquad (7.9)$$

where we have identified the *apparent force*

$$\mathbf{F}_a \equiv m\frac{d\mathbf{v}}{dt}$$

by analogy with the Second Law. The quantity $\mathbf{F}_a$ *cannot* be a force in the strict Newtonian sense because changes in $\mathbf{v}$ are being measured relative to a noninertial frame, in which the First Law cannot be valid. Equation 7.9 tells us the "forces" that a particle in a rotating frame of reference will experience. The first one, $\mathbf{F}_0$, is simply a true Newtonian force that is applied in the inertial frame (e.g., an electric force). The second quantity on the right-hand side of (7.9) comes about solely because of the rotational motion. The magnitude of this apparent force is $mr\omega^2 \sin\alpha = mr\dot{\phi}^2 \sin\alpha$ and, as shown in Figure 7.3, its direction is along a line outward and perpendicular to the axis of rotation. Because of the latter property this force is known as the *centrifugal force*.[1] For the case of motion in a plane about an axis, the magnitude of the centrifugal force reduces to the familiar $mr\dot{\phi}^2$ mentioned in footnote 3 of Chapter 4. It should be understood clearly that the centrifugal force is an "apparent force" only because it does not arise in an inertial

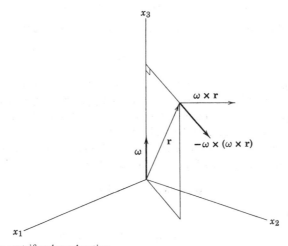

FIGURE 7.3. The centrifugal acceleration.

[1] The name is derived from latin words meaning "center fleeing."

frame of reference, where all true forces *by definition* must exist. But the effects of this apparent force are quite real, as the ubiquitous laboratory centrifuge amply demonstrates.

The second apparent force in Equation 7.9 can occur only if the particle is in motion relative to the rotating frame. This quantity,

$$\mathbf{F}_{Cor} \equiv 2m(\mathbf{v} \times \boldsymbol{\omega}) \tag{7.10}$$

is known as the *Coriolis force.*[2] The Coriolis force produces a motion which is directed at a right angle to the plane containing the axis of rotation and the line of motion. In this sense it resembles the well-known Lorentz force, $\mathbf{F} = (q/c)(\mathbf{v} \times \mathbf{H})$, with $\boldsymbol{\omega}$ playing the mathematical role of the applied magnetic field, $\mathbf{H}$.

With Equation 7.9 in hand we may proceed to construct the "total energy" for the motion of a single particle in a rotating frame. We begin by using the vector identity

$$\mathbf{A} \times (\mathbf{B} \times \mathbf{C}) \equiv (\mathbf{A} \cdot \mathbf{C})\mathbf{B} - (\mathbf{A} \cdot \mathbf{B})\mathbf{C}$$

to write

$$m\boldsymbol{\omega} \times (\mathbf{r} \times \boldsymbol{\omega}) = m[\omega^2 \mathbf{r} - (\boldsymbol{\omega} \cdot \mathbf{r})\boldsymbol{\omega}]$$

This expression, the centrifugal force in Equation 7.9, can be verified to be the negative gradient of the "centrifugal potential function"

$$V_{cen} = \tfrac{1}{2}m[(\boldsymbol{\omega} \cdot \mathbf{r})^2 - \omega^2 r^2] \tag{7.11}$$

Similarly, the Coriolis force in (7.9) can be shown to be the negative gradient of the "Coriolis potential function"

$$V_{Cor} = 2m(\boldsymbol{\omega} \times \mathbf{v}) \cdot \mathbf{r} = 2m(\mathbf{r} \times \boldsymbol{\omega}) \cdot \mathbf{v} \tag{7.12}$$

The apparent total energy of the particle in the rotating frame now may be written

$$E_a = \tfrac{1}{2}mv^2 + V_0(\mathbf{r}) + V_{cen} \tag{7.13}$$

where $V_0(\mathbf{r})$ is the potential function for the true forces acting on the particle. The quantity E_a looks just like a legitimate total energy. Even though it is a constant of the motion (because $\boldsymbol{\omega}$ is a constant vector), it is not a total energy for the same reason that the centrifugal and Coriolis forces are not Newtonian forces. The form of Equation 7.13 deemphasizes the velocity-dependent Coriolis force. This apparent omission disappears, however, if we define

$$\mathbf{p}_c \equiv m\mathbf{v} - m(\mathbf{r} \times \boldsymbol{\omega}) \tag{7.14}$$

and employ the vector identity

$$(\mathbf{r} \times \boldsymbol{\omega}) \cdot (\mathbf{r} \times \boldsymbol{\omega}) \equiv r^2 \omega^2 - (\mathbf{r} \cdot \boldsymbol{\omega})^2$$

---

[2] Equation 7.9 was derived by G. G. Coriolis in 1835.

to rewrite Equation 7.13 in the remarkably simple form

$$E_a = \frac{p_c^2}{2m} + V_0(\mathbf{r}) + (\mathbf{r} \times \boldsymbol{\omega}) \cdot \mathbf{p}_c \qquad (7.15)$$

The apparent total energy is now expressed as the sum of a "kinetic energy" term in the "Coriolis momentum" $\mathbf{p}_c$, the true potential function, and a term proportional to a Coriolis potential expressed in $\mathbf{p}_c$ rather than mv as in Equation 7.12. It follows that we may formally regard the motion of a particle with momentum mv subject to Newtonian, centrifugal, and Coriolis forces as equivalent to that of a hypothetical particle with momentum $\mathbf{p}_c$ subject only to Newtonian potentials and the "potential" $\mathbf{p}_c \cdot (\mathbf{r} \times \boldsymbol{\omega})$. This fact can be of great help in applications of Equations 7.9 and 7.15.

7.2 MOTION RELATIVE TO A ROTATING PLANET

The most interesting example of a rotating frame of reference is that which turns with the Earth throughout each day. In this frame, which is that of our common experience, a number of important geophysical phenomena occur whose causes can be understood quite well on the basis of Equation 7.9. We shall consider four of these phenomena to make our point clear: that Equation 7.9 is a powerful tool for analyzing motion relative to a rotating planet.

(a) *Apparent gravitational acceleration and the general shape of the Earth.* Suppose a particle is situated at the surface of the Earth a distance R from its center, as shown in Figure 7.4. If the particle is initially at rest on the Earth and the Earth itself is rotating with angular velocity $\boldsymbol{\omega}_E$ relative to an inertial frame,

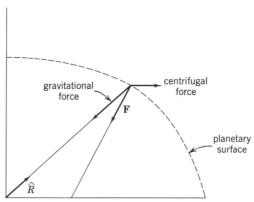

FIGURE 7.4. Addition of centrifugal and gravitational forces at the surface of a molten planet. The magnitude of the centrifugal force has been exaggerated for clarity of illustration.

then the initial force on the particle in the frame of reference of the planet is

$$\mathbf{F} = \frac{-GM_E m}{R^2} \hat{\mathbf{R}} - m\boldsymbol{\omega}_E \times (\boldsymbol{\omega}_E \times \mathbf{R}) \tag{7.16}$$

according to Equation 7.9, where the first term is just the Newtonian gravitational force. (We are assuming here that $\boldsymbol{\omega}_E$ is not time dependent, which is an excellent first approximation.[3]) Equation 7.16 and Figure 7.4 show that, in general, $\mathbf{F}$ will not point along $\mathbf{R}$, but will be directed along a line that intersects the equatorial plane of the Earth at a spot removed from the center of the planet. Thus $\mathbf{F}$ can be resolved into a component along $\mathbf{R}$ whose magnitude is *less* than that of the gravitational force (because of the centrifugal force) and a component along a line perpendicular to the vector $\mathbf{R}$ that will accelerate the particle toward the equator. Thus the initial state of rest of the particle cannot be maintained on the rotating planet, if no retarding frictional forces are present. In fact, if the particle in question were just a bit of the surface of the Earth itself, we would conclude from this analysis that the general shape of the Earth would be distorted, over a long period of time, from that of a perfect sphere (formed, for example, in the molten state) to that of an oblate spheroid (sphere flattened at the poles).

Equation 7.16, after a rather simple analysis, has shown us that the effects of the centrifugal force in the frame of reference of the Earth are: (1) a decrease in the apparent gravitational acceleration relative to its static value and (2) an increase in the equatorial radius of the Earth. Both of these effects depend essentially on the magnitude of $\boldsymbol{\omega}_E$. The measured value[4] of this quantity, $\omega_E = 7.29211515 \times 10^{-5}$ rad/s, is quite small and, therefore, the effects themselves are small. In particular, the apparent gravitational acceleration, the coordinate of $\mathbf{F}/m$ along $-\hat{\mathbf{R}}$, is

$$g = g_0 - \omega_E^2 R \sin^2 \theta \tag{7.17}$$

where g_0 is the static value and θ is the angle between $\boldsymbol{\omega}_E$ and $\mathbf{R}$. At the equator $(\theta = \pi/2)$, g differs from g_0 only by 3.392 cm/s^2, or about 0.35 percent of the static value. In the same vein we find that the polar and equatorial radii of the Earth are 6.3568×10^3 km and 6.3782×10^3 km, respectively, which represents a difference of just one part in 300. Therefore, we shall be on quite safe ground in what follows to make the approximation of considering the surface of the Earth to be perfectly spherical, but with the local value of g expressed by Equation 7.17.

(b) *Deflection of a freely falling particle.* Now let us take up the problem of a particle falling freely from a point near the surface of the Earth. A convenient frame of reference for this problem is one whose origin is at the planetary surface and whose x_3 axis is along the direction of the gravitational acceleration (see

---

[3] The rotational period of the Earth varies by *milliseconds* during the year. See C. W. Allen, *Astrophysical Quantities*, The Athlone Press, London, 1973, §11.
[4] C. W. Allen, op. cit., §48.

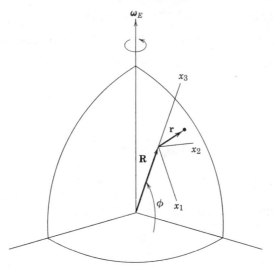

FIGURE 7.5. A rotating frame at the surface of the Earth.

Figure 7.5). The Second Law for this particle follows from Equation 7.9 as

$$\frac{d\mathbf{v}}{dt} = \mathbf{g} + 2(\mathbf{v} \times \boldsymbol{\omega}_E) \tag{7.18}$$

where $\mathbf{v}$ is measured in the new frame of reference. (The transformation of co-ordinates implied by the change in reference frame does not affect the applicability of Equation 7.9 because $\boldsymbol{\omega}_E$ remains the same and the centrifugal force on the particle, which effectively depends only on $\mathbf{R}$, due to the nearness of the particle to the Earth, can be lumped into $m\mathbf{g}$.) Written out in coordinate form, Equation 7.18 becomes

$$\frac{dv_1}{dt} = 2\omega_E \sin \phi v_2(t) \tag{7.19}$$

$$\frac{dv_2}{dt} = -2\omega_E \sin \phi v_1(t) - 2\omega_E \cos \phi v_3(t) \tag{7.20}$$

$$\frac{dv_3}{dt} = -g + 2\omega_E \cos \phi v_2(t) \tag{7.21}$$

where we have introduced ϕ, the *geographic latitude*, the complement of the colatitudinal angle θ in Equation 7.17. In getting Equations 7.19 to 7.21 we have resolved $\boldsymbol{\omega}_E$ along the coordinate axes as

$$\boldsymbol{\omega}_E = -\omega_E \cos \phi \hat{\mathbf{x}}_1 + \omega_E \sin \phi \hat{\mathbf{x}}_3 \tag{7.22}$$

According to Equations 7.19 and 7.21, we can write

$$v_1(t) = 2\omega_E \sin \phi x_2(t) \qquad v_3(t) = -gt + 2\omega_E \cos \phi x_2(t) \qquad (7.23)$$

Upon putting these two equations into (7.20), we get

$$\frac{dv_2}{dt} = -4\omega_E{}^2 x_2(t) + (2\omega_E g \cos \phi)t$$

or

$$\frac{d^2 x_2}{dt^2} + 4\omega_E{}^2 x_2(t) = (2\omega_E g \cos \phi)t \qquad (7.24)$$

Equation 7.24 has the mathematical form of the equation of motion for a linear oscillator driven by a force proportional to the time. Thus it is comparable with the prototype driven oscillator, described in Equation 3.52, if we set

$$\omega_0 \equiv 2\omega_E \qquad \tau = +\infty \qquad a(t) \equiv (2\omega_E g \cos \phi)t$$

in that equation. Moreover, we can put to work the theory developed in Section 3.5 to write the general solution

$$x_2(t) = \left(\frac{g \cos \phi}{2\omega_E}\right) t + C \cos (2\omega_E t - \delta) \qquad (7.25)$$

of Equation 7.24, where C and δ are the usual amplitude and phase constants. Under the initial conditions

$$x_1(0) = x_2(0) = 0 \qquad x_3(0) = h \qquad v_1(0) = v_2(0) = v_3(0) = 0 \qquad (7.26)$$

Equation 7.25 reduces to

$$x_2(t) = \left(\frac{g \cos \phi}{2\omega_E}\right) t - \frac{g \cos \phi}{4\omega_E{}^2} \sin (2\omega_E t) \qquad (7.27)$$

Now we can solve the remaining equations of motion. With (7.27) inserted, the first of Equations 7.23 is the differential equation

$$v_1(t) = \frac{dx_1}{dt} = \left(\frac{g \sin 2\phi}{2}\right) t - \frac{g \sin 2\phi}{4\omega_E} \sin (2\omega_E t)$$

The solution of this equation is

$$x_1(t) = (\tfrac{1}{4} g \sin 2\phi) t^2 - \frac{g \sin 2\phi}{8\omega_E{}^2} (1 - \cos (2\omega_E t)) \qquad (7.28)$$

according to Equation 1.16, where we have noted the initial conditions in doing the integration. The solution for $x_3(t)$ follows in an exactly analogous way from the second of Equations 7.23:

$$x_3(t) = h - (\tfrac{1}{2} g \sin^2 \phi) t^2 - \frac{g \cos^2 \phi}{4\omega_E{}^2} (1 - \cos (2\omega_E t)) \qquad (7.29)$$

Equations 7.27 to 7.29 combine to give the position of the particle during its free fall at any time. We note that, by contrast with free fall in an inertial frame of reference, *the particle deviates from a purely vertical trajectory.* Because of the small magnitude of ω_E and our assumption that the particle is near the surface of the Earth, the quantity $\omega_E t$ should be a small parameter, in terms of which Equations 7.27 to 7.29 can be expanded. If we develop these expressions in a Maclaurin series, we find in third order

$$x_1(t) \simeq 0 \tag{7.30a}$$

$$x_2(t) \simeq (\tfrac{1}{3} g \omega_E \cos \phi) t^3 \tag{7.30b}$$

$$x_3(t) \simeq h - \tfrac{1}{2} g t^2 \tag{7.30c}$$

which shows that the deflection along the x_2 axis is the greater of the two deviations. This deflection has been measured a number of times in careful experiments. That along the x_1 axis has also been detected by sensitive observation over long flight paths.[5]

It should be noted that all of the anomalous effects discussed here derive essentially from the presence of the Coriolis acceleration in Equation 7.18, and that the characteristic period associated with these effects is exactly one-half the period of rotation of the Earth $(2\pi/\omega_E)$. Both of these observations can be seen qualitatively from Equation 7.18 alone. If $\boldsymbol{\omega}_E$ points along the positive *fixed* x_3 axis (Northern Hemisphere) and **v** is directed along the negative *rotating* x_3 axis, then Equation 7.18 tells us to expect a deflection eastward (i.e., along the rotating x_2 axis). But the Coriolis force can act on this eastward component of the velocity as well to produce a secondary deflection southward (along the x_1 axis). The latter deflection clearly will be smaller than the former. Neither deflection will be very large at any rate (unless the particle momentum is enormous) because the deflecting force depends upon ω_E, which, as noted earlier, is quite small. Moreover, the characteristic frequency for any of these effects should be $2\omega_E$, rather than ω_E alone, because the Coriolis acceleration is an axial vector comprising **v** and $2\boldsymbol{\omega}_E$.

(c) *Inertia oscillations of moving fluids.* An interesting phenomenon governed wholly by the Coriolis force may be predicted as a special case of Equation 7.18. Let us suppose that a particle moves in the rotating x_1-x_2 plane shown in Figure 7.5 and, therefore, is governed by the equations of motion

$$\frac{dv_1}{dt} = 2\omega_E \sin \phi v_2(t) \tag{7.31}$$

$$\frac{dv_2}{dt} = -2\omega_E \sin \phi v_1(t) \tag{7.32}$$

These equations can be solved in the same way as were Equations 7.19 to 7.21.

---

[5] A short historical note on these measurements is given in J. B. Marion, *Classical Dynamics of Particles and Systems*, Academic Press, New York, 1970, p. 350.

The results are

$$x_1(t) = C \sin [(2\omega_E \sin \phi)t - \delta]$$ (7.33)

and

$$x_2(t) = C \cos [(2\omega_E \sin \phi)t - \delta]$$ (7.34)

respectively. Equations 7.33 and 7.34 are formally identical with the expressions for the displacement coordinates of an isotropic harmonic oscillator whose natural frequency is $2\omega_E \sin \phi$. Because the two coordinates have the same amplitude of oscillation but differ by $\pi/2$ in their phase factors, the motion in the x_1-x_2 plane will be in a *circle* of radius C. Indeed,

$$[x_1(t)]^2 + [x_2(t)]^2 = C^2 \sin^2 [(2\omega_E \sin \phi)t - \delta]$$
$$+ C^2 \cos^2 [(2\omega_E \sin \phi)t - \delta] = C^2$$

shows this fact quite conclusively. By differentiating Equations 7.33 and 7.34, we find that the radius C is expressed in terms of dynamical quantities as

$$C = \frac{v}{2\omega_E \sin \phi}$$ (7.35)

where $v = (v_1{}^2 + v_2{}^2)^{1/2}$ is the speed of the particle. Since v is a constant of the motion,[6] we conclude that the radius varies with the geographic latitude. The period of the circular motion is just 2π divided by the natural frequency of oscillations. Thus

$$T_C = \frac{\pi}{\omega_E \sin \phi}$$ (7.36)

which also depends on the geographic latitude. Our results here tell us that a particle moving horizontally in a rotating frame of reference will appear to undergo a circular motion (known as an *inertia oscillation* or *inertia current* because no true forces are acting on the particle) whose period depends on the period of rotation of the Earth and the location on its surface. Usually this motion will be strongly perturbed by true forces, especially dissipative ones, and will be difficult to observe. However, in what apparently was a classic instance, a clockwise circulating current in the Baltic Sea was discovered in a well-isolated layer beneath the ocean surface. The period of rotation was 14 hours; that predicted by Equation 7.36 for the appropriate latitude was 14 hours 8 minutes.[7]

(*d*) *Geostrophic and cyclonic air flow.* The horizontal motion of a parcel of air subjected to a gradient in pressure is affected significantly by the Coriolis force. To see this fact qualitatively we need only add to Equations 7.31 and 7.32 the

---

[6] This can be seen by calculating v using Equations 7.33 and 7.34.
[7] Cited in S. L. Hess, *Introduction to Theoretical Meteorology*, Henry Holt, New York, 1959, p. 172. See also G. Neumann, *Ocean Currents*, Elsevier, Amsterdam, 1968, Chapter IV.

coordinates of the "pressure-gradient acceleration"

$$\mathbf{a}_p = -\frac{1}{\rho}\nabla P \tag{7.37}$$

where ρ is the density of the fluid and P is the hydrostatic pressure.[8] Thus we have

$$\frac{dv_1}{dt} = 2\omega_E \sin \phi v_2(t) - \frac{1}{\rho}\frac{\partial P}{\partial x_1}$$

$$\frac{dv_2}{dt} = -2\omega_E \sin \phi v_1(t) - \frac{1}{\rho}\frac{\partial P}{\partial x_2} \tag{7.38}$$

as the equations of motion for a fluid parcel small enough to meet our definition of "particle." Without solving these equations explicitly, we can see quite readily from their vector form,

$$\frac{d\mathbf{v}}{dt} = 2(\mathbf{v} \times \boldsymbol{\omega}_E) - \frac{1}{\rho}\nabla P$$

that the motion of a parcel initially at rest will commence along the direction of the pressure gradient, then will be deflected to the *right* by the Coriolis force if the motion occurs in the Northern Hemisphere. (If the motion is in the Southern Hemisphere, the direction of $\boldsymbol{\omega}_E$ relative to "up" is reversed and the Coriolis force deflects to the *left*.) At some point in time the Coriolis force will have deflected the parcel far enough to the right to make it move perpendicularly to its initial direction. Then the parcel can no longer gain speed from the pressure gradient, the Coriolis force cannot change, and the two forces remain balanced with uniform motion the result (see Figure 7.6). In this equilibrium situation the components

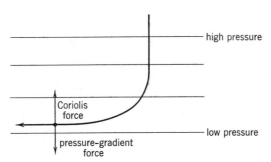

FIGURE 7.6. Geostrophic flow in the Northern Hemisphere. The horizontal lines denote isobars.

[8] For a simple derivation of Equation 7.37, see S. L. Hess, op. cit., §11.4.

of the velocity satisfy

$$v_2 = (2\rho\omega_E \sin \phi)^{-1} \frac{\partial P}{\partial x_1}$$

$$v_1 = -(2\rho\omega_E \sin \phi)^{-1} \frac{\partial P}{\partial x_2}$$

(7.39)

respectively. Equations 7.39 are of practical value because they permit a computation of part of the velocity distribution in the Earth's atmosphere based solely on measurements of the pressure distribution. The motion described by Equations 7.39 is known as *geostrophic flow*. This kind of flow, characterized by a constant wind velocity directed parallel with the pressure isobars, with low pressure to the left (right) along the direction of motion in the Northern (Southern) Hemisphere, is well known to meteorologists as an example of *Buys-Ballot's law*.

When the pressure isobars in a nonrotating fluid form circles about a given point, the motion of a fluid parcel will be along radial lines emanating from that point, as shown in Figure 7.7(a). When the central point is in a region of low pressure and the fluid parcel is in a rotating frame of reference, Equation 7.10 tells us to expect a deflection of the parcel to its right (left) if it is in the Northern (Southern) Hemisphere. This deflection results in a circulating motion known as *cyclonic flow* (see Figure 7.7b). Cyclonic flow, then, is counterclockwise in the Northern Hemisphere. If the central point is in a region of high rather than low pressure, the motion of a fluid parcel is away from it and the resulting circulation

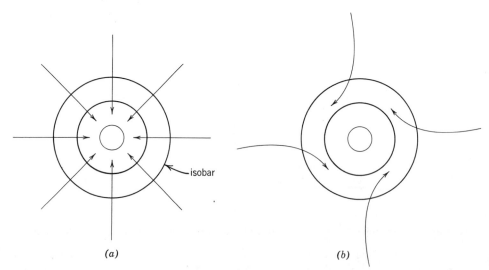

(a)   (b)

FIGURE 7.7. (a) Motion in a fluid toward a point of low pressure. (b) The same motion in a rotating frame (cyclonic flow).

is known as *anticyclonic flow*. This flow, of course, is clockwise in the Northern Hemisphere.

7.3 THE FOUCAULT PENDULUM

One of the classic problems in the study of motion relative to a rotating frame of reference is the theory of the *Foucault pendulum*.[9] This device was the first to provide direct evidence for the rotation of the Earth. For the purpose of our discussion, the Foucault pendulum may be defined simply as a plane pendulum acted on by centrifugal and Coriolis forces. Strictly, the Foucault pendulum moves in *three* dimensions and is acted upon by these "forces." But, with the swing no longer restricted to a plane, the gravitational force can induce a precession which masks that due to the Coriolis force. We shall omit a discussion of this complication here. In order to simplify the mathematics even further we shall consider initially a plane pendulum located at the north geographic pole of the Earth. Then the frame of reference of the pendulum and the inertial frame with respect to which it rotates are in the same relation as are the frames of reference shown in Figure 7.1. If the radius of the semicircle upon which the pendulum is constrained to move is equal to l, then, by Equation 7.15, the "total energy" may be written

$$E_a = \frac{P_c{}^2}{2m} + mg_0(l + x_3) - \omega_E J_{c3} \tag{7.40}$$

where

$$P_c = m\mathbf{v} + m\omega_E(x_2\hat{\mathbf{x}}_1 - x_1\hat{\mathbf{x}}_2) \tag{7.41}$$

and

$$J_{c3} \equiv x_1 P_{c2} - x_2 P_{c1} = m(x_1 v_2 - x_2 v_1) + m\omega_E(x_1{}^2 + x_2{}^2) \tag{7.42}$$

The angular velocity ω_E has been chosen to point along the x_3 axis and the gravitational potential energy of the pendulum (the second term on the right-hand side of Equation 7.40) has been set equal to zero when $x_3 = -l$. Since the potential energy of the hypothetical particle with momentum $\mathbf{p}_c$ (but with mass m) depends only on x_3, the arguments of Section 4.2 tell us to expect cylindrical symmetry in the motion. This conclusion in turn suggests that a transformation to cylindrical polar coordinates should be useful (see Figure 7.8.):

$$x_1 = \rho \cos \theta \qquad x_2 = \rho \sin \theta \qquad x_3 = z \tag{7.43}$$

Under this transformation, for example, Equation 7.42 becomes

$$J_{c3} = m\rho^2(\dot{\theta} - \omega_E) \tag{7.44}$$

[9] The pendulum was constructed by Leon Foucault in 1851. A complete theory of its motion was not available until 1869, when the thesis of H. Kamerlingh Onnes appeared. Onnes went on to become one of the foremost low-temperature physicists.

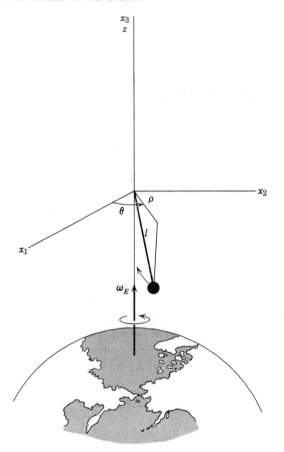

FIGURE 7.8. A Foucault pendulum at the North Pole.

Equation 7.44 makes it very clear that J_{c3} differs from the z coordinate of the angular momentum of the real particle only through the term $m\rho^2\omega_E$. Now J_{c3} represents the conserved angular momentum generated by the symmetry imposed on the problem through the rotating coordinate frame. When there is no rotation it is evident that J_{c3} is just the z coordinate of the ordinary angular momentum, $J_3 = m\rho^2\dot{\theta}$. Thus it appears that the effect of rotation in this instance is merely the augmentation of $\dot{\theta}$ by the angular speed ω_E. To substantiate this idea, we write down the transformation of coordinates

$$\rho' = \rho \qquad \theta' = \theta + \omega_E t \qquad z' = z \qquad (7.45)$$

which represents a set of cylindrical polar coordinates whose axis system is rotating at the constant rate $-\omega_E$ relative to the unprimed coordinate system. With the help of Equations 7.43 and 7.45, the expression for E_a can be transformed as follows:

$$E_a = \tfrac{1}{2}mv^2 + \omega_E J_3 + \tfrac{1}{2}m\omega_E^2(x_1^2 + x_2^2) + mg_0(l + x_3) - \omega_E J_{c3}$$
$$= \tfrac{1}{2}m(\dot{\rho}^2 + \rho^2\dot{\theta}^2 + \dot{z}^2) - \tfrac{1}{2}m\rho^2\omega_E^2 + mg_0(l + z)$$

and

$$E_a' = \tfrac{1}{2}m(\dot{\rho}'^2 + \rho'^2\dot{\theta}'^2 + \dot{z}'^2) + mg_0(l + z') + m\rho'^2\dot{\theta}'\omega_E \tag{7.46}$$

Since the unprimed coordinate system rotates with the angular speed ω_E relative to the inertial frame shown in Figure 7.8, the set of coordinates $\{\rho', \theta', z'\}$ must refer to that frame. Because the motion of the pendulum is by hypothesis in a *fixed* plane in the inertial frame, we can fix the value of θ' and can thereby reduce Equation 7.46 to

$$E_a' = \tfrac{1}{2}m(\dot{\rho}'^2 + \dot{z}'^2) + mg_0(l + z') \tag{7.47}$$

This expression simply gives the total energy of a pendulum restricted in its motion to the ρ'-z' plane. Therefore, noting again Equations 7.45, we may conclude that *the effect of the Coriolis force is to set the plane of swing of the Foucault pendulum into rotation at the rate* $-\omega_E$.

At the north geographic pole of the Earth, the plane of swing of a Foucault pendulum rotates, relative to the Earth, at a rate equal but opposite to the rate of planetary rotation, ω_E. At a latitude ϕ different from $\pi/2$ we do not expect to see the same rate of rotation because the Coriolis force diminishes in strength as ϕ decreases, according to Equations 7.31 and 7.32. The relevant parameter at an arbitrary latitude should be the component of $\boldsymbol{\omega}_E$ along the x_3 axis in the frame of reference of the pendulum, since $\mathbf{F}_{\text{Cor}}$ is determined only by that part of $\boldsymbol{\omega}_E$ that is perpendicular to the pendulum velocity. According to the decomposition of $\boldsymbol{\omega}_E$ given in Equation 7.22, the x_3 coordinate of $\boldsymbol{\omega}_E$ in the rotating frame is just $\omega_E \sin\phi$. Therefore, the period of rotation of the plane of swing at latitude ϕ must be

$$T_E = \frac{2\pi}{\omega_E \sin\phi} \tag{7.48}$$

At the equator this period goes to infinity (no rotation) because $\mathbf{F}_{\text{Cor}} = \mathbf{0}$.

FOR FURTHER READING

K. R. Symon   *Mechanics*, Addison-Wesley, Reading, Mass., 1971. Chapter 7 of this book presents a thorough description of noninertial frames at a level comparable to that of Section 7.1.

P. S. Eagleson   *Dynamic Hydrology*, McGraw-Hill, New York, 1970. Noninertial frames are discussed in a geophysical context in Chapter 7 of this fine introduction to planetary fluid mechanics.

S. L. Hess   *Introduction to Theoretical Meteorology*, Holt, Rinehart and Winston, New York, 1959. Chapters 11 and 12 present a good discussion of the meteorological applications of the physics of apparent forces.

G. Neumann   *Ocean Currents*, Elsevier, Amsterdam, 1968. Chapter IV of this text
on physical oceanography discusses the influence of apparent forces
on ocean currents.

PROBLEMS

1. Consider a frame of reference that moves with a constant linear acceleration $\mathbf{a}_0$ relative to some inertial frame and rotates with a fixed angular velocity ω about the origin of that frame as well. Derive the generalization of Equation 7.9 for this situation.

2. Generalize Equation 7.9 to the case where ω is not a constant vector.

3. Show that Equation 7.5 applies to any vector $\mathbf{A} = A_1\hat{\mathbf{x}}_1 + A_2\hat{\mathbf{x}}_2 + A_3\hat{\mathbf{x}}_3$.

4. A space station is to be built in the shape of a doughnut and spun about an axis through the "hole of the doughnut" such that the outer wall of the station becomes the floor: if the spinning is at a constant rate, a static gravitational field directed toward the floor will be simulated. The space station is thus a doughnut-shaped centrifuge.

   (a) Calculate the period of rotation necessary to simulate the gravitational field at the Earth's surface for any point on the outer periphery of a station with outer diameter 200 m.

   (b) Calculate the simulated gravitational acceleration as a function of distance along a "vertical" radial line, beginning at the inner periphery of the station and ending at the outer periphery.

5. A railroad car of mass 10^4 kg moves on a northbound route at latitude 40°. If its speed is 60 mph, what lateral force does it exert on the railroad track?

6. A river in northern Argentina, at latitude 32°S, has a width of 1 km and flows from west to east with a speed of 2 m/s. If the flow is steady, what is the difference in surface elevation across the river and on which bank will the surface be higher? (*Hint.* The slope of a fluid surface subject to a constant acceleration $\mathbf{a}$ can be determined from the equation $a_1\,dx_1 + a_2\,dx_2 + a_3\,dx_3 = 0$, which expresses the constancy of the potential energy per unit mass along that surface.)

7. Calculate the magnitude of the Coriolis force on an object of mass 1 kg, which is "dropping" with a velocity of 6 m/s at a point 3 m above the outer periphery of the space station described in Problem 4. What fraction of the "gravitational force" is the Coriolis force?

8. Show that

   (a) $\mathbf{F}_{cen} = -\nabla V_{cen}$

   (b) $\mathbf{F}_{Cor} = -\nabla V_{Cor}$

as stated in Section 7.1.

9. Repeat the analysis leading from Equation 7.9 to Equation 7.15 for a particle of charge q subject to the Lorentz force

$$\mathbf{F} = q\mathbf{E}_0 + \frac{q}{c}(\mathbf{v} \times \mathbf{H}_0)$$

where $\mathbf{E}_0$ and $\mathbf{H}_0$ are static electric and magnetic field intensities, respectively, and c is the speed of light in vacuum. What is the analog of the Coriolis force here? Of the centrifugal force?

10. In the table below are listed some data on the planets Jupiter, Saturn, and Uranus. Calculate the value of g_0 at the equator in each case.

Planet	g	Equatorial Radius	Rotation Period
	(cm/s²)	(km)	(hr)
Jupiter	2600	71,400	9.84
Saturn	1120	60,400	10.23
Uranus	940	23,800	10.82

11. The crab pulsar is a rotating neutron star with a radius of about 10 km and a mass of about one solar mass. If the rotation period is 0.033 s, what is the ratio of centrifugal to gravitational forces at the equator?

12. Expand Equations 7.27 through 7.29 to fifth order in powers of $\omega_E t$ and show that

$$x_1(t) \simeq (\tfrac{1}{12}g\omega_E{}^2 \sin 2\phi)t^4$$

$$x_2(t) \simeq (\tfrac{1}{3}g\omega_E \cos \phi)t^3[1 - \tfrac{2}{5}(\omega_E t)^2]$$

$$x_3(t) \simeq h - \tfrac{1}{2}gt^2[1 - \tfrac{1}{3}\cos^2 \phi(\omega_E t)^2]$$

How sensitively would g have to be measured to deduce ω_E from data on a 100-m fall?

13. An astronaut drops an object from the ceiling of the rotating space station described in Problem 4. The distance of fall to the floor is 3 m. How far away from the spot on the floor directly "below" the point of release does the object land? Can you neglect the variation of the apparent gravitational acceleration with position?

14. How much deflection can one expect for a particle dropped 100 m at latitude 45°N on Earth? What assumptions did you make in doing the calculation?

15. Calculate the period and radius of the inertia currents to be expected at latitude 30°N in the sea if the current speed is 50 cm/s.

16. The following data were taken during observations of the motions of a neutrally buoyant float at a depth of 2500 m in the Atlantic Ocean (latitude 28°N): diameter of circular path = 2.38 ± 0.27 km; period of the motion = 25.2 hr; mean speed = 7.97 cm/s. Calculate the period of rotation of the Earth in *two* ways using these data, thereby checking the theory presented in Section 7.2c for both accuracy and self-consistency.

17. In the sea we expect that the inertia currents described in Section 7.2c actually will decay because of fluid friction. If the frictional force per unit mass is given by $f = -v/\tau$ (an hypothesis known in physical oceanography as the Guldbert-Mohn assumption), what will be the time dependence of the radius C, expressed in the absence of friction by Equation 7.35?

18. The speed of the inertia current described in Problem 16 was observed to drop to 10 percent of its initial value after 4 days. What is the value of the time constant τ, given in Problem 17?

19. Calculate the geostrophic wind speed for a pressure gradient of 0.2 millibar/100 km at latitude 35°N. Take the density of air to be 1.29 kg/m³.

20. Demonstrate that cyclonic flow is predicted by Equations 7.38 in the case of a constant, positive pressure gradient that has equal x_1 and x_2 coordinates.

8

The Special Theory of Relativity

8.1 GALILEAN AND LORENTZ INVARIANCE

The Laws of Motion as stated in Chapter 1 contain a tacit but very important assumption about dynamical behavior that we now shall examine in detail. Recall that an essential feature of the First and Second Laws was the necessary existence of a preferred class of reference frames called *inertial frames*. In these sets of three mutually perpendicular cartesian axes, a particle free of influence from matter and radiation moves with a constant velocity. As we have seen through the discussion in Chapter 7, the inertial frames are the only ones for which this observation can be made, and, therefore, they form the required setting in which to interpret dynamical phenomena through the Laws of Motion. But this conclusion suggests a problem of communication that we have not yet discussed. We know already that a particle moving at constant velocity with respect to one inertial frame will be observed to do so with respect to any other inertial frame: the frames are equivalent in that respect. What we do not know is if the same is true for *accelerated* motion. It is not immediately obvious nor have we considered whether changes in the velocity of a particle observed in an inertial frame depend in any way on the (constant) velocity of that frame. (On the other hand, we do know, from the discussion in Chapter 7, that observed changes in the velocity of a particle do depend on the *acceleration* of a *non*inertial frame.) This rather deep question motivates what is generally called a Principle of Relativity. An analysis is needed that will determine unambiguously the relation between observations of dynamical phenomena carried out in an arbitrary pair of inertial frames. In particular, we would like to know the physical quantities that remain invariant under this relation. Since we shall not be concerned with the observations made in the most general, noninertial frame of reference, the equations establishing the relationship we seek and the attendant invariant quantities are said to constitute a *special* Principle of Relativity. Naturally enough, the equations of relation between accelerated frames of reference constitute part of a *general* Principle of Relativity.

As we saw in Chapter 1, the most basic observations made in dynamics are those of position and time. Derived from them are velocity and acceleration, and it is the latter quantity, along with the mass, that appears as the decisive physical element in the Laws of Motion. Therefore, a relativity principle ought to describe the effect of motion, at constant velocity, upon measurements of mass and acceleration. It is implicit in the way the Laws of Motion were stated in Chapter 1 that the particular choice of inertial frame with respect to which one measures these quantities is *not* important to making calculations using the Second Law. We shall accept this suggestion for the moment and codify it in the following principle.

Galilean-Newtonian Principle of Relativity   The mass and the acceleration of a particle, measured in an inertial frame of reference, are not affected by the velocity of that frame relative to any other inertial frame.

This Principle of Relativity gives us the freedom to choose any convenient inertial frame from which to base measurements of acceleration or deductions from the Second Law. Experimental support for this possibility was cited by Galileo[1] even before the Laws of Motion were formulated, but we shall concern ourselves instead with the equations of relation or *transforms* that follow from it.

Regardless of which quantities are deemed invariants, the transforms between two inertial observers in motion with respect to one another must comply with the physical requirement that a particle moving with constant velocity relative to one of them does so as well relative to the other. This means that the transforms of position and time must be *linear* and, if both frames of reference are assumed to be congruent at "time zero" for each, the transforms must have the mathematical

---

[1] Galileo Galilei, *Dialogue Concerning the Two Chief World Systems—Ptolemaic and Copernican*, translation by Stillman Drake, University of California Press, Berkeley, 1953, p. 144f.

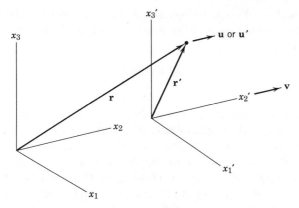

FIGURE 8.1. A pair of inertial frames, the primed one moving with the constant velocity **v** relative to the unprimed one, in which a moving particle is located.

form[2]

$$\mathbf{r}' = \alpha(\mathbf{v})\mathbf{r} + \boldsymbol{\beta}(\mathbf{v})t \qquad (8.1)$$

$$\mathbf{r} = a(\mathbf{v})\mathbf{r}' + \mathbf{b}(\mathbf{v})t' \qquad (8.2)$$

where $\alpha(\mathbf{v})$, $\boldsymbol{\beta}(\mathbf{v})$, $a(\mathbf{v})$, and $\mathbf{b}(\mathbf{v})$ are as yet arbitrary functions of the velocity **v** of one frame relative to the other (see Figure 8.1). Two special cases of the linearity condition on the transforms can be used now to place some general constraints on the unknown functions in Equations 8.1 and 8.2. First, the origin of the primed coordinate system must move with velocity **v** relative to that of the unprimed system and the origin of the unprimed system must move with velocity $-\mathbf{v}$ relative to that the of the primed system. Therefore

$$\frac{\mathbf{r}}{t} = \mathbf{v} \qquad (\mathbf{r}' = \mathbf{0})$$

$$\frac{\mathbf{r}'}{t'} = -\mathbf{v} \qquad (\mathbf{r} = \mathbf{0})$$

Equations 8.1 and 8.2 then lead to the relations

$$\boldsymbol{\beta}(\mathbf{v}) = -\alpha(\mathbf{v})\mathbf{v} \qquad \mathbf{b}(\mathbf{v}) = a(\mathbf{v})\mathbf{v} \qquad (8.3)$$

With the help of Equations 8.3 we can immediately rewrite the transforms as

$$\mathbf{r}' = \alpha(\mathbf{v})(\mathbf{r} - \mathbf{v}t) \qquad (8.4)$$

$$\mathbf{r} = a(\mathbf{v})(\mathbf{r}' + \mathbf{v}t') \qquad (8.5)$$

---

[2] The time transforms need not be given separately just now, as they are deduced directly from the position transforms.

and only the scalar-valued functions α and a remain undetermined. Second, we note that measurements of position must coincide in both frames of reference when $\mathbf{v} = \mathbf{0}$. Thus we deduce

$$\alpha(0) = a(0) = 1 \tag{8.6}$$

upon putting $\mathbf{v} = \mathbf{0}$ in Equations 8.4 and 8.5. Equations 8.3 and 8.6 are general conditions that the transforms corresponding to *any* Principle of Relativity ought to meet. Since a special Principle of Relativity also implies a certain symmetry for inertial frames, we can add in the present case that Equations 8.4 and 8.5 must remain invariant under a change in the designation of an inertial frame from "moving" to "resting" and *vice versa*. This change in designation may be accomplished by: (a) permuting the prime label and changing the sign of $\mathbf{v}$, or (b) permuting the label and writing $t \rightarrow -t$, $t' \rightarrow -t'$. If we carry out (a) on Equation 8.4, we conclude

$$\alpha(-\mathbf{v}) = a(\mathbf{v}) \tag{8.7}$$

while (b) yields simply

$$\alpha(\mathbf{v}) = a(\mathbf{v}) \tag{8.8}$$

Equations 8.6 to 8.8 are the general properties that the transforms generated by a special relativity principle should have.

Now we must impose the invariance requirement of the Galilean-Newtonian Principle of Relativity: that the acceleration measured in the primed frame of reference be identical with that measured in the unprimed frame. For the sake of simplicity we shall restrict the problem at this time to one spatial dimension, whereupon Equations 8.4 and 8.5 become

$$x' = \alpha(v)(x - vt) \tag{8.9}$$

$$x = \alpha(v)(x' + vt) \tag{8.10}$$

respectively. If we eliminate x' from these two equations, we get the time transform

$$t' = \alpha(v)\left[t - \left(1 - \frac{1}{\alpha^2(v)} \right)\frac{x}{v} \right] \tag{8.11}$$

which we need in calculating the relation of velocity between the two frames of reference. Equations 8.9 and 8.11 then can be made to yield

$$
\begin{aligned}
u'(t') \equiv \frac{dx'}{dt'} &= \frac{\alpha(v)(dx - v\,dt)}{\alpha(v)\{dt - [1 - (1/\alpha^2(v))]\,dx/v\}} \\[2mm]
&= \frac{(dx/dt) - v}{1 - [1 - (1/\alpha^2(v))](1/v)(dx/dt)} \\[2mm]
&\equiv \frac{u(t) - v}{1 - [1 - (1/\alpha^2(v))](u(t)/v)}
\end{aligned}
\tag{8.12}
$$

The transformation of acceleration follows as

$$a'(t') = \frac{du'}{dt'} = \left\{ \alpha(v) \left[1 - \left(1 - \frac{1}{\alpha^2(v)} \right) \frac{u(t)}{v} \right] \right\}^{-3} a(t)$$

The Galilean-Newtonian relativity principle asserts that $a'(t') \equiv a(t)$. This condition, when introduced into the above expression, leads to

$$\alpha^2(v) - \frac{1}{1 - (u/v)} \alpha(v) + \frac{u/v}{1 - (u/v)} = 0$$

The only solution of this equation that satisfies Equation 8.6 is $\alpha(v) \equiv 1$. Therefore, Equations 8.9 to 8.11 become:

$$x' = x - vt \qquad x = x' + vt' \tag{8.13}$$

$$t' = t \qquad t = t' \tag{8.14}$$

which are known as the *Galileo Transforms*. Equations 8.13 are the same as what the intuitive geometric picture in Figure 8.1 would suggest. Equations 8.14, the time transforms, are quite consistent with the Third Law, which requires influences of one particle on another to propagate instantaneously, whatever the interparticle separation. Thus, the Galileo transforms give full mathematical expression to the Galilean-Newtonian Principle of Relativity and may be regarded as representatives of the fundamental symmetries underlying the Laws of Motion.

The implications of Equations 8.13 and 8.14 run into serious difficulties when phenomena involving electromagnetic interactions are considered. The essential unity one might hope to see among descriptions of these phenomena and what can be deduced solely from the Laws of Motion is challenged severely by the fact that the interactions between charged particles are known to be propagated by fields. These quantities are described by Maxwell's Equations, not by the Second Law. The difficulty we allude to is presented very forcefully if one considers a pure electromagnetic field propagating through otherwise empty space. The electric and magnetic field intensities can be shown[3] in this case to satisfy an equation of the form

$$\frac{\partial^2 E}{\partial x_1^2} + \frac{\partial^2 E}{\partial x_2^2} + \frac{\partial^2 E}{\partial x_3^2} - \frac{1}{c^2}\frac{\partial^2 E}{\partial t^2} = 0 \tag{8.15}$$

where E is, for example, the electric field intensity and c is the speed of light in vacuum,

$$c = 2.997925 \pm 0.000003 \times 10^{10} \text{ cm/s}$$

The same kind of equation holds for H, the magnetic field intensity. Equation 8.15 has the form of a *wave equation* and leads to the conclusion, first drawn by Maxwell

[3] Equation 8.15 is derived and discussed fully in, for example, R. Becker, *Electromagnetic Fields and Interactions*, Blaisdell, New York, 1964, Vol. I, Chapters DI and DII.

himself, that an electromagnetic field in empty space propagates as a (transverse) wave whose phase speed is equal to c. The fundamental point we wish to make here is that the wave Equation 8.15 contains a *velocity* rather than an acceleration and, therefore, *cannot* be invariant under a Galileo transform, which relates velocities according to an equation like (8.12). On the other hand, there must be a set of transforms between inertial observers under which Equation 8.15 does remain invariant. But this set apparently would be consistent with the condition

$$c' = c$$

which is inherent in the wave equation, and thus would strongly contradict Equation 8.12 as applied to the speed of light propagation. The dichotomy we are faced with, then, runs approximately as follows: either the Maxwell equations and their concomitant invariance property $c' = c$ are not correct, because they are inconsistent with the Galilean-Newtonian Principle of Relativity, or the equations are correct and the Newtonian formulation of dynamics is faulty. It took nothing less than the courage of genius to decide, as did Albert Einstein in a paper published in 1905[4], that the second alternative was the sound one. He epitomized his views by stating a special Principle of Relativity that would be valid for all dynamical phenomena.

Einsteinian Principle of Relativity   The laws of physics contain an essential symmetry with respect to inertial frames of reference in that:

(a) Observations of dynamical and electromagnetic phenomena made entirely in one inertial frame are unaffected by the velocity of that frame relative to any other frame.
(b) The speed of light in vacuum has the same numerical value when measured in any inertial frame.

It is obvious that the Einsteinian relativity principle is of a much more comprehensive character than is the Galilean-Newtonian principle and that its central invariant quantity is the speed of light in vacuum, not the acceleration of a particle. Indeed, the set of transforms derivable from this principle will not be expected to preserve the Laws of Motion as stated by Newton, but rather will leave the Maxwell equations and equations like (8.15) unchanged. The physical basis for this momentous difference, $c' = c$, it should be mentioned, has been supported by a number of important experiments beginning with the celebrated effort of Michelson and Morley, published in 1887.[5] Our interest now lies with the development of the

---

[4] Einstein's fundamental papers on the theory of relativity are reprinted in English in *The Principle of Relativity*, translation by W. Perrett and G. B. Jeffery, Dover Publications, New York, 1958.
[5] A good discussion of the experimental foundation of the Einsteinian relativity principle is given by A. P. French, *Special Relativity*, W. W. Norton, New York, 1968, p. 72ff. A presentation of the key experiments from the point of view of the self-consistency of electromagnetic theory is in P. Lorrain and D. Corson, *Electromagnetic Fields and Waves*, W. H. Freeman, San Francisco, 1970, Chapter 5.

Einsteinian principle into a set of transforms whose effect on Newtonian dynamics we can investigate in the subsequent sections of this chapter.

Let us once again, for simplicity, restrict our consideration of the relative motion of two inertial observers to one spatial dimension. If the primed frame of reference moves with velocity v along the positive x direction relative to the unprimed frame, then the transforms between the two observers must be of the form of Equations 8.9 and 8.10 with $\alpha(v)$ subject to Equation 8.7. The invariance requirement of the Einsteinian principle is $c' = c$, which implies that

$$x'(t') = ct' \qquad x(t) = ct \qquad (8.16)$$

are the equations describing the path of a beam of light in the primed and unprimed frames of reference, respectively. (We assume that the beam moves in the positive x direction relative to the unprimed frame.) When Equations 8.16 are introduced into Equations 8.9 and 8.10 and the latter are multiplied, we find without difficulty

$$\alpha^2(v) = \left(1 - \frac{v^2}{c^2}\right)^{-1} \qquad (8.17)$$

Therefore

$$\alpha(v) = a(v) = \left(1 - \frac{v^2}{c^2}\right)^{-1/2} \qquad (8.18)$$

and Equations 8.9 to 8.11 become

$$x' = \left(1 - \frac{v^2}{c^2}\right)^{-1/2}(x - vt) \qquad x = \left(1 - \frac{v^2}{c^2}\right)^{-1/2}(x' + vt') \qquad (8.19)$$

$$t' = \left(1 - \frac{v^2}{c^2}\right)^{-1/2}\left(t - \frac{v}{c^2}x\right) \qquad t = \left(1 - \frac{v^2}{c^2}\right)^{-1/2}\left(t' + \frac{v}{c^2}x'\right) \qquad (8.20)$$

which are known as the *Lorentz transforms*.[6] We note that Equations 8.19 and 8.20 reduce to Equations 8.13 and 8.14, respectively, in the limit $c \to \infty$. This behavior is fully consistent with the Newtonian assumption of instantaneous propagation of particle-particle influences, mentioned in Chapter 1. More importantly, perhaps, we also notice that $\alpha(v)$ in the Lorentz transforms differs from the comparable function in the Galileo transforms, to lowest order, by terms of order $(v/c)^2$. This fact explains the apparent consistency of many dynamical experiments with the Newtonian precepts. The velocities typically encountered in them are remarkably smaller than the speed of light in vacuum. For example, the speed of the Earth in its orbit about the sun is only about 10^{-4} c. To complete our initial acquaintance with the Lorentz transforms we remark that the Lorentz invariance

---

[6] The transform equations are named in honor of H. A. Lorentz, the Dutch theoretical physicist who initiated their serious use in electromagnetic theory in 1904. The equations were first derived by Albert Einstein in 1905 using his special Principle of Relativity.

of the wave equation (8.15) can be demonstrated,[7] and that the almost universal notation

$$\boldsymbol{\beta} \equiv \frac{\mathbf{v}}{c} \tag{8.21}$$

$$\gamma \equiv (1 - \beta^2)^{-1/2} \tag{8.22}$$

will be inserted into (8.19) and (8.20) to facilitate the construction of a similarly Lorentz-invariant particle dynamics.

8.2 SPACE-TIME COORDINATES

The dynamical aspects of the Einsteinian Principle of Relativity are seen most directly when a careful survey is made to deduce all the relevant invariants of the principle. Unlike the situation that existed for the Galilean-Newtonian relativity principle, the primary invariant quantities we seek are not dictated by a previously established set of dynamical laws. Instead, the basic invariant—the speed of light in vacuum—has come to us from electromagnetic theory and we have yet to discover the nature of the equation of motion that is consistent with it. Therefore, we ought to begin by examining those most basic ingredients of any physical theory, position and time.

For an arbitrary pair of inertial observers moving with a relative velocity $\mathbf{v}$, the Lorentz transforms are

$$\mathbf{r}' = \mathbf{r} + \frac{(\mathbf{r} \cdot \mathbf{v})}{(\beta c)^2} (\gamma - 1)\mathbf{v} - \gamma \mathbf{v} t \tag{8.23}$$

$$t' = \gamma \left(t - \frac{(\mathbf{r} \cdot \boldsymbol{\beta})}{c} \right) \tag{8.24}$$

In the special case $\mathbf{v} = v\hat{\mathbf{x}}_1$ the transforms (8.23) and (8.24) reduce to the first of Equations 8.19 and 8.20, respectively, along with a pair of identity transforms for x_2 and x_3. The appearance of the transform equations already makes it evident that a certain symmetry exists in the relations between position and time that they represent. This symmetry is brought out most clearly if we define

$$x_0 = ct \tag{8.25}$$

and rewrite Equations 8.23 and 8.24 as the set

$$x_0' = \gamma x_0 - \gamma \beta_1 x_1 - \gamma \beta_2 x_2 - \gamma \beta_3 x_3$$

$$x_1' = -\gamma \beta_1 x_0 + \left[1 + \frac{(\gamma - 1)}{\beta^2} \beta_1{}^2 \right] x_1 + \frac{(\gamma - 1)}{\beta^2} \beta_1 \beta_2 x_2 + \frac{(\gamma - 1)}{\beta^2} \beta_1 \beta_3 x_3$$

[7] See, for example, R. Resnick, *Introduction to Special Relativity*, Wiley, New York, 1968, §4.3.

$$x_2' = -\gamma\beta_2 x_0 + \frac{(\gamma - 1)}{\beta^2}\beta_2\beta_1 x_1 + \left[1 + \frac{(\gamma - 1)}{\beta^2}\beta_2{}^2\right]x_2 + \frac{(\gamma - 1)}{\beta^2}\beta_2\beta_3 x_3$$

$$x_3' = -\gamma\beta_3 x_0 + \frac{(\gamma - 1)}{\beta^2}\beta_3\beta_1 x_1 + \frac{(\gamma - 1)}{\beta^2}\beta_3\beta_2 x_2 + \left[1 + \frac{(\gamma - 1)}{\beta^2}\beta_3{}^2\right]x_3$$

The unwieldy character of these equations is improved significantly when they are written in the matrix form

$$(\mathbf{X}') = (\mathscr{L}(\boldsymbol{\beta})) \times (\mathbf{X}) \tag{8.26}$$

where

$$(\mathbf{X}') = \begin{pmatrix} x_0' \\ x_1' \\ x_2' \\ x_3' \end{pmatrix} \qquad (\mathbf{X}) = \begin{pmatrix} x_0 \\ x_1 \\ x_2 \\ x_3 \end{pmatrix}$$

$$\mathscr{L}(\boldsymbol{\beta}) = \begin{bmatrix} \gamma & -\gamma\beta_1 & -\gamma\beta_2 & -\gamma\beta_3 \\ -\gamma\beta_1 & 1 + \frac{(\gamma - 1)}{\beta^2}\beta_1{}^2 & \frac{(\gamma - 1)}{\beta^2}\beta_1\beta_2 & \frac{(\gamma - 1)}{\beta^2}\beta_1\beta_3 \\ -\gamma\beta_2 & \frac{(\gamma - 1)}{\beta^2}\beta_2\beta_1 & 1 + \frac{(\gamma - 1)}{\beta^2}\beta_2{}^2 & \frac{(\gamma - 1)}{\beta^2}\beta_2\beta_3 \\ -\gamma\beta_3 & \frac{(\gamma - 1)}{\beta^2}\beta_3\beta_1 & \frac{(\gamma - 1)}{\beta^2}\beta_3\beta_2 & 1 + \frac{(\gamma - 1)}{\beta^2}\beta_3{}^2 \end{bmatrix} \tag{8.27}$$

and $\times$ refers to matrix multiplication. The column matrices $(\mathbf{X}')$ and $(\mathbf{X})$ represent the space and time coordinates of a single particle as observed in primed and unprimed inertial frames of reference, respectively. Alternatively, we may imagine them to represent the *four-vectors*

$$\mathbf{X}' = \{x_0', x_1', x_2', x_3'\} \qquad \mathbf{X} = \{x_0, x_1, x_2, x_3\}$$

that describe the motion of a particle in terms of four cartesian coordinates. Even though the coordinate x_0 is not a position coordinate, we can suppose that it, along with the space coordinates, determines a point in a four-dimensional space that is in correspondence with the space containing the vector $\mathbf{X}$. A moving particle, then, will trace out a curve in this four-dimensional space according to the positions in ordinary space that it occupies at different times. In Figure 8.2 we have a geometric picture of this idea for the simple case of one spatial dimension. The coordinate x_0 is conventionally marked off along a vertical axis, while x_1 is measured along a horizontal one. The sets of coordinates $\{x_0, x_1\}$ then denote the endpoints of the vector $\mathbf{X}$ as it locates the moving particle. The curve generated in this way is known as the *world-line* of the particle. The space in which world-lines are created is called the space-time continuum or *Minkowski space*.[8] It is important

---

[8] The name is in honor of Hermann Minkowski, who proposed the use of four-vectors in 1908.

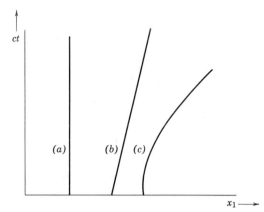

FIGURE 8.2. A geometric picture of two-dimensional Minkowski space, illustrating the world-lines of: (*a*) a particle at rest, (*b*) a particle in uniform motion, and (*c*) a particle with constant acceleration.

to realize at the outset that world-lines differ radically from the trajectories of particles in ordinary three-dimensional space. For example, the path of a single particle, as observed in a frame of reference in which it is at rest, is merely a *point* at some (one-dimensional) position x_1, whereas in the corresponding two-dimensional Minkowski space,[9] the path is a *straight line*, parallel with the x_0 axis, which emanates from the point x_1.

Having established the possibility of describing the motion of a particle in terms of space-time coordinates, we should turn to the problem of interpreting Equation 8.26. Evidently the Lorentz transforms can be represented by an operator matrix $(\mathscr{L}(\beta))$ which takes the four-vector $(\mathbb{X})$ into $(\mathbb{X}')$. We note in particular that

$$(\mathscr{L}(0)) = \begin{pmatrix} 1 & 0 & 0 & 0 \\ 0 & 1 & 0 & 0 \\ 0 & 0 & 1 & 0 \\ 0 & 0 & 0 & 1 \end{pmatrix}$$

is simply an identity transformation, as it should be, since it corresponds to $\mathbf{v} = \mathbf{0}$.

Now we can state in a precise way what is meant by an invariant of the Einsteinian Principle of Relativity. For the moment let us restrict our attention to the special case $\mathbf{v} = v\hat{\mathbf{x}}_1$. Then

$$(\mathscr{L}(\beta)) = \begin{pmatrix} \gamma & -\gamma\beta & 0 & 0 \\ -\gamma\beta & \gamma & 0 & 0 \\ 0 & 0 & 1 & 0 \\ 0 & 0 & 0 & 1 \end{pmatrix}$$

[9] Introductory discussions of two-dimensional Minkowski space are given, for example, in the reference in Footnote 7 and in A. P. French, *Special Relativity*, W. W. Norton, New York, 1968.

and we may consider the matrix defined by

$$(\mathscr{L}(\boldsymbol{\beta}))^{-1} \equiv (g) \times (\widetilde{\mathscr{L}(\boldsymbol{\beta})}) \times (g) = \begin{pmatrix} 1 & 0 & 0 & 0 \\ 0 & -1 & 0 & 0 \\ 0 & 0 & -1 & 0 \\ 0 & 0 & 0 & -1 \end{pmatrix} \times \begin{pmatrix} \gamma & -\gamma\beta & 0 & 0 \\ -\gamma\beta & \gamma & 0 & 0 \\ 0 & 0 & 1 & 0 \\ 0 & 0 & 0 & 1 \end{pmatrix}$$

$$\times \begin{pmatrix} 1 & 0 & 0 & 0 \\ 0 & -1 & 0 & 0 \\ 0 & 0 & -1 & 0 \\ 0 & 0 & 0 & -1 \end{pmatrix} = \begin{pmatrix} \gamma & \gamma\beta & 0 & 0 \\ \gamma\beta & \gamma & 0 & 0 \\ 0 & 0 & 1 & 0 \\ 0 & 0 & 0 & 1 \end{pmatrix}$$

It is straightforward to show that $(\mathscr{L}(\boldsymbol{\beta}))^{-1} \times (\mathscr{L}(\boldsymbol{\beta})) = (\mathscr{L}(0))$. Therefore, $(\mathscr{L}(\boldsymbol{\beta}))^{-1}$ is the matrix that represents the transformation inverse to $(\mathscr{L}(\boldsymbol{\beta}))$. It is equal to the product of (g), known as the metric tensor for Minkowski space,[10] and the transpose of $(\mathscr{L}(\boldsymbol{\beta}))$. If $(\mathscr{L}(\boldsymbol{\beta}))^{-1}$ were simply equal to the transpose, then $(\mathscr{L}(\boldsymbol{\beta}))$ would be an *orthogonal* matrix, as discussed in Section A.2 of the Appendix. Orthogonal matrices have the important property of preserving the scalar product. Obviously, since $(\mathscr{L}(\boldsymbol{\beta}))$ is not orthogonal, we cannot say that the usual scalar product of the four-vectors $\mathbb{X}$ is invariant under $\mathscr{L}(\boldsymbol{\beta})$. However, if we define a *modified* scalar product by

$$\mathbb{X} \cdot \mathbb{X} \equiv \sum_{m=0}^{3} \sum_{n=0}^{3} g_{mn} x_m x_n$$
$$= x_0{}^2 - x_1{}^2 - x_2{}^2 - x_3{}^2 = (ct)^2 - x_1{}^2 - x_2{}^2 - x_3{}^2 \quad (8.28)$$

we do indeed have a quantity that is invariant under Lorentz transformations. (See Problem 2 at the end of this chapter.)

We shall define, generally, as a four-vector *any* quantity $\mathbb{A} = \{A_0, A_1, A_2, A_3\}$ that transforms between inertial frames of reference according to Equation 8.26. Then every scalar product, defined as in Equation 8.28, of whatever four-vectors we might create is a Lorentz-invariant quantity, that is, an invariant of the Einsteinian Principle of Relativity. Also, the quantities that are scalars with respect to four-vectors will be automatically invariant. To make these notions concrete, let us consider several examples.

The scalar product of $d\mathbb{X}$ with itself

$$d\mathbb{X} \cdot d\mathbb{X} = -(dx_1)^2 - (dx_2)^2 - (dx_3)^2 + (c\,dt)^2 \quad (8.29)$$

is the square of the infinitesimal "distance" between two points in Minkowski space and is invariant under $\mathscr{L}(\boldsymbol{\beta})$. Because of the minus signs, the right-hand side of Equation 8.29 need not be positive, in contrast with the lengths of intervals in

---

[10] The concept of the metric tensor is introduced in Section A.3 of the Appendix. A nontrivial metric tensor arises whenever the element of length in a space is not expressed by the square root of a simple sum of squared coordinates.

ordinary space. In fact, we can distinguish three cases:

$$dX \cdot dX \begin{cases} >0 & \text{(timelike)} \\ =0 & \text{(lightlike)} \\ <0 & \text{(spacelike)} \end{cases}$$

From Equation 8.29 we can deduce that the displacement of one point in Minkowski space from another is *timelike* if the rate of transit between the two points is less than the speed of light in vacuum, c. (Remember that $dx_i = u_i\, dt$; $i = 1, 2, 3$.) For a reason to be given shortly, the speed c represents the upper limit for the speed of any known material particle. Therefore we may conclude that *points along the world-line of any particle always correspond to timelike displacements.* Moreover, these points can be put into communication with one another by a signal traveling with speed c and so always display a causal relationship.

It follows from this fact that the purely temporal part of the distance between two points separated by a timelike displacement can never be zero, although the purely spatial part can be made so by an appropriate choice of frame of reference.

If the space-time displacement between two points in Minkowski space is zero, it is said to be *lightlike*. This name derives from the fact that Equation 8.29, when set equal to zero, describes the path of a beam of light whose position coincided with the origin at time zero. Thus lightlike displacements separate the points on the world-lines of any kind of electromagnetic radiation (and of neutrinos) traveling in vacuum.

A *spacelike* displacement exists between two points when they cannot be put into communication through a light signal and, therefore, when they show no particular causal relation with one another. For this reason, the purely temporal part of a spacelike displacement between two points in Minkowski space can be reduced to zero in a suitably chosen inertial frame with no loss of physical significance: the two points in question are merely simultaneous in that frame. On the other hand, the purely spatial separation of these two points can *never* be zero. This fact leads to the conclusion that the question of temporal order can be entirely relative for events (space-time points) that are not occurring at the same place. It was this remarkable aspect of the relativity of time intervals that in fact prompted Einstein's investigation into the self-consistency of Newtonian dynamics.

The infinitesimal element of timelike path length,

$$\|dX\| = [(c\,dt)^2 - (dx_1)^2 - (dx_2)^2 - (dx_3)^2]^{1/2} \qquad (8.30)$$

has an interesting physical meaning that is brought out more clearly if we define the real-valued element

$$c\,d\tau \equiv \|dX\| \qquad (8.31)$$

where $d\tau$ is called the *proper time* between two points separated by a timelike displacement. The significance of $d\tau$ is seen if we evaluate it along the world-line of a particle that is instantaneously at rest in some inertial frame. Then we have

$c\,d\tau = [(c\,dt)^2]^{1/2} = c\,dt$ or $d\tau = dt$, which means that the proper time is the time interval *as measured in the particle's own frame of reference*. Since $d\tau$ is an invariant, we know that it retains the same physical interpretation regardless of the inertial frame in which the particle is observed. Therefore, in the general case,

$$
\begin{aligned}
d\tau &= \frac{1}{c}\left[-(dx_1)^2 - (dx_2)^2 - (dx_3)^2 + (c\,dt)^2\right]^{1/2} \\
&= \frac{1}{c}\left(-u_1{}^2 - u_2{}^2 - u_3{}^2 + c^2\right)^{1/2} dt \\
&= \left(1 - \frac{u^2}{c^2}\right)^{1/2} dt
\end{aligned}
\tag{8.32}
$$

we have a relation between the time interval measured in the frame of reference of a particle and that measured in a frame wherein the particle moves with the velocity **u**. Equation 8.32 shows that time intervals measured in a "moving" frame of reference will be *shorter* than those measured in a "resting" frame. This conclusion, it should be noted, does not require that the velocity of the particle, **u**, be constant, although the measurement of the time interval dt must take place in an inertial frame. The relativity in time implied by Equation 8.32 is, of course, a direct result of the inadequacy of the Galileo transform (8.14) to account for the invariance relation $c' = c$. This failure of the Galilean-Newtonian relativity principle has been documented experimentally in a most dramatic way through measurements of the radioactive decay times for muons produced by cosmic rays.[11]

As an example of a purely scalar quantity, with respect to four-vectors, we may consider the *proper mass* of a particle. This mass is denoted by m_0 and is the result—like proper time—of a measurement carried out entirely in the frame of reference of the particle. The proper mass is also known as the *rest mass* of the particle for obvious reasons. With m_0 we may define the *four-vector momentum*

$$
\mathbb{P} \equiv m_0 \frac{d\mathbb{X}}{d\tau}
\tag{8.33}
$$

which has the coordinates

$$
\frac{m_0 c}{[1 - (u^2/c^2)]^{1/2}} \qquad \frac{m_0 u_1}{[1 - (u^2/c^2)]^{1/2}} \qquad \frac{m_0 u_2}{[1 - (u^2/c^2)]^{1/2}} \qquad \frac{m_0 u_3}{[1 - (u^2/c^2)]^{1/2}}
$$

according to Equations 8.30 and 8.32. If we rewrite (8.33) in the form

$$
\mathbb{P} = \left\{ \frac{m_0 c}{[1 - (u^2/c^2)]^{1/2}}, \quad \frac{m_0 \mathbf{u}}{[1 - (u^2/c^2)]^{1/2}} \right\}
\tag{8.34}
$$

we see that $\mathbb{P}$ consists of a "spatial part" that resembles the Newtonian linear momentum, except for the factor in the denominator. If we wish to preserve the

---

[11] See D. H. Frisch and J. H. Smith, Measurement of the relativistic time dilation using μ-mesons, *American Journal of Physics* 31:342–355 (1963).

convention that the "spatial part" of a four-vector is to be interpreted as a vector in ordinary space, we must give up the Newtonian idea that mass is an invariant quantity with respect to inertial observers. Instead we write

$$m \equiv \frac{m_0}{[1 - (u^2/c^2)]^{1/2}} \tag{8.35}$$

The quantity m is a measure of the inertia of a particle in a frame relative to which it is observed to move with the velocity $\mathbf{u}$. Then we have

$$\mathbb{P} = \{mc, \mathbf{p}\} \tag{8.36}$$

for the four-vector momentum, where $\mathbf{p} = m\mathbf{u}$ in complete analogy with the Newtonian definition. Equation 8.35 indicates that the inertia of a moving particle *increases* with its speed relative to a given reference frame.[12] We note that the infinite upper limit of this increase occurs when $u = c$. Therefore, in order that the inertia of a particle may remain finite and real valued, we must stipulate that the speed with which the particle travels be always less than the speed of light in vacuum.

The scalar product $\mathbb{P} \cdot \mathbb{P}$ has an interesting physical interpretation whose full significance will be seen in the next section. For now, we can write

$$\begin{aligned} \mathbb{P} \cdot \mathbb{P} &= \frac{m_0{}^2 u^2}{1 - (u^2/c^2)} - \frac{m_0{}^2 c^2}{1 - (u^2/c^2)} = -(m_0 c)^2 \\ &= p^2 - (mc)^2 \end{aligned}$$

or

$$(mc)^2 = p^2 + m_0{}^2 c^2 \tag{8.37}$$

upon using Equations 8.34 and 8.36 in that order. Equation 8.37 is an important relation between the momentum of a particle and its inertia. We note in passing that $\mathbb{P}$ qualifies as a spacelike four-vector, since $\mathbb{P} \cdot \mathbb{P} < 0$.

Before going on to consider the Einsteinian versions of the conservation theorems, we note that the natural analog of the Second Law may be formed by writing

$$\mathbf{F} = \frac{d\mathbf{p}}{dt} = \frac{d}{dt} \left\{ \frac{m_0 \mathbf{u}}{[1 - (u^2/c^2)]^{1/2}} \right\} \tag{8.38}$$

The first part of Equation 8.38 is constructed to be formally identical with what appears in Newtonian dynamics. The second part shows clearly the effect of the requirement of Lorentz invariance: The time derivative of the particle mass is no longer zero and, therefore, $\mathbf{F} = m\mathbf{a}$ is patently incorrect.

---

[12] Equation (8.35) has been verified in a number of experiments carried out since 1909. See, for example, A. P. French, op. cit., p. 23.

8.3 RELATIVISTIC ENERGY

If we continue the process of reasoning by analogy, we expect that the change in kinetic energy for a single particle should derive from the integral

$$\int_{\mathbf{r}_1}^{\mathbf{r}_2} \mathbf{F(r)} \cdot d\mathbf{r} = \int_{u_1}^{u_2} \mathbf{u} \cdot d \left\{ \frac{m_0 \mathbf{u}}{[1 - (u/c)^2]^{1/2}} \right\}$$

$$= m_0 \int_{u_1}^{u_2} \frac{\mathbf{u} \cdot d\mathbf{u}}{[1 - (u^2/c^2)]^{3/2}} = \frac{m_0 c^2}{[1 - (u^2/c^2)]^{1/2}} \Big|_{u_1}^{u_2}$$

according to Equation 8.38. Following the usual procedure, we would regard the quantity

$$\frac{m_0 c^2}{[1 - (u^2/c^2)]^{1/2}} = mc^2$$

as the kinetic energy measured relative to its value when $u = 0$. But mc^2 certainly does not look much like a kinetic energy. Besides the unusual appearance, we have

$$\lim_{u/c \to 0} mc^2 = m_0 c^2 \qquad (8.39)$$

which is not equal to zero for any material particle. Equation 8.39 states that a particle at rest in an inertial frame still possesses energy. This energy is

$$E_0 = m_0 c^2 \qquad (8.40)$$

Einstein considered E_0 to represent the energy a particle has by virtue of its inertia. Today we would say that Equation 8.40 expresses the energy equivalent of the mass of a particle and would call E_0 simply the *rest energy*. The physical significance of the rest energy has been demonstrated amply since 1905 in observations of the creation of matter from gamma rays and in the production of radiant energy through nuclear fission and fusion.

Therefore, we are led to define

$$T(\mathbf{u}) \equiv mc^2 - m_0 c^2 \qquad (8.41)$$

to be the *relativistic kinetic energy* of a single particle. The basic soundness of this definition is supported by the fact that

$$\lim_{c \to \infty} T(\mathbf{u}) = \lim_{c \to \infty} m_0 c^2 \left\{ \left[1 - \frac{u^2}{c^2} \right]^{-1/2} - 1 \right\}$$

$$= \lim_{c \to \infty} m_0 c^2 \left[1 + \frac{1}{2} \left(\frac{u}{c} \right)^2 + \cdots - 1 \right] = \tfrac{1}{2} m_0 u^2 \qquad (8.42)$$

which is in perfect agreement with Newtonian dynamics. With Equations 8.40 and 8.41 we can express the total energy of a free particle in the form

$$E = T(\mathbf{u}) + E_0 \tag{8.43}$$

The Energy Theorem as presented here is not the same in meaning as what we met in Chapters 1 or 4. Instead, the constancy of E now refers to the *conservation of mass-energy*, since any process wherein the conversion of matter to energy, or the reverse, occurs is compatible with Equation 8.43. This enlarged concept of the physical significance of energy is one of the most remarkable aspects of the Einsteinian Principle of Relativity and undoubtedly has been the effect of that principle most strongly felt in the fields of physics outside classical dynamics.

8.4 MOTION IN A UNIFORM FIELD

Because of Equation 8.35, the relativistic motion of a single particle that is sub-ject to the force $\mathbf{F} = m\mathbf{a}_0$ will depend sensitively upon whether $\mathbf{a}_0$ is a function of the mass. For this reason we shall divide our consideration of the uniform field into two special cases of general importance, each of which in Newtonian me-chanics would have received the *same* mathematical description.

(a) *A uniform, constant gravitational field.* In this case we shall write

$$\mathbf{a}_0 = -g\hat{\mathbf{x}}_3$$

into Equation 8.38 to produce the three differential equations

$$\frac{dp_1}{dt} = 0 \qquad \frac{dp_2}{dt} = 0 \qquad \frac{dp_3}{dt} = -mg \tag{8.44}$$

The first two of these expressions lead to the expected result that p_1 and p_2 are constants of the motion. We shall set those two momentum coordinates equal to zero in order to simplify matters. The third equation in (8.44) may be expressed

$$\frac{d}{dt}\left\{\frac{m_0 u_3}{[1 - (u_3{}^2/c^2)]^{1/2}}\right\} = \frac{m_0}{[1 - (u_3{}^2/c^2)]^{3/2}}\frac{du_3}{dt} = -\frac{m_0 g}{[1 - (u_3{}^2/c^2)]^{1/2}}$$

or

$$\frac{du_3}{dt} + g - \frac{g}{c^2}u_3{}^2(t) = 0 \tag{8.45}$$

The influence of the relativity principle appears in the third term on the left-hand side of Equation 8.45. In the limit $c \to \infty$ the expression reduces to Equation 1.13, as it should. Otherwise (8.45) is a differential equation of the same mathematical form as is (3.14) if we set $u_3(t) = y(t) + c$ and write $a = -2g/c$, $b = g/c^2$, $n = 2$. It then follows that the solution for $y(t)$ is

$$y(t) = -2c(1 - C^{-1}e^{-2gt/c})^{-1}$$

where C is a constant of integration. Finally

$$u_3(t) = -c[-1 + 2(1 - C^{-1}e^{-2gt/c})^{-1}]$$
$$= -c \tanh\left[\frac{g}{c}(t - t_0)\right] \tag{8.46}$$

upon writing in the definition of $y(t)$ and putting

$$C \equiv e^{-2gt_0/c}$$

Equation 8.46 is graphed in Figure 8.3 for the case $t_0 = 0$. There we see that the

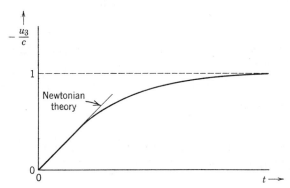

FIGURE 8.3. The downward velocity of a particle accelerated to relativistic speed by a uniform gravitational field.

velocity of the particle is initially linear in the time, but approaches the value $-c$ asymptotically. The early time behavior is not difficult to understand, since

$$\tanh\left[\frac{g}{c}(t - t_0)\right] = \frac{g}{c}(t - t_0) - \frac{1}{3}\left[\frac{g}{c}(t - t_0)\right]^3 + \cdots.$$

and, accordingly,

$$u_3(t) \simeq -g(t - t_0) \qquad (c \to \infty) \tag{8.47}$$

to the lowest order in the time difference $(t - t_0)$. Equation 8.47 is in perfect agreement with the Newtonian result and with the expectation that, for small elapsed times, the field cannot accelerate the particle to relativistically important velocities. On the other hand, the behavior of the particle for large values of t is consistent with the Einsteinian relativity principle. The degree of the contradiction between this behavior and that predicted on the basis of Newtonian mechanics is made very apparent if we note that, even though the velocity approaches a *finite* limiting value, the relativistic momentum of the particle,

$$p_3(t) = \frac{m_0}{[1 - (u_3(t)/c)^2]^{1/2}} u_3(t)$$

becomes asymptotically *infinite*. At large elapsed times most of the energy transferred by the field to the particle is converted to mass instead of being used to promote acceleration.

The parameter t that appears in Equation 8.46 is, of course, the elapsed time as measured in an inertial frame of reference wherein the particle is observed to move with the velocity $u_3(t)\hat{\mathbf{x}}_3$. The elapsed time in the particle's own frame of reference, the proper time, is given by Equation 8.32:

$$d\tau = \left[1 - \left(\frac{u_3(t)}{c}\right)^2\right]^{1/2} dt$$

With the help of Equation 8.46, we can rewrite this expression as

$$d\tau = \text{sech}\left[\frac{g}{c}(t - t_0)\right] dt \tag{8.48}$$

upon noting the identity

$$\tanh^2 x + \text{sech}^2 x = 1$$

Equation 8.48 is, in effect, a differential equation for $\tau(t)$ whose solution can be obtained directly from a table of integrals. The solution is

$$\tau(t) = \frac{c}{g} \tan^{-1}\left\{\sinh\left[\frac{g}{c}(t - t_0)\right]\right\} \tag{8.49}$$

if we assume that the proper time is synchronous with the observer time at $t = t_0$. Equation 8.49 is plotted in Figure 8.4. We note that the proper time is always less than the observer time and, in fact, is finite when the latter is infinitely large.

(b) *A uniform, constant electric field.* In this case we have

$$\mathbf{a}_0 = \frac{q}{m} \mathbf{E}_0$$

where q is the charge on the particle and $\mathbf{E}_0$ is the electric field intensity. Equation 8.38 becomes

$$\frac{m_0}{[1 - (u/c)^2]^{3/2}} \frac{d\mathbf{u}}{dt} = q\mathbf{E}_0 \tag{8.50}$$

which is, as usual, a set of three differential equations for the velocity u. The most interesting solutions of these equations are obtained when we impose the conditions

$$\mathbf{E}_0 = E_{02}\hat{\mathbf{x}}_2$$

$$p_1(0) = p_0 > 0 \qquad p_2(0) = 0 \qquad p_3(0) = 0$$

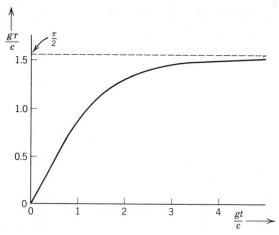

FIGURE 8.4. The relation between the proper time of a particle moving in a uniform gravitational field and the time of an observer at rest with respect to the field.

These expressions describe a particle moving initially along the positive x_1 direction into a transverse electric field directed along the positive x_2 axis. It is clear that, in this case, the motion in the x_3 direction is of no physical consequence. The two remaining equations of motion in (8.50) are

$$\frac{dp_1}{dt} = 0 \tag{8.51}$$

$$\frac{dp_2}{dt} = qE_{02} \tag{8.52}$$

The solution of (8.51) is, of course, ·

$$p_1(t) = p_0 \tag{8.53}$$

For Equation 8.52 we have quite as simply

$$p_2(t) = qE_{02}t \tag{8.54}$$

Equations 8.63 and 8.54 have an appearance deceptively similar to the Newtonian solutions for the problem of motion in a transverse electric field. But the momentum coordinates in these equations are expressed in Equation 8.36 and this fact makes the observed motion of the particle generally quite different from that in the non-relativistic regime. To see this we rearrange (8.53) to read

$$u_1(t) = \frac{p_0}{m} = \frac{p_0 c^2}{[\varepsilon^2 + (qE_{02}ct)^2]^{1/2}} \tag{8.55}$$

where Equations 8.37 and 8.54 have been used to write

$$m = \left(\frac{p^2}{c^2} + m_0{}^2\right)^{1/2}$$

$$= \frac{1}{c^2}\left[p_0{}^2c^2 + (qE_{02}ct)^2 + m_0{}^2c^4\right]^{1/2}$$

$$\equiv \frac{1}{c^2}\left[\varepsilon^2 + (qE_{02}ct)^2\right]^{1/2} \tag{8.56}$$

$\varepsilon \equiv (p_0{}^2c^2 + m_0{}^2c^4)^{1/2}$ being the initial total energy of the particle. Equation 8.53 leads us to the remarkable conclusion that the component of velocity along the x_1 direction actually *decreases* in magnitude as time passes, even though there is no electric field along that direction. This result is just another reminder that a conserved relativistic momentum (Equation 8.51) does not necessarily imply a constant velocity, as it does in the Newtonian program.

With regard to Equation 8.54, the velocity coordinate along the x_2 axis is

$$u_2(t) = \frac{qE_{02}t}{m} = \frac{qE_{02}c^2t}{\left[\varepsilon^2 + (qE_{02}ct)^2\right]^{1/2}} \tag{8.57}$$

This coordinate increases in response to the field, but has a smaller value at every

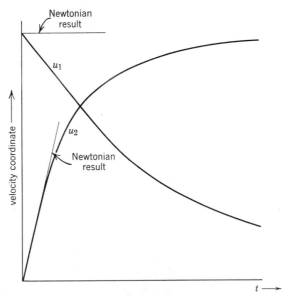

FIGURE 8.5. The velocity coordinates of a particle moving in a uniform electric field directed along the x_2 axis.

point in time than the Newtonian result

$$u_2(t) = \frac{q}{m_0} E_{02}t \qquad \text{(Newton)} \tag{8.58}$$

The reduction in value, of course, comes about for the same reason it did in Equation 8.55: the kinetic energy gained by the particle goes partly to increase its inertia. We note in passing that the asymptotic values of the velocity coordinates,

$$u_1(t) \sim 0 \qquad u_2(t) \sim c$$

are in marked contradiction with the Newtonian theory (see Figure 8.5).

8.5 RELATIVISTIC COLLISION THEORY

The special Principle of Relativity exerts a profound influence on the theory of elastic collisions because of the radical changes the relativity principle brings about in the concept of mass. Since mass and energy are equivalent entities, in the sense of Equation 8.40, the expression for the conservation of energy in a two-particle scattering process, such as portrayed in Figure 8.6, can no longer be

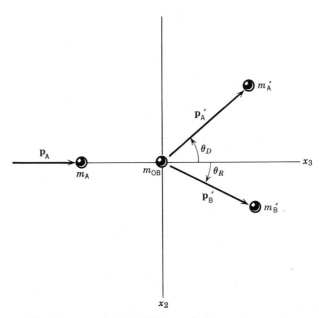

FIGURE 8.6. Relativistic scattering in the LAB frame of reference.

Equation 5.1 but instead is[13]

$$E_T \equiv m_A c^2 + m_{0B} c^2 = m'_A c^2 + m'_B c^2 \tag{8.59}$$

where the masses are to be evaluated under the same conditions as were the *velocities* in (5.1). In a similar way the relativistic increase of the mass with velocity alters the equations of momentum conservation to the extent that they must be written

$$p_A = p'_A \cos \theta_D + p'_B \cos \theta_R \tag{8.60}$$

$$0 = p'_A \sin \theta_D - p'_B \sin \theta_R \tag{8.61}$$

rather than as are Equations 5.3 and 5.4. It is evident that relativistic statements of the conservation theorems avoid explicit representation of the particle velocities.

In the LAB frame of reference we can proceed as in Chapter 5 to define the recoil energy

$$\Delta E_R \equiv m'_B c^2 - m_{0B} c^2 = (m_A - m'_A) c^2 \tag{8.62}$$

and to inquire as to its relation with θ_R. Using Equations 8.60 and 8.61, we can write the analog of Equation 5.5:

$$(p'_A)^2 = p_A{}^2 + (p'_B)^2 - 2 p_A p'_B \cos \theta_R \tag{8.63}$$

The momentum p'_A is eliminated from Equation 8.63 by introducing Equations 8.37 and 8.59. Ultimately we find

$$(m_A + m_{0B})(m'_B - m_{0B}) = [(m_A{}^2 - m_{0A}{}^2)(m'_B{}^2 - m_{0B}{}^2)]^{1/2} \cos \theta_R \tag{8.64}$$

after expressing all of the momenta in terms of masses. Equation 8.64 may be rearranged, on squaring both sides, to produce a quadratic equation for m'_B, which is the quantity analogous to v'_B in Equation 5.6. The solution of the quadratic equation is

$$m'_B = \frac{(m_A + m_{0B})^2 m_{0B} \pm m_{0B}(m_A{}^2 - m_{0A}{}^2) \cos^2 \theta_R}{(m_A + m_{0B})^2 - (m_A{}^2 - m_{0A}{}^2) \cos^2 \theta_R}$$

or, in a more compact form,

$$m'_B = \left[\frac{E_T{}^2 \pm p_A{}^2 c^2 \cos^2 \theta_R}{E_T{}^2 - p_A{}^2 c^2 \cos^2 \theta_R} \right] m_{0B} \tag{8.65}$$

We keep only the plus sign in the numerator of Equation 8.65 in order to discard the trivial solution $m'_B = m_{0B}$. With this choice for m'_B, Equation 8.62 becomes

$$\Delta E_R = \left[\frac{2 p_A{}^2 c^2 \cos^2 \theta_R}{E_T{}^2 - p_A{}^2 c^2 \cos^2 \theta_R} \right] m_{0B} c^2 \tag{8.66}$$

---

[13] In agreement with the notation of Chapter 5, a *prime* will be used to denote a quantity measured after a collision has occurred. A *tilde* over a symbol, rather than the prime used in the previous sections of this chapter, will denote measurement in a "moving" frame of reference.

This expression is the relativistic generalization of Equation 5.7. We see that now the recoil energy is determined by the momentum of the incident particle instead of its rest mass and that the relativistic total energy E_T supercedes T_{LAB}. In the nonrelativistic limit,

$$p_A^2 \simeq (m_{0A}u)^2 \qquad E_T^2 \simeq (m_{0A}c^2 + m_{0B}c^2) \gg p_A^2 c^2$$

and we recover Equation 5.7:

$$\Delta E_R \simeq \left[\frac{2(m_{0A}u)^2 c^2 \cos^2 \theta_R}{(m_{0A}c^2 + m_{0B}c^2)^2} \right] m_{0B}c^2$$

$$= 4 \frac{(m_{0A}/m_{0B})}{[1 + (m_{0A}/m_{0B})]^2} T_{LAB} \cos^2 \theta_R$$

To find the relation between θ_R and θ_D that is analogous to Equation 5.8 we combine Equations 8.60 and 8.61. The result is

$$\tan \theta_D = \frac{\sin \theta_R}{p_A/p_B' - \cos \theta_R}$$

Now, Equations 8.37, 8.59, 8.62, and 8.63 lead to

$$\Delta E_R E_T = p_A p_B' c^2 \cos \theta_R$$

which permits us to write

$$\frac{p_A}{p_B'} = \frac{\Delta E_R E_T}{(p_B')^2 c^2 \cos \theta_R} = \frac{E_T}{(m_B' + m_{0B})c^2 \cos \theta_R}$$

$$= \frac{1 + (m_A/m_{0B})}{[1 + (m_B'/m_{0B})]\cos \theta_R}$$

In reducing $p_B'^2$ we have used the identities

$$p_B'^2 = (m_B'c)^2 - (m_{0B}c)^2 \equiv (m_B' - m_{0B})c^2(m_B' + m_{0B})$$
$$\equiv \Delta E_R(m_B' + m_{0B})$$

With these results we can deduce finally

$$\tan \theta_D = \frac{\sin 2\theta_R}{\left[\dfrac{1 - (m_B'/m_{0B}) + (2m_A/m_{0B})}{1 + (m_B'/m_{0B})} \right] - \cos 2\theta_R} \qquad (8.67)$$

as the generalization of Equation 5.8. The most interesting aspect of Equation 8.83 is that it does *not* lead to $\theta_D + \theta_R = \pi/2$ when the rest masses of the incident and target particles are equal. Thus the scattering of identical particles at high energies will deviate from the well known "billiard ball" condition of right-angle collisions.

The CM coordinate system is not a viable construction in the relativistic domain because the particle mass is not an invariant quantity. Therefore, instead of generalizing the picture of scattering in the CM frame as described in Section 5.1, we shall follow the customary procedure of defining a new frame of reference which will be of value in theoretical discussions. This frame is defined so as to move with the constant velocity

$$\mathbf{V} = \frac{\mathbf{p}_A c^2}{E_T} \tag{8.68}$$

relative to the LAB frame.[14] For reasons to be made clear soon, the new frame is called the *center-of-momentum frame of reference* or *CP frame*. Since the CP frame moves at constant velocity with respect to the LAB frame, the two are related through the Lorentz transforms. Therefore

$$\tilde{\mathbf{v}}_B = -\mathbf{V} = \frac{-\mathbf{p}_A c^2}{E_T} \tag{8.69}$$

and

$$
\begin{aligned}
\tilde{\mathbf{v}}_A = \frac{d\tilde{\mathbf{r}}_A}{dt} &= \frac{\mathbf{v}_A + [(\mathbf{v}_A \cdot \mathbf{V})/v^2](\gamma - 1)\mathbf{V} - \gamma \mathbf{V}}{\gamma[1 - (\mathbf{v}_A \cdot \mathbf{V})/c^2]} \\
&= \frac{\mathbf{v}_A - \mathbf{V}}{1 - (v_A V/c^2)} = \frac{\mathbf{p}_A[(1/m_A) - (c^2/E_T)]}{1 - (p_A^2/m_A E_T)} \\
&= \frac{\mathbf{p}_A(E_T - m_A c^2)}{m_A E_T - p_A^2} = \frac{m_{0B} c^2}{m_{0B} E_A + (m_{0A} c)^2} \mathbf{p}_A
\end{aligned}
\tag{8.70}
$$

are the velocities of the target and incident particles, respectively, in the CP frame. In getting Equation 8.70 we have employed Equations 8.23 and 8.24 in differential form and have noted that $\mathbf{V}$ and $\mathbf{v}_A$ are parallel vectors. The quantity E_A is just $m_A c^2$. The relativistic particle energies in the CP frame are

$$
\begin{aligned}
\tilde{E}_A = \tilde{m}_A c^2 &= \frac{m_{0A} c^2}{[1 - (\tilde{v}_A^2/c^2)]^{1/2}} \\
&= m_{0A} c^2 \left[1 - \frac{m_{0B}^2 p_A^2}{(m_{0B} E_A + (m_{0A} c)^2)^2} \right]^{-1/2} = \frac{E_A E_T - p_A^2 c^2}{(E_T^2 - p_A^2 c^2)^{1/2}}
\end{aligned}
\tag{8.71}
$$

$$
\tilde{E}_B = \tilde{m}_B c^2 = \frac{m_{0B} c^2}{[1 - (v^2/c^2)]^{1/2}} = \frac{m_{0B} c^2 E_T}{(E_T^2 - p_A^2 c^2)^{1/2}}
\tag{8.72}
$$

With these expressions we deduce

$$
\begin{aligned}
\tilde{E}_T = \tilde{E}_A + \tilde{E}_B &= \frac{E_A E_T - p_A^2 c^2 + m_{0B} c^2 E_T}{(E_T^2 - p_A^2 c^2)^{1/2}} \\
&= (E_T^2 - p_A^2 c^2)^{1/2} = E_T \left(1 - \frac{V^2}{c^2} \right)^{1/2}
\end{aligned}
\tag{8.73}
$$

[14] In the nonrelativistic limit, $\mathbf{V}$ in (8.68) reduces to $\mathbf{V}$ given in Section 5.1. For this reason the practice of viewing a collision from the CP frame is often carried out in the nonrelativistic limit as well. However, in adopting that procedure one automatically loses the opportunity to describe the scattering process in terms of relative coordinates. Since the latter are the simplest in which to represent the particle interactions, we have chosen to retain them and, therefore, to avoid the CP coordinate system in Chapter 5.

and

$$\tilde{\mathbf{p}}_A = \left(\frac{\tilde{E}_A}{c^2}\right)\tilde{\mathbf{v}}_A$$

$$= \frac{m_{0B}c^2}{(E_T{}^2 - p_A{}^2 c^2)^{1/2}}\mathbf{p}_A \tag{8.74}$$

$$\tilde{\mathbf{p}}_B = \left(\frac{\tilde{E}_B}{c^2}\right)\tilde{\mathbf{v}}_B$$

$$= \frac{-m_{0B}c^2}{(E_T{}^2 - p_A{}^2 c^2)^{1/2}}\mathbf{p}_A \tag{8.75}$$

Equations 8.74 and 8.75 lead us to the important conclusion that *the total momentum of the two-particle system is zero in the CP frame*. Because of this fact and the Momentum Theorem, we infer that the effect of a scattering process in the CP frame is simply to rotate the line of centers of the two particles (see Figure 8.7).

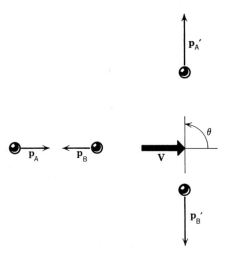

FIGURE 8.7. Relativistic scattering in the CP frame of reference. (Compare with Figure 5.4.)

The relativistic extension of Equation 5.20 is found by writing

$$v'_{A3} = \frac{\tilde{v}_{AB} + V}{1 + (\tilde{v}_{A3}V/c_2)} \qquad v'_{A2} = \frac{1}{\gamma}\frac{\tilde{v}_{A2}}{1 + (\tilde{v}_{A3}V/c^2)}$$

for the x_3 and x_2 coordinates of the velocity of the scattered particle, respectively. These equations are the inverses of the general velocity transform

$$\tilde{\mathbf{v}} = \frac{\mathbf{v} + [(\mathbf{v} \cdot \mathbf{V})/v^2](\gamma - 1)\mathbf{V} - \gamma\mathbf{V}}{\gamma\{1 - [(\mathbf{v} \cdot \mathbf{V})/c^2]\}} \tag{8.76}$$

with $\mathbf{V}$ taken to point along the x_3 direction. Since

$$v'_{AB} = v'_A \cos \theta_D \qquad v'_{A2} = v'_A \sin \theta_D \qquad \tilde{v}_{AB} = \tilde{v}_A \cos \theta \qquad \tilde{v}_{A2} = \tilde{v}_A \sin \theta$$

we have

$$\tan \theta_D = \frac{\tilde{v}'_{A2}}{\gamma(\tilde{v}_{AB} + V)} = \frac{\sin \theta}{\gamma[(V/\tilde{v}_A) + \cos \theta]} \tag{8.77}$$

But

$$\frac{V}{\tilde{v}_A} = \frac{V}{\tilde{v}_A} = \frac{p_A c^2/E_T}{m_{0B}c^2 p_A/[m_{0B}E_A + (m_{0A}c)^2]} = \frac{m_A + (m_{0A}/m_{0B})m_{0A}}{m_A + m_{0B}} \tag{8.78}$$

which allows us to express (8.77) in terms of the masses alone. It is evident that Equation 8.77 reduces to Equation 5.20 in the nonrelativistic limit. We note also that, in the case of equal rest masses,

$$\tan \theta_D = \frac{1}{\gamma} \frac{\sin \theta}{1 + \cos \theta} = \frac{1}{\gamma} \tan \frac{\theta}{2} \qquad (m_{0A} = m_{0B}) \tag{8.79}$$

The proportionality between the deflection angles in the LAB and CP systems is lost at high energies.

FOR FURTHER READING

K. R. Symon — *Mechanics*, Addison-Wesley, Reading, Mass., 1971. Chapters 13 and 14 present a lengthy introduction to special relativity at the mathematical level of this chapter.

A. P. French — *Special Relativity*, W. W. Norton, New York, 1968. This short book is a fine, basic introduction to relativity, with an emphasis on the experimental confirmation of the theory.

D. Bohm — *The Special Theory of Relativity*, W. A. Benjamin, New York, 1965. For the reader wishing insight into the physical meaning of relativity and a discussion of some of the apparent paradoxes which arise in connection with the relativity of time, this book is highly recommended.

W. G. V. Rosser — *Introductory Relativity*, Plenum Press, New York, 1967. This book gives a sound introduction to all aspects of special relativity. It is recommended for the reader wishing an extensive physical discussion with many examples.

A. Baldin, V. I. Goldanskii, and I. L. Rozental' — *Kinematics of Nuclear Reactions*, Oxford University Press, London, 1961. The first three chapters of this technical discussion of nuclear reactions provide a complete account of relativistic collision theory.

PROBLEMS

1. Derive the general result, valid for any relativity principle, that

$$\alpha(v) \simeq 1 + kv^2$$

for sufficiently small v and give the value of the constant k according to Einsteinian relativity.

2. The length of a rod may be defined as the length of the difference between the position vectors specifying its endpoints at one instant in a given inertial frame of reference.

   (a) Show that the length is an invariant of the Galilean-Newtonian relativity principle.

   (b) Show that this length is not an invariant of the Einsteinian relativity principle. What length *is* an invariant in this case?

3. Show that the relativistic transform for the velocity of a single particle is

$$\mathbf{u}' = \frac{\mathbf{u} + (\mathbf{u} \cdot \mathbf{v}/c^2)(\gamma - 1)\mathbf{v} - \gamma\mathbf{v}}{\gamma[1 - (\mathbf{u} \cdot \mathbf{v}/c^2)]}$$

4. Show that the elapsed time $\Delta t'$ measured on a clock moving at the constant velocity $\mathbf{v}$ with respect to another clock will always be related to the elapsed time Δt shown on the latter according to

$$\Delta t = \gamma \, \Delta t' \qquad \text{(Einstein time dilation)}$$

Note that Δt refers to a macroscopic time interval, not an infinitesimal one.

5. Define mathematically the term "simultaneous" and show that two events that are simultaneous in one inertial frame generally cannot be so in another inertial frame. What case brings an exception to this conclusion?

6. Show that a clock on Earth loses 0.16 s per year relative to a hypothetical clock on the sun. Ignore the accelerations of the Earthbound clock in doing the calculation.

7. A beam of monochromatic light is sent along the positive x direction through a pipe filled with water moving in the same direction. The phase speed of this beam, measured relative to the laboratory, is found to be represented by the empirical equation

$$u_{\text{LAB}} = 2.26 \times 10^{10} \text{ cm/s} + 0.432 \, v$$

where v is the speed of the water in cm/s. Derive this equation using the Lorentz transform equations. The index of refraction of water is 1.33. Why does the Einsteinian relativity principle not apply *directly* in this case?

8. Show that the Galileo transform equations, when applied to Problem 7, lead to the result

$$u_{\text{LAB}} = 2.26 \times 10^{10} \text{ cm/s} + v$$

Is there any condition under which this equation could be correct?

9. The mean proper lifetime of the π^+ meson is 2.55×10^{-8} s. Calculate the mean distance such particles will travel, at the speed $0.9995 \, c$, in the laboratory before decaying into muons and neutrinos.

10. Sketch a graph of the angle θ that a meter stick makes with the x_1 axis as a function of the speed along the x_1 direction of the observer measuring the angle. Take $\theta = 60°$ when the observer is at rest relative to the meter stick.

11. In the list below are some data on the charge-to-mass ratio of the electron as a function of electron speed. Calculate the charge-to-mass ratios expected on the basis of the u/c data. What assumption *must* be made?

$e/m(10^{11}C/kg)$	u/c
1.661	0.3173
1.630	0.3787
1.590	0.4281
1.511	0.5154

12. Consider carefully the essential "relativistic invariance assumption" alluded to in Problem 11. Can you think of a critical experiment which could verify or refute the assumption?

13. Demonstrate that

$$\frac{d\mathbf{p}}{dt} = m\mathbf{a}_\perp + m\gamma^2\mathbf{a}_\|$$

where m is the mass-energy of a particle and $\mathbf{a}_\perp$ and $\mathbf{a}_\|$ are the components of the particle's acceleration that are perpendicular to and parallel with the direction of its instantaneous motion, respectively. At high speeds, which should be easier—to accelerate a particle perpendicularly to or along its direction of motion?

14. Calculate the relativistic kinetic energy of a single particle to terms of order u^6. What value of u/c would be necessary in order to observe a 10 percent deviation in the kinetic energy from its nonrelativistic value?

15. Compute the motion of a charged particle in a transverse electric field, relative to an observer at rest with respect to the field source, by integrating Equations 8.55 and 8.57. Examine your result for its consistency with the nonrelativistic limit.

16. Find the relation between the proper time for a charged particle in a uniform, constant electric field and that for an observer at rest with respect to the field source. This relationship is the analog of Equation 8.49.

17. Show that $T_{LAB} = \Delta E_D + \Delta E_R$ regardless of magnitudes of the velocities of the particles involved in an elastic scattering process.

18. Demonstrate that Equations 8.74 and 8.75 may be rewritten

$$\tilde{p}_A = \tilde{m}_A\tilde{v}_A \qquad \tilde{p}_B = \tilde{m}_B\tilde{v}_B$$

respectively.

19. An electron of incident momentum 2 BeV/c scatters elastically through a head-on collision with a resting proton. What percent of the initial energy of the electron is lost in the scattering process? The rest energy of the electron is 0.511 MeV and that of the proton is 938 MeV.

20. A theorist wishes to have experimental data on relativistic proton-proton scattering at a CP angle of 30°. Given that the LAB momentum of the incident proton is to be 800 MeV/c, for what LAB scattering angle should the empirical data be sought?

SPECIAL TOPIC 4

The Relativistic Kepler Problem

It is not possible to describe completely the Kepler problem within the context of the special Principle of Relativity because the relative, bounded motion of two gravitating bodies must be observed from an accelerated frame, and therefore, is outside the province of the theory we have understood to this point. What is necessary in order to do full justice to the problem is Einstein's *general* relativity principle, a subject beyond the scope of our discussion.[1] On the other hand, a consideration of the simplest relativistic aspects of the Kepler problem, employing only the *special* relativity principle, can be quite enlightening because the results one obtains from the theory turn out to be in perfect qualitative agreement with those achieved through the rigorous formulation. Thus we shall not at all be wasting time in extending the discussion of Section 4.6 to permit an accounting of planetary motion when the velocity dependence of the mass cannot be neglected.

To avoid the rather complex situation encountered[2] (because of the finite speed of propagation for gravitational influences) when both the sun and the planet are moving relative to some frame of reference, we shall assume that the sun is at rest and is stationed at the origin of the coordinate system in which the position of the planet is specified. The relativistic form of the Second Law for the motion of the planet, Equation 8.38, is accordingly

$$\frac{d}{dt}(\bar{\gamma}M_p\mathbf{u}) = -\frac{GM_s\bar{\gamma}M_p}{r^2}\,\hat{\mathbf{r}}$$

or, in a more compact form,

$$\frac{d\mathbf{p}}{dt} = -\frac{K}{r^2}\,\hat{\mathbf{r}} \tag{S.1}$$

where M_s and M_p are rest masses and $\hat{\mathbf{r}}$ is a unit vector lying along the line of centers of the sun and planet. It should be noted that the relativistic factor, $\bar{\gamma} = [1 - (u^2/c^2)]^{-1/2}$, appears in the expression for the gravitational force because of the presumed equivalence of the inertial mass with the gravitational mass of the planet.[3] The form of Equation S.1 indicates that the axial vector $\mathbf{r} \times d\mathbf{p}/dt$ vanishes and, therefore, that the relativistic angular momentum $\mathbf{J} = \mathbf{r} \times \mathbf{p}$ is a constant of the motion, in complete analogy with the result found in Section 4.6. It follows that the motion of the planet takes place entirely in a fixed plane and that the problem reduces, as before, to a two-dimensional one. This fact, in turn, suggests that it should be profitable to consider the relativistic version of the Runge-Lenz vector, which is defined to have the coordinates

$$A_1 = \frac{x_1}{r} - \frac{Jp_2}{K_0M_p}$$

$$A_2 = \frac{x_2}{r} + \frac{Jp_1}{K_0M_p} \tag{S.2}$$

[1] For a good introductory discussion of the general relativity principle, see M. Born, *Einstein's Theory of Relativity*, Dover Publications, New York, 1965, Chapter VII.
[2] The difficulties encountered when the sun moves are described in J. L. Synge, *Relativity: The Special Theory*, North Holland, Amsterdam, 1965, p. 395.
[3] Sometimes K in Equation S.1 is replaced by its nonrelativistic limit $K_0 \equiv GM_sM_p$, whereupon the factor of 2 appearing in Equation S.5 disappears. The possible inconsistency in this approach has been discussed by R. Engelke and C. Chandler, Planetary perihelion precession with velocity-dependent gravitational mass, *American Journal of Physics* 38:90–93 (1970).

In Equations S.2, J is the (constant) magnitude of the relativistic angular momentum, p_i ($i = 1, 2$) is a coordinate of the relativistic momentum, and $K_0 = GM_sM_p$. Proceeding as in Section 4.6, we find

$$\frac{dA_1}{dt} = \frac{Jx_2}{r^3}\left(\frac{K}{K_0M_p} - \frac{1}{\bar{M}_p}\right)$$

$$\frac{dA_2}{dt} = -\frac{Jx_1}{r^3}\left(\frac{K}{K_0M_p} - \frac{1}{\bar{M}_p}\right)$$

where $\bar{M}_p \equiv \bar{\gamma}M_p$ is the relativistic mass of the planet. It is rather clear that $\mathbf{A}$ is *not* a constant of the motion except in the nonrelativistic limit, $K \to K_0, \bar{M}_p \to M_p$. Evidently, the velocity dependence of the planetary mass is sufficient to destroy the "accidental degeneracy" of the total energy, and with it, the constancy of $\mathbf{A}$. Since the length of $\mathbf{A}$ is the orbital eccentricity and its direction is along the semimajor axis, we infer that the planetary ellipses are distorted in some way in the relativistic domain. The exact expression for the length of $\mathbf{A}$ is in fact

$$\|\mathbf{A}\| = (A_1{}^2 + A_2{}^2)^{1/2} = \left[1 + \frac{2}{K_0{}^2}\frac{J^2}{M_p}\left(\frac{p^2}{2M_p} - \frac{K_0}{r}\right)\right]^{1/2} \tag{S.3}$$

The quantity in parentheses on the right-hand side of (S.3) is not the total energy

$$E = mc^2 - \frac{K}{r}$$

and, therefore, is not generally a constant of the motion. But, in practical terms, $p^2/2M_p$ and K_0 differ from their relativistic counterparts only by quantities the order of $(u/c)^2$ or less. At most the value of $(u/c)^2 \simeq 10^{-8}$ for any of the known planets, and the difference is ignorable. Therefore $\|\mathbf{A}\| \simeq e$, even in the practical relativistic case. Given this conclusion, we suggest further that the primary effect of the velocity dependence of the planetary mass must be the change in *direction* of $\mathbf{A}$ as time passes. Because $\mathbf{A}$ points from the sun to the orbital aphelion (see again Figure 4.10), we see that *the planetary ellipse precesses in the plane of motion,* as shown in Figure 8.8. The magnitude of this precession can be calculated approximately in the following way. First we note that

$$\left\|\frac{d\mathbf{A}}{dt}\right\| = \left[\left(\frac{dA_1}{dt}\right)^2 + \left(\frac{dA_2}{dt}\right)^2\right]^{1/2}$$

$$= \left[\frac{J^2(x_1{}^2 + x_2{}^2)}{r^6}\left(\frac{K}{K_0M_p} - \frac{1}{\bar{M}_p}\right)^2\right]^{1/2}$$

$$= \frac{J}{r^2}\left(\frac{K}{K_0M_p} - \frac{1}{\bar{M}_p}\right) = \frac{J}{M_pr^2}\left(\bar{\gamma} - \frac{1}{\bar{\gamma}}\right) \simeq \frac{J}{M_pc^2}\left(\frac{u}{r}\right)^2 \tag{S.4}$$

upon expanding $\bar{\gamma}$ to first order in $(u/c)^2$. With $\|\mathbf{A}\|$ fixed, Equation S.4 must express the rate of rotation of $\mathbf{A}$ in the orbital plane. To calculate the average rate of

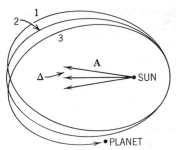

FIGURE 8.8. Relativistic precession of a planetary orbit.

rotation we write

$$\left\langle \left\| \frac{d\mathbf{A}}{dt} \right\| \right\rangle = \frac{1}{T} \int_0^T \left\| \frac{d\mathbf{A}}{dt} \right\| dt \simeq [a^2(1 - e^2)^{1/2}]^{-1} \frac{J}{2\pi M_p c^2} \int_0^{2\pi} u^2 \, d\varphi$$

where T is the orbital period. Here we are assuming that the precession is small enough to justify averaging over the period of an undistorted and, therefore, closed, orbit. (The last equality is obtained from $J = \mu r^2 \dot{\varphi}$ and Equations 4.76 and 4.81.) Now,

$$\frac{1}{2\pi} \int_0^{2\pi} u^2 \, d\varphi \simeq \frac{1}{2\pi} \int_0^{2\pi} \frac{2}{M_p} \left(E + \frac{K_0}{r(\varphi)} \right) d\varphi = \frac{2E}{M_p} + \frac{K_0}{\pi M_p} \int_0^{2\pi} r^{-1}(\varphi) \, d\varphi$$

$$= \frac{2E}{M_p} + \frac{2K_0^2}{J^2} = -\frac{2E}{M_p}(1 - 2e^2) \simeq -\frac{2E}{M_p}$$

if we approximate

$$E \simeq \tfrac{1}{2} M_p u^2 - \frac{K_0}{r}$$

and put Equations 4.76 and 4.81 to work. Therefore

$$\left\langle \left\| \frac{d\mathbf{A}}{dt} \right\| \right\rangle \simeq \frac{-2JE}{M_p^2 c^2 a^2 (1 - e^2)^{1/2}} = \left(\frac{K_0}{M_p} \right)^{3/2} a^{-5/2} c^{-2}$$

upon using Equations 4.81 and 4.83. During a single revolution the aphelion (or, as is more customarily measured, the perihelion) is displaced by the angle

$$\Delta_s \equiv T \left\langle \left\| \frac{d\mathbf{A}}{dt} \right\| \right\rangle \simeq 2\pi \left(\frac{M_p}{K_0} \right)^{1/2} a^{3/2} \cdot \left(\frac{K_0}{M_p} \right)^{3/2} a^{-5/2} c^{-2}$$

$$= \frac{2\pi G M_S}{a c^2} \tag{S.5}$$

where Equation 4.83 has been employed to express T. The only dependence of Δ_s on the properties of the planet is through the semimajor axis a. For the case of the planet mercury, whose orbit exhibits the smallest value of a, we find, using the data in Table 4.1, $\Delta_s \simeq 14$ seconds per century. This value is about one third that of the observed relativistic precession of the orbit. Indeed, Equation S.5 differs almost exactly by a factor of 3 from the accurate, general relativistic expression[4]:

$$\Delta_G \simeq \frac{6\pi G M_s}{ac^2(1 - e^2)} \tag{S.6}$$

A comparison of (S.6) with (S.5) indicates that the special relativistic calculation at least contains most of the essential physics that appears in the more rigorous theory. In terms of the primary qualitative effect of the relative motion—precession—the two calculations are in complete accord.

---

[4] See J. B. Marion, *Classical Dynamics of Particles and Systems*, Academic Press, New York, 1970, §8.10, for a heuristic derivation of Equation S.6.

9

Hamilton's Principle

9.1 MINIMUM PRINCIPLE FOR DYNAMICS

In the preceding chapters the motions of particles subject to a variety of forces have been discussed according to the prescription of the Second Law. The results obtained, in addition to their obvious practical value as approximate descriptions of the real world, have provided a fairly precise understanding of the relation between the mathematical forms of the position and velocity as functions of the time and the character of the force that generates them. Thus we know, for example, that in the nonrelativistic limit a uniform field causes the position of a particle moving within it to be at most a quadratic function of the time, since higher powers of t are not compatible with a constant acceleration. On the other hand, the position of a harmonic oscillator exhibits periodic behavior as a function of time because the force to which it is subject always acts in a direction opposed to that of the displacement from the point of equilibrium. Although these well-known examples are evidence for the efficacy of the Second Law in understanding the causes of motion, they should not be regarded as demonstrating that the Newtonian Program is unique. For, as we well know through the discussion in previous chapters, an alternative formulation of dynamics is sometimes possible in terms of the concept of energy. If the motion is restricted to a single spatial dimension and the total energy

$$E = \tfrac{1}{2}m(v(t))^2 + V(x(t))$$ (9.1)

is defined, the position $x(t)$ may be determined without recourse to the Second Law by employing quadrature:

$$t = \int_{x(0)}^{x(t)} \left[\frac{2}{m} (E - V(x)) \right]^{-1/2} dx$$ (9.2)

In addition to the purely manipulative value of Equation 9.1, there is the physical value derived from that expression through its use to portray the motion of a single particle as a continual exchange of kinetic for potential energy, and the reverse. Considering once again the motion in a uniform field, we may describe the situation in which a turning point is observed in terms of a kinetic energy exhausted at first to provide the maximum possible potential energy, then replenished quickly as the potential energy diminishes in response to the pull of the applied force. It follows from this method of description and from the explicit mathematical forms of $T(v)$ and $V(x)$ in this example that the velocity will go to zero as the position takes on an extremum value, in agreement with the definition of a turning point, then will increase as the position returns to its initial value. It is evident that we have here a somewhat more tangible account of the behavior, expressed in terms of a property of the moving particle—its energy, than that to be obtained through a direct mathematical examination of $x(t)$ calculated using the Second Law.

The physical meaning of the energy concept was, of course, discussed in Chapters 2 and 4. What we should like to point out at this time is that the importance of the concept is far greater than what might be imagined solely on the basis of expressions such as (9.2). In particular, we shall endeavor to show now that the energy concept is of sufficient generality to be employed in the mathematical structure of classical dynamics not only as an alternative to the Second Law, *but even as the foundation of dynamics itself*. The truth of this statement will be demonstrated rigorously in Section 9.2. In anticipation of that proof we shall direct our attention to some physical arguments which can further illuminate the Energy Theorem and provide a motivation for the new principle soon to be uncovered.

Consider a particle initially thrown straight upward in the Earth's gravitational field under conditions in which friction is unimportant. The total energy of the particle is given by

$$E = \tfrac{1}{2}mv^2 + mgx$$

and is a constant of the motion. For the purpose of illustration we shall assume that the zero of potential energy has been chosen to be at the initial position of the particle. Then the total energy of the particle at the outset is wholly kinetic energy. As the motion proceeds, this kinetic energy is converted to potential energy until it is completely exhausted. Then the particle suddenly "turns around" and begins to increase its kinetic energy at the expense of potential energy, the increase of the former and depletion of the latter continuing until the particle abruptly collides with the ground. If we analyze these observations and try to remain entirely within the scope of the energy concept, we might be led to entertain the figurative notion that the particle is acting at all times to prevent, on the average, its energy from being "too much" kinetic or "too much" potential while yet maintaining the sum of these two kinds of energy absolutely fixed. At the start of its flight, the particle moves to deprive the kinetic energy from what is evidently an excessive proportion of the total energy. But the particle overdoes the job and reduces that proportion to zero. Now it is the potential energy which is excessive and must be diminished. The particle accordingly turns around and falls in order to try to restore the balance. Once more, however, it exceeds the mark and loses *all* of the potential energy it had. But this time it does not bring about parity by traveling on a bit and turning around, because more falling serves to *decrease* the potential energy. Instead, the particle takes the only alternative left to it and continually increases the kinetic energy in order to compensate the reduction in its potential energy and thereby maintain balance, on the average.

The behavior of the harmonic oscillator can be analyzed in the same way. If we assume that initially the particle is moving and is at the equilibrium point, we have again the situation wherein the total energy is entirely kinetic energy. This condition evidently is an anathema to the particle, and it quickly attempts to reduce the kinetic energy and gather potential energy. It gathers too much, however, and achieves a reduction of its velocity to zero. This is also unwanted, so the particle turns around to promote a return to a more equable distribution of energy. Overcompensation occurs still another time, but now, because of the form of $V(x)$, there is the possibility of rectifying matters by moving to the side of the equilibrium point opposite to that from which it has just come. Doing this, the particle increases its potential energy to a maximum again, turns around, and returns to the equilibrium point to begin the motion anew. Thus we are led to view the behavior of this particle, and any whose total energy is conserved, as the result of a competition between kinetic and potential energy wherein the motion is always such as to prevent either kind of energy from permanently achieving a dominant share of what is available. Indeed, we may regard the periodic motion of the oscillator to be simply that particle's successful if frenetic attempt to maintain its kinetic and potential energies, at least on average, equal to one another.

If we suppose that the foregoing illustrative remarks are generally true, then it appears that we have a new insight as to the basis of at least a part of classical dynamics. It may be reasonable to propose that *the conservative motion of a particle tends to keep the kinetic and potential energies equal to one another on the average.* Put another way, the motion attempts to give the quantity

$$\tfrac{1}{2}m(v(t))^2 - V(x(t))$$

a *minimum* value when this difference in energy is averaged over the time the particle takes to go between two chosen points on its path. Our new principle, then, says something like: "the particle moves so as to keep the difference between its kinetic and potential energies the least possible, on the average." We can translate this statement into mathematical terms by insisting that the actual path traveled by a particle between times t_1 and t_2 give the integral

$$S \equiv \int_{t_2}^{t_1} \left[\tfrac{1}{2}m(v(t))^2 - V(x(t))\right] dt \qquad (9.3)$$

which, for a given time interval, is proportional to the mean difference of $T(v)$ and $V(x)$, its smallest possible value. Hypothetically, if the particle were to travel along some path different from the physically permitted one, that is, if its hypothetical position $\hat{x}(t)$ differed to any extent from the "true" position $x(t)$, then the integral would presumably have a value greater than its least possible value (see Figure 9.1). But if this is the case, then we have a means for selecting the true position function, for a chosen $V(x)$, from out of the multitudes of possible mathematical forms. For, if $S_{\min}$ is the least possible value of the integral S for a given

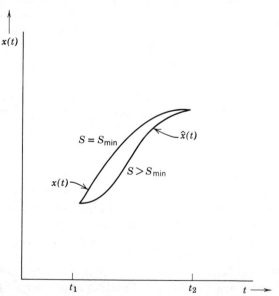

FIGURE 9.1. The "true" path of a particle $x(t)$ compared with a hypothetical path $\hat{x}(t)$. The Action Integral S is least for the true path.

$V(x)$, then the value of the integral computed with a trial $\hat{x}(t)$ should not show a difference from $S_{\min}$, to some first order of approximation, if $\hat{x}(t)$ is the right choice. This statement is just a generalization of the simplest (necessary) condition for a minimum: A quantity like S can have a minimum value for the position function that makes some kind of first-order difference vanish.

Let us try to be a little more precise about this. Suppose we define a trial position function by

$$\hat{x}(t) \equiv x(t) + \eta(t) \qquad (9.4)$$

where $\eta(t)$ is a function of small value that vanishes at the endpoints of S, but which is otherwise arbitrary. We are not saying that $\hat{x}(t)$ has the most general form conceivable for a trial position function. Instead, we have prescribed it to differ from the true position function $x(t)$ by a small amount in order to see clearly the implication of minimizing S. Thus we expect to find a condition on $x(t)$ that derives from making the difference between $S_{\min}$ and S computed with $\hat{x}(t)$ vanishingly small, in the first order of approximation, for small $\eta(t)$. To accomplish this task we shall "expand" S about $x(t)$ to first order in $\eta(t)$ and require that the first-order term be identically zero. Thus we write

$$\left(\frac{d\hat{x}}{dt}\right)^2 = \left(\frac{dx}{dt} + \frac{d\eta}{dt}\right)^2 \simeq \left(\frac{dx}{dt}\right)^2 + 2\frac{dx}{dt}\frac{d\eta}{dt}$$

and

$$V(\hat{x}) \simeq V(x) + \left(\frac{dV}{d\hat{x}}\right)_{\hat{x}=x} \eta(t)$$

to terms of first order in $\eta(t)$. If we put these expressions into Equation 9.3 we find

$$S \simeq \int_{t_1}^{t_2} \left[\frac{1}{2}m\left(\frac{dx}{dt}\right)^2 - V(x) + m\frac{dx}{dt}\frac{d\eta}{dt} - \left(\frac{dV}{d\hat{x}}\right)_{\hat{x}=x}\eta(t)\right] dt$$

or

$$\delta S \equiv S - S_{\min} \simeq \int_{t_1}^{t_2} \left[m\frac{dx}{dt}\frac{d\eta}{dt} - \left(\frac{dV}{d\hat{x}}\right)_{\hat{x}=x}\eta(t)\right] dt \qquad (9.5)$$

The quantity δS is defined to be the difference between S and its minimum value, to a first order of approximation in $\eta(t)$. We can greatly simplify the expression (9.5) by a partial integration of the first term on the right-hand side:

$$\int_{t_1}^{t_2} \frac{dx}{dt}\frac{d\eta}{dt} dt = \int_{t_1}^{t_2} \left[\frac{d}{dt}\left(\frac{dx}{dt}\eta(t)\right) - \frac{d^2x}{dt^2}\eta(t)\right] dt$$

$$= \left(\frac{dx}{dt}\right)\eta(t)\Big|_{t_1}^{t_2} - \int_{t_1}^{t_1} \frac{d^2x}{dt^2}\eta(t)\, dt$$

$$= -\int_{t_1}^{t_2} \frac{d^2x}{dt^2}\eta(t)\, dt$$

the last step coming from the condition $\eta(t_2) = \eta(t_1) = 0$. With this result put in, Equation 9.5 becomes

$$\delta S \simeq -\int_{t_1}^{t_2} \left[m\frac{d^2x}{dt^2} + \left(\frac{dV}{d\hat{x}}\right)_{\hat{x}=x} \right] \eta(t)\, dt \tag{9.6}$$

If we are going to insist that δS vanish, then we are saying that the integral of the quantity in square brackets above multiplied by the more or less arbitrary function $\eta(t)$ must be zero identically. In general we have absolutely no guarantee that $\eta(t)$ will have just the right mathematical form to produce a zero value for the integral, even though it is always very small. It appears as if we are forced to require that the terms in the brackets be identically zero if δS is to disappear in all circumstances. But this requirement means that $x(t)$ is the solution of the differential equation

$$m\frac{d^2x}{dt^2} = -\left(\frac{dV}{d\hat{x}}\right)_{\hat{x}=x} = F(x) \tag{9.7}$$

the last equality coming from Equation 1.29. Equation 9.7 leads to the remarkable conclusion that *the position function that produces the least difference between the kinetic and potential energies on the average is the one that solves the Second Law.* Thus the energy concept has provided us with a minimum principle for dynamics that is not only in agreement with the Newtonian Second Law of Motion, but can be its foundation.

9.2 HAMILTON'S PRINCIPLE FOR CONSERVATIVE MOTION

Our discussion thus far has served to give a basic, qualitative idea of what form a minimum principle for dynamics ought to have. What is lacking now is the mathematical sharpness that would derive from possessing a rigorous understanding of what it means to insist that an integral like S in Equation 9.3 be an extremum with respect to a chosen position function. The argument that has been given already is suggestive, but it is not clear that a Taylor expansion of an integrand about a function has a precise meaning or that it is equivalent to some kind of expansion of an integral. Therefore, we should do well at this time to present a sound mathematical description of the minimum principle by restating and extending the results of the previous section in a set of theorems. The first of these is a pair of rather general lemmas whose value in proving our main result will soon be apparent.

Lemma 1 Let $f(x(\alpha, t), v(\alpha, t))$ be a continuous and differentiable function of its arguments, where

$$v(\alpha, t) = \left(\frac{\partial x}{\partial t}\right)_{\alpha}$$

and α is an arbitrary parameter. If $x(\alpha, t)$ and $v(\alpha, t)$ are also continuous and differentiable functions of α, and if the integral

$$I(\alpha) = \int_{t_1}^{t_2} f(x(\alpha, t), v(\alpha, t)) \, dt$$

exists, then

$$\frac{dI}{d\alpha} = \int_{t_1}^{t_2} \left[\frac{\partial f}{\partial x} \frac{\partial x}{\partial \alpha} + \frac{\partial f}{\partial v} \frac{\partial v}{\partial \alpha} \right] dt$$

The proof of this lemma generally can be found in textbooks on elementary calculus.[1] Its essential ingredients are the continuity of $I(\alpha)$ and the chain rule for partial derivatives.

Lemma 2 Let $f(x)$ be a continuous function defined on the closed interval $[x_1, x_2]$ and let

$$\int_{x_1}^{x_2} f(x)\varepsilon(x) \, dx = 0$$

where $\varepsilon(x)$ is any function that is continuous and twice-differentiable on $[x_1, x_2]$ and that vanishes at the endpoints. Then $f(x) \equiv 0$.

Proof Let us suppose that the lemma is false, that $f(x)$ does *not* vanish at some point x_0 in $[x_1, x_2]$, and that at this point $f(x_0) > 0$. Since $f(x)$ is continuous, this stipulation means there is some open interval (a, b), where $x_1 \leqslant a < b \leqslant x_2$, which contains x_0 and in which $f(x) > 0$. Now let us put

$$\varepsilon(x) = \begin{cases} [(b - x)(x - a)]^3 & a < x < b \\ 0 & x_1 \leqslant x < a, b < x \leqslant x_2 \end{cases}$$

We have chosen here a mathematical representation of $\varepsilon(x)$ that meets all of the conditions that function is supposed to satisfy. Thus we can write

$$\int_{x_1}^{x_2} f(x)\varepsilon(x) \, dx = \int_a^b f(x)[(b - x)(x - a)]^3 \, dx > 0$$

the inequality coming from the fact that both $f(x)$ and the chosen $\varepsilon(x)$ are positive-valued functions. But this inequality contradicts our hypothesis that the integral is zero. It follows that we cannot have $f(x) > 0$ anywhere on $[x_1, x_2]$. In a similar way we can show that $f(x) < 0$ is also impossible and the lemma is proved.

These lemmas make it possible to state and prove the theorem which epitomizes the results in Section 9.1. In its most general form this theorem is called *Hamilton's Principle*, after W. R. Hamilton, the mathematician who first proposed it in 1834.[2]

---

[1] See, for example, J. M. H. Olmstead, *Calculus with Analytic Geometry*, Appleton-Century-Crofts, New York, 1966, Vol. II, pp. 406f, 387f.
[2] W. R. Hamilton, on a general method in dynamics, *Philosophical Transactions of the Royal Society* (1834), II, pp. 247–308.

We shall give here a restricted version of Hamilton's Principle, which is applicable to the conservative motion of a single particle. The more general statements of the theorem will be developed in later sections.

Hamilton's Principle The motion of a single particle whose total energy is conserved always traces the path $x(t)$ that makes the *Action Integral*

$$S = \int_{t_1}^{t_2} L(x(t), v(t))\, dt$$

have a stationary value, where

$$L(x(t), v(t)) \equiv T(v(t)) - V(x(t)) \tag{9.8}$$

Proof We shall prove the theorem by giving a precise definition of the term "stationary" and then showing that this definition leads directly to the condition that $x(t)$ satisfy the Second Law. To accomplish the first task we shall appeal to the branch of mathematics known as the calculus of variations,[3] which was conceived to solve problems of just the kind before us.

Following the precepts of variation theory, we begin by defining the one-parameter family of *comparison functions*

$$x(\alpha, t) = x(t) + \alpha \varepsilon(t)$$

$$v(\alpha, t) = v(t) + \alpha \frac{d\varepsilon}{dt} \tag{9.9}$$

where $x(t)$ and $v(t)$ are the actual position and velocity, respectively, of a particle subject to the potential function $V(x(t))$, α is a positive parameter, and $\varepsilon(t)$ is an arbitrary function of the time that is continuous and twice differentiable and that vanishes at $t = t_1$ and $t = t_2$. According to Equations 9.9, the comparison function $x(\alpha, t)$ reduces to the true position function when $\alpha = 0$. If α is very small, $x(\alpha, t)$ is the same as $\hat{x}(t)$ given in Equation 9.4, with $\eta(t) = \alpha \varepsilon(t)$. By partitioning η into α and $\varepsilon(t)$ we shall be able to make a rigorous expansion of the action integral without restricting the mathematical form of the difference $x(\alpha, t) - x(t)$ other than to stipulate its vanishing at the endpoints t_1 and t_2. Now we form the integral

$$S(\alpha) \equiv \int_{t_1}^{t_2} L(x(\alpha, t), v(\alpha, t))\, dt$$

It is evident that $S(0)$ is just the integral we wish to be stationary. Moreover, because $L(x(\alpha, t), v(\alpha, t))$ depends continuously on the parameter α, we can employ Lemma 1 to give meaning to the expansion

$$S(\alpha) = S(0) + \left(\frac{dS}{d\alpha}\right)_{\alpha=0} \alpha + \cdots \tag{9.10}$$

---

[3] Those wishing a fuller discussion of this subject than is given here might consult L. A. Pars, *An Introduction to the Calculus of Variations*, Wiley, New York, 1962.

where

$$\left(\frac{dS}{d\alpha}\right)_{\alpha=0} = \int_{t_1}^{t_2} \left[\frac{\partial L}{\partial x}\,\varepsilon(t) + \frac{\partial L}{\partial v}\,\frac{d\varepsilon}{dt}\right] dt \qquad (9.11)$$

Equation 9.11 comes directly from Lemma 1, the definition of the function L, and Equations 9.9. The form of the latter when $\alpha = 0$ should make it clear that the integrand in (9.11) contains L as defined in Equation 9.8 with the *actual* position and velocity of the particle, respectively. Equation 9.10 has the form of a Maclaurin expansion of $S(\alpha)$. The general coefficient

$$\left(\frac{d^n S}{d\alpha^n}\right)_{\alpha=0} \equiv \delta^n S$$

is called the *nth variation of the integral* $S(\alpha)$. By analogy with the theory of ordinary differentiable functions of a single variable, we can say that $S(\alpha)$ possesses an extremum at $\alpha = 0$ or is *stationary* if its first variation is zero. The mathematical expression of this condition is, therefore,

$$\left(\frac{dS}{d\alpha}\right)_{\alpha=0} = 0 \qquad (9.12)$$

or

$$\delta S = \int_{t_1}^{t_2} \left[\frac{\partial L}{\partial x}\,\varepsilon(t) + \frac{\partial L}{\partial v}\,\frac{d\varepsilon}{dt}\right] dt = 0 \qquad (9.13)$$

according to Equation 9.11. Equation 9.13 is the mathematically rigorous version of Equation 9.5. To see that it will lead to a differential equation for $x(t)$, we must first integrate the term in $d\varepsilon/dt$ by parts:

$$\int_{t_1}^{t_2} \frac{\partial L}{\partial v}\,\frac{d\varepsilon}{dt}\,dt = \frac{\partial L}{\partial v}\,\varepsilon(t)\,\bigg|_{t_1}^{t_2} - \int_{t_1}^{t_2} \frac{d}{dt}\left(\frac{\partial L}{\partial v}\right)\varepsilon(t)\,dt$$

$$= -\int_{t_1}^{t_2} \frac{d}{dt}\left(\frac{\partial L}{\partial v}\right)\varepsilon(t)\,dt$$

and rewrite (9.13) as

$$\int_{t_1}^{t_2} \left[\frac{d}{dt}\left(\frac{\partial L}{\partial v}\right) - \frac{\partial L}{\partial x}\right]\varepsilon(t)\,dt = 0 \qquad (9.14)$$

Since $\varepsilon(t)$ is an arbitrary function of the time, not restricted to being small in value, and so on, an application of Lemma 2 can be made. Equation 9.14 can be satisfied only if the term in square brackets in the integrand equals zero *identically*. (We have assumed that all of the derivatives shown in the integrand actually exist, as they must for a physically meaningful problem.) Therefore, we have

$$\frac{d}{dt}\left(\frac{\partial L}{\partial v}\right) - \frac{\partial L}{\partial x} \equiv 0 \qquad (9.15)$$

as the differential equation for $x(t)$ which must be satisfied if $S(0)$ is to be stationary.

If we note, finally, that Equation 9.8 implies

$$p = \frac{\partial L}{\partial v} \tag{9.16}$$

and

$$\frac{\partial L}{\partial x} = -\frac{dV}{dx} = F(x)$$

Equation 9.15 reduces to the Second Law and the theorem is proved.

Equation 9.15 is the general condition that Hamilton's Principle imposes on $x(t)$ through the function $L(x(t), v(t))$ defined in Equation 9.8. This latter function is called the *Lagrangian* of the particle, after the mathematician Joseph Louis Lagrange who, in 1788, presented a comprehensive treatise on mechanics[4] based on a general form of Equation 9.15. Thus Equation 9.15 has come to be called the *Lagrangian equation of motion*, for a single particle whose total energy is conserved, and Equation 9.16 is the Lagrangian definition of the linear momentum.

To get a better feeling for the physical meaning of Equation 9.15 and its relation to Hamilton's Principle, let us consider the motion of a particle in a uniform field. The Lagrangian in this instance is

$$L(x, v) = \tfrac{1}{2}mv^2 + ma_0 x \tag{9.17}$$

where a_0 is the (constant) acceleration due to the field. According to (9.15), the equation of motion is

$$\frac{d}{dt}(mv) - ma_0 = 0$$

or

$$\frac{dp}{dt} = ma_0 \tag{9.18}$$

which, of course, we recognize as the appropriate form of the Second Law. The physical consequences of (9.18) as regards the form of $x(t)$ are well known to us through the discussion in Section 2.1. More important at present is the implication of the equation for the integral S. If we introduce the comparison functions $x(\alpha, t)$ and $v(\alpha, t)$ into the integral $S(\alpha)$, we find

$$S(\alpha) = \int_{t_1}^{t_2} (\tfrac{1}{2}mv^2 + ma_0 x)\, dt + \alpha \int_{t_1}^{t_2} \left(mv\frac{d\varepsilon}{dt} + ma_0 \varepsilon(t) \right) dt + \tfrac{1}{2}m\alpha^2 \int_{t_1}^{t_2} \left(\frac{d\varepsilon}{dt}\right)^2 dt$$

$$= S(0) + \alpha mv(t)\varepsilon(t) \Big|_{t_1}^{t_2} - \alpha \int_{t_1}^{t_2} \left(\frac{dp}{dt} - ma_0\right) \varepsilon(t)\, dt + \tfrac{1}{2}m\alpha^2 \int_{t_1}^{t_2} \left(\frac{d\varepsilon}{dt}\right)^2 dt$$

[4] J. L. Lagrange, *Mecanique Analytique*, A. Blanchard, Paris, France, 1965. Lagrange, of course, did not achieve his results through Hamilton's Principle, but instead derived them from a version of D'Alembert's formulation of the Second Law. For a discussion of this work, see R. Dugas, *A History of Mechanics*, Éditions du Griffon, Switzerland, 1955, Chapter XI.

where, in order to get the second step, we have done a partial integration. Since $\varepsilon(t)$ vanishes identically at t_1 and t_2, and because of Equation 9.18, we may rewrite the expression above as

$$S(\alpha) - S(0) = \tfrac{1}{2}m\alpha^2 \int_{t_1}^{t_2} \left(\frac{d\varepsilon}{dt}\right)^2 dt \geqslant 0 \tag{9.19}$$

This inequality demonstrates that the actual path taken by the particle in a uniform field always imparts a *minimum* value to the integral $S(\alpha)$. Any other hypothetical position function defined by means of comparison functions would provide a greater value of $S(\alpha)$ than does the "true" one. Thus Equation 9.19 substantiates directly the extremum character of the Action Integral.

We can see in a more concrete way the minimum character of S if we do not limit ourselves to trial functions in the form of the comparison function $x(\alpha, t)$ in Equations 9.9. For example, consider the three possibilities

$$x_1(t) = \tfrac{1}{2}a_0 t$$

$$x_2(t) = \tfrac{1}{2}a_0 t^2$$

$$x_3(t) = \tfrac{1}{2}a_0 t^3$$

subject to $x(t = 1\ \text{sec}) = \tfrac{1}{2}a_0$ meters and $0 \leqslant t \leqslant 1$ sec. Now,

$$L_1(x_1, v_1) = \frac{ma_0{}^2}{2}(\tfrac{1}{4} + t)$$

$$L_2(x_2, v_2) = ma_0{}^2 t^2$$

$$L_3(x_3, v_3) = \frac{ma_0{}^2}{2}(\tfrac{9}{2}t^4 + t^3)$$

upon applying Equation 9.17, and

$$S_1 = \int_0^1 L_1(x_1, v_1)\, dt = \tfrac{3}{8}ma_0{}^2$$

$$S_2 = \int_0^1 L_2(x_2, v_2)\, dt = \tfrac{1}{3}ma_0{}^2$$

$$S_3 = \int_0^1 L_3(x_3, v_3)\, dt = \tfrac{37}{30}ma_0{}^2$$

It is obvious that the correct result, $x_2(t)$, yields in this case the least value of S.

The example discussed here brings up a point that has been overlooked until now. Our statement of Hamilton's Principle makes it evident that we are not necessarily imparting a *minimum* value to the integral S when we determine $x(t)$ from (9.15) but, instead, that we can be assured of only an extremum value for the integral. This is as it should be, since the vanishing of the first variation of S is only a necessary condition for a minimum, not a sufficient one. On the other hand, the case of motion in a uniform field suggests that S may very well be minimized, rather than maximized, by the "true" position function. The question of a sufficient condition for Hamilton's Principle to produce a minimum value of S is a complex one that involves a careful study of the second variation, $\delta^2 S$. We shall only quote the result of the analysis here, leaving the details for interested readers to seek elsewhere.[5] The necessary and sufficient conditions for the integral

$$\int_{t_1}^{t_2} F(t, x, v)\, dt$$

to be a relative minimum are that it be stationary and that

$$\frac{\partial^2 F}{\partial v^2} \geqslant 0 \tag{9.20}$$

for all values of v along the minimizing $x(t)$. The function $F(t, x, v)$ is presumed to be continuous and to have continuous partial derivatives with respect to x and v. Thus we can translate (9.20) very easily into a physical expression by writing $F(t, x, v) = L(x, v)$ and looking at (9.8). The result is the simple condition

$$m \geqslant 0$$

where m is the mass of the particle. It follows that, for any single particle moving under the influence of a potential function $V(x)$, the integral S must have a minimum value when the Lagrangian equation of motion is satisfied.

9.3 CONSERVED QUANTITIES AND LAGRANGE'S EQUATIONS

The conservation theorems for the motion of a single particle in one dimension are direct and simple results of the Lagrangian definition of momentum (Equation 9.16) and of the equation of motion (9.15). But the physical meaning of these theorems deepens when they are placed into the context of the Lagrangian formalism because they are then seen to present, in addition to computationally useful expressions, an association between the conserved quantities and the *symmetry properties* of motion. Thus we shall come to see in this section how the invariance of the behavior of a particle with respect to certain transformations of the space and time coordinates can lead necessarily to the conservation theorems.

The simplest conserved quantity is the linear momentum, the first theorem for which was stated in Section 1.3. In the present context it takes the form:

The Linear Momentum Theorem   *If the Lagrangian for a single particle is not a function of the position, then the linear momentum of the particle is a constant of the motion.*

---

[5] See, for example, S. G. Mikhlin, *Mathematical Physics, An Advanced Course*, American Elsevier, New York, 1970, Chapter 4.

Proof If $L(x, v)$ does not in fact depend on x, then the equation of motion (9.15) must reduce to

$$\frac{d}{dt}\left(\frac{\partial L}{\partial v}\right) \equiv 0$$

or

$$\frac{dp}{dt} \equiv 0$$

by Equation 9.16.

It is evident that the essential part of this rather simple proof is the condition

$$\frac{\partial L}{\partial x} \equiv 0 \tag{9.21}$$

which states that the Lagrangian is not sensitive to changes in the position of the particle. We can manufacture this condition into an invariance property by considering the *space-displacement transformation*

$$x' = x + a \tag{9.22}$$

where a is a constant of arbitrary magnitude. If we solve Equation 9.22 for x and introduce the result into any Lagrangian $L(x, v)$, a Lagrangian $L'(x', v)$ is created that, in general, has an appearance different from $L(x, v)$. However, the numerical values of the two functions must be the same:

$$L'(x', v) = L(x, v) \qquad (x' = x + a) \tag{9.23}$$

Now consider the Taylor expansion

$$L(x, v) = L(x', v) + \left(\frac{\partial L}{\partial x}\right)_{x=x'} a + \cdots + \frac{1}{n!}\left(\frac{\partial^n L}{\partial x^n}\right)_{x=x'} a^n + \cdots.$$

If $L(x, v)$ satisfies the condition (9.21), this expansion reduces at once to $L(x, v) = L(x', v)$. It then follows that

$$L'(x', v) = L(x', v) \tag{9.24}$$

according to Equation 9.23. Equation 9.24 states that, if we replace x in $L(x, v)$ *directly* with x', we shall find a Lagrangian that is identical with that created by replacing x by $x' - a$. Hence the Lagrangian $L(x, v)$ is invariant under the space displacement (9.22), if $\partial L/\partial x$ vanishes identically.

The foregoing argument suggests that the conservation of linear momentum is related closely to the invariance of the Lagrangian under space displacements. This relationship may be stated:

If the Lagrangian of a single particle is invariant under arbitrary space displacements, the linear momentum of the particle is a constant of the motion.

In physical terms this conclusion tells us that, should the frame of reference in which we observe the motion of a particle be displaced by $-a$ along the negative x direction, changing x to $x + a$, and should we then discover that this action makes no change in the Lagrangian that describes what is observed, we may surely expect to find that the linear momentum of the particle is conserved. The motion we are examining, then, may be said to display *space-displacement symmetry* and this is tantamount to momentum conservation.

The second conservation theorem we shall consider, the Energy Theorem, we know already to be assured because the Lagrangian is defined in Equation 9.8 to contain a potential energy function. Therefore, we have to look at energy conservation in the Lagrangian formalism a little more carefully than we did momentum conservation in order to discover the relation to invariance and symmetry. In some way we shall need to "get behind" the notion of a Lagrangian to see how the symmetry of a conservative motion may be broken to preclude the constancy of the total energy.

We begin our investigation by stating:

The Total Energy Theorem   *If the motion of a single particle is described by the Lagrangian (9.8), then the quantity*

$$p(t)v(t) - L(x(t), v(t))$$

is a constant of the motion and is equal to the total energy of the particle.

Proof   We show the constancy of the quantity by calculating its total time derivative. Thus

$$\frac{d}{dt}\left[p(t)v(t) - L(x(t), v(t))\right] = \frac{dp}{dt}v(t) + p(t)\frac{dv}{dt} - \frac{\partial L}{\partial x}\frac{dx}{dt} - \frac{\partial L}{\partial v}\frac{dv}{dt}$$

$$= \left[\frac{dp}{dt} - \frac{\partial L}{\partial x}\right]v(t) + \left[p(t) - \frac{\partial L}{\partial v}\right]\frac{dv}{dt} \qquad (9.25)$$

where we have employed the chain rule to compute dL/dt. The coefficient of $v(t)$ is just the left-hand side of the equation of motion (9.15) and, therefore, it vanishes identically. The same fate befalls the coefficient of dv/dt, according to Equation 9.16. Thus the total time derivative is zero and the first part of the theorem is proved. Now

$$p(t)v(t) - L(x(t), v(t)) = (mv)v - \tfrac{1}{2}mv^2 + V(x)$$
$$= \tfrac{1}{2}mv^2 + V(x) = E$$

according to Equations 9.8 and 1.30. This proves the Energy Theorem.

The symmetry aspect of energy conservation is not terribly obvious in the proof we have just given. However, in anticipation of the more general discussion to be given in the following section, we can point out that an important ingredient in the chain of equalities (9.25) is the absence of an *explicit* time dependence of the Lagrangian. For, if it had shown such a time dependence, we would have found

$$\frac{d}{dt}\left[p(t)v(t) - L(x(t), v(t), t)\right] = -\frac{\partial L}{\partial t}$$

according to the chain rule, instead of zero. This fact suggests at least that a *necessary* condition for energy conservation is

$$\frac{\partial L}{\partial t} = 0 \qquad\qquad (9.26)$$

Given Equation 9.26, we may develop a logical parallel with our examination of the basis in symmetry of linear momentum conservation. Evidently we can argue that 9.26 implies a relation of the form

$$L'(x'(t'), v'(t')) = L(x'(t'), v'(t'))$$

where $t' = t + \tau$ and τ is an arbitrary constant. This relationship leads us to speculate that energy conservation is allied with the invariance of the Lagrangian under arbitrary *time* displacements. Although we are not fully justified in doing so at present, we shall accept this speculation and write:

> If the Lagrangian of a single particle is invariant under arbitrary time displacements, the total energy of the particle is a constant of the motion.

The physical meaning of this statement is that we shall find the behavior of a particle whose total energy is conserved to be insensitive to the absolute value of the time. In order words, if we measure the position of the particle, for example, it would not matter to our measurement where we set "time zero." The behavior will have the same character, relative to the initial conditions, regardless of the lack of synchronization among the many possible time frames of reference we may choose. This *time-displacement symmetry* we find to be equivalent to energy conservation.

9.4 HAMILTON'S PRINCIPLE FOR NONCONSERVATIVE MOTION

When the motion of a single particle is not conservative we must expect (in light of our experience in Chapter 3) that the dynamical analysis will necessarily depend more on the concept of force than on that of energy. In principle the total energy of a particle is not even a defined quantity whenever the force cannot be said to derive from a potential function. Within the Newtonian program this fact is sufficient to preclude any description in terms of an energic equation [such as (2.10)] and we are compelled to cast all problems involving nonconservative motion into the formalism of the Second Law. On the face of it, a similar difficulty would appear to confront any attempt to extend the Lagrangian formalism, as contained in Hamilton's Principle, to the description of dissipative or driven motion. For, in Equation 9.8, we have made the very definition of the Lagrangian subject to the existence of a potential energy and, therefore, we have made it dependent on the requirement of energy conservation. It would indeed appear from this that the generalization of Hamilton's Principle to the problem of non-conservative forces will not be a trivial matter.

Fortunately the difficulty posed here does not prove to be insuperable. It is possible, as we shall discover, to reformulate Hamilton's Principle in one dimension in a way that provides a basis for the analysis of any motion that falls within the province of the Second Law. In order to accomplish this end, however, we shall have to examine and restate very carefully the method of proof given in Section 9.2, with the aid of sifting from it what is not inextricably bound to the assumption of energy conservation. We shall commence this investigation by considering the notion of a *virtual displacement*: a particle is said to undergo a virtual displacement if its position changes by an arbitrary infinitesimal amount at the time t in a way that is consistent with the forces acting on it *at that instant*. A virtual displacement, then, is an instantaneous one. It differs from a real displacement of a particle in that any real displacement always takes place in a nonzero time interval during which the forces on a particle, in general, *are changing*. Thus the virtual displacement does not correspond directly to reality, although it is a well-defined mathematical entity.

Now let us look once again at the first of Equations 9.9, which relates the true position of a particle, $x(t)$, to a hypothetical position of the particle distinguished by the parameter α:

$$x(\alpha, t) = x(t) + \alpha\varepsilon(t) \tag{9.9}$$

where $\varepsilon(t)$ is the arbitrary function of the time defined previously. We shall assume for the moment that α is an infinitesimally small quantity in order that $x(\alpha, t)$ may be regarded as a virtual displacement of the particle from its true position $x(t)$. In other words, we shall now consider the succession of positions $x(\alpha, t)$ as time passes to be the result of virtually displacing the set of true positions $x(t)$ by $\alpha\varepsilon(t)$ at each instant t (see Figure 9.2). This is a much more precise description of the possible path of a particle than was given in Section 9.2, but in no way does it contradict what was done there in proving Hamilton's Principle.

Given the new interpretation of Equation 9.9 we examine the quantity

$$T(v(t)) + W(x(t))$$

where $T(v(t))$ is *defined* by Equation 1.27 and

$$W(x(t)) \equiv Fx(t) \tag{9.27}$$

the quantity F being just the total force acting on a particle, as described in Chapter 1. If we should impose a virtual displacement on the position (and, consequently, the velocity) the quantity $(T + W)$ becomes

$$T(v(\alpha, t)) + W(x(\alpha, t)) = \tfrac{1}{2}m(v(\alpha, t))^2 + Fx(\alpha, t) \tag{9.28}$$

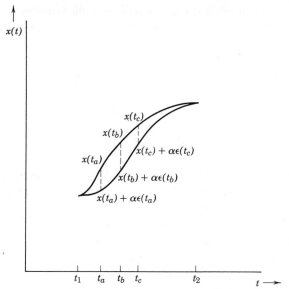

FIGURE 9.2. The hypothetical path $x(\alpha, t)$ as a set of virtual displacements from the true path $x(t)$.

In W, only the position factor changes because a virtual displacement is carried out instantaneously with the total force on the particle held *fixed*, as stated earlier. Bearing this very important point in mind, we can now prove a general version of:

Hamilton's Principle   The motion of a single particle in one dimension always traces out the path $x(t)$ that makes the integral

$$S = \int_{t_1}^{t_2} [T(v) + W(x)] \, dt \tag{9.29}$$

take on a stationary value.

Proof   We begin by forming the integral

$$S(\alpha) = \int_{t_1}^{t_2} [T(v(\alpha, t)) + W(x(\alpha, t))] \, dt$$

and noting that, as before, $S = S(0)$. The requirement that S be stationary implies through Equations 9.9, 9.12, 9.28, and 9.29,

$$\int_{t_1}^{t_2} \left[\frac{\partial T}{\partial v} \frac{d\varepsilon}{dt} + F\varepsilon(t) \right] dt \equiv 0 \tag{9.30}$$

Upon integrating by parts the first term in the integrand of (9.30) we get

$$\int_{t_1}^{t_2} dt \, \varepsilon(t) \left[\frac{d}{dt} \left(\frac{\partial T}{\partial v} \right) - F \right] \equiv 0$$

which, because $\varepsilon(t)$ is an arbitrary function of the time, requires

$$\frac{d}{dt}\left(\frac{\partial T}{\partial v}\right) - F \equiv 0 \qquad (9.31)$$

A glance at Equation 1.27 shows that (9.31) is the same as

$$\frac{dp}{dt} = F(x, v, t) \qquad (9.32)$$

which is the Second Law for an arbitrary force.

It should not be construed from the method of proof presented here that our generalization of Hamilton's Principle has cost us the Lagrangian formulation of dynamics. To see the error in that idea, we need only remark that, in general, $F(x, v, t)$ may be thought of as comprising two parts: one that derives from a "generalized potential function" and one that can never be related to a potential function. Thus we write

$$F(x, v, t) \equiv -\frac{\partial U}{\partial x} + \frac{d}{dt}\left(\frac{\partial U}{\partial v}\right) + Q(x, v, t) \qquad (9.33)$$

where $U(x, v, t)$ is called the *generalized potential function* and is defined by Equation 9.33, and $Q(x, v, t)$ represents all of the forces that cannot be associated with a function such as $U(x, v, t)$. A special case of the generalized potential function is, of course, the potential energy $V(x)$. A velocity-dependent function of the general form of $U(x, v, t)$ arises in the description of the motion of a single charged particle through an electromagnetic field, in three dimensions.[6] Forces that are represented by $Q(x, v, t)$ include the damping forces discussed in Section 3.1. Now we exploit Equation 9.33 by redefining the Lagrangian for a single particle to be

$$L(x, v, t) \equiv T(v) - U(x, v, t) \qquad (9.34)$$

Then Equation 9.31 becomes

$$\frac{d}{dt}\left(\frac{\partial L}{\partial v}\right) - \frac{\partial L}{\partial x} = Q(x, v, t) \qquad (9.35)$$

As a concrete example of Equation 9.35, consider the motion of a single particle in a uniform field subject to Stokes damping. In that case we have

$$U(x, v, t) = V(x) = -ma_0 x$$

$$Q(x, v, t) = F_s(v) = -\beta v$$

$$L(x, v) = T(v) - ma_0 x$$

---

[6] For an introductory discussion of this problem, see T. C. Bradbury, *Theoretical Mechanics*, Wiley, New York, 1968, §7.6.

and the Lagrangian equation of motion is

$$\frac{dp}{dt} - ma_0 = -\beta v(t)$$

which is identical with Equation 3.20.

The general form of the Lagrangian equation of motion (9.35) brings up some very important questions concerning the conservation theorems. First of all, Equation 9.16, which we have called the Lagrangian definition of linear momentum, tells us that, in general, *the Lagrangian momentum of a particle will not be the same as the Newtonian momentum.* To see this clearly, note that

$$\frac{\partial L}{\partial v} = \frac{\partial T}{\partial v} - \frac{\partial U}{\partial v} = mv - \frac{\partial U}{\partial v}$$

according to Equation 9.34. Therefore, in the Lagrangian scheme,

$$p_L = mv - \frac{\partial U}{\partial v} \tag{9.36}$$

while in Newtonian theory, of course, $p = mv$. The appearance of a generalized potential in the Lagrangian makes it possible that the linear momentum will no longer be the simple product mv. We can, of course, remove this difference in concept by renormalizing p_L relative to $\partial U/\partial v$. However, to do so would effectively obscure the significant advance in generality obtained by replacing $V(x)$ by $U(x, v, t)$ in the Lagrangian, since Equation 9.35 would have the awkward form

$$\frac{dp}{dt} - \frac{d}{dt}\left(\frac{\partial U}{\partial v}\right) - \frac{\partial L}{\partial x} = Q(x, v, t)$$

with the associated complication of the Momentum Theorem. If we choose instead to accept the definition (9.16) (and we shall!), Equation 9.35 may be written

$$\frac{dp_L}{dt} - \frac{\partial L}{\partial x} = Q(x, v, t) \tag{9.37}$$

and we may state in generalized form:

The Linear Momentum Theorem *If the Lagrangian for a single particle is not a function of position, and if forces not derivable from a generalized potential function do not act on the particle, then the Lagrangian momentum of the particle is a constant of the motion.*

The proof of this theorem follows directly from its hypotheses and from Equation 9.37. We note that space displacement invariance is now only a *necessary* condition for linear momentum conservation, the complete requirement stipulating the absence of dissipative and driving forces. Moreover, the conserved momentum is not necessarily the Newtonian mv.

In Section 9.2 we derived the Energy Theorem by calculating the first time derivative of the quantity

$$p_L(t)v(t) - L(x(t), v(t), t)$$

here expressed in terms of the generalized momentum and Lagrangian. If we repeat that calculation, we find

$$\frac{d}{dt}\left[p_L(t)v(t) - L(x(t), v(t), t)\right]$$

$$= \left[\frac{dp_L}{dt} - \frac{\partial L}{\partial x}\right]v(t) + \left[p_L(t) - \frac{\partial L}{\partial v}\right]\frac{dv}{dt} - \frac{\partial L}{\partial t}$$

$$= Q(x, v, t)v(t) - \frac{\partial L}{\partial t} \tag{9.38}$$

according to Equations 9.37 and 9.16. It is evident that the total time derivative will not vanish unless the dissipative forces are absent and the Lagrangian is explicitly time independent. But, even when these conditions *are* met,

$$p_L v - L(x, v, t) = mv^2 - \frac{\partial U}{\partial v}v - \tfrac{1}{2}mv^2 + U(x, v)$$

$$= \tfrac{1}{2}mv^2 + U(x, v) - \frac{\partial U}{\partial v}v$$

which is certainly not the same as the total energy of the particle, in general. This is a most interesting situation! It appears that we have uncovered an extension of the Energy Theorem that reduces to the usual conservation statement whenever the potential function $U(x, v, t)$ is simply a potential energy. Otherwise we have found "something else" to be conserved. We can summarize our conclusions in the following:

The Generalized Energy Theorem   *If the Lagrangian of a single particle is not an explicit function of the time, and if no forces act on the particle that are not derivable from a generalized potential function, then the quantity*

$$p_L(t)v(t) - L(x(t), v(t))$$

is a constant of the motion. If the generalized potential under whose influence the particle moves is a potential energy, the quantity above is equal to the total energy of the particle.

The proof of this theorem should be obvious from our previous discussion. Notice here that time-displacement invariance appears as a necessary condition for energy conservation but is by no means a sufficient one. Just as we discovered in Chapter 3, the forces acting on a particle must be explicitly independent of the velocity and the time in order that the total energy be conserved.

9.5 HAMILTON'S EQUATIONS

In our discussion of Hamilton's Principle we have come to regard the Lagrangian as an important quantity in the dynamics of a single particle. However, it has also become apparent that the quantity

$$p_L v - L(x, v, t)$$

is of comparable or even greater significance when the question of energy conservation is raised. This function, we have seen, is the same as the total energy for a conservative motion and otherwise bears an important relation to whatever dissipative or absorptive process may be occurring. For this reason we should like now to examine the function more carefully. In particular, we shall attempt to sharpen the mathematical relation between it and the Lagrangian and shall seek to clarify its role as a conserved quantity. A successful conclusion of our efforts will not only make the physical meaning of the function transparent, but, in fact, will provide us with a new dynamical formulation alternate and equivalent to that of the Lagrangian equation of motion. This unexpected and remarkable by-product will emerge naturally from the discussion which follows.

Suppose we have at hand a function $y(x)$ of a single variable and we wish to interpret it geometrically. One obvious way is to represent the function as a set of points in the cartesian plane, as shown in Figure 9.3. On the other hand, an alternative (if little-used) portrayal is to present the set of straight lines that are

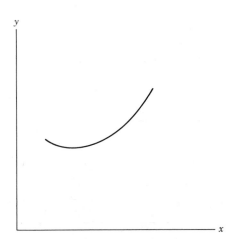

FIGURE 9.3. A function $y(x)$ represented as a locus of points.

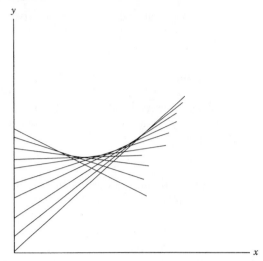

FIGURE 9.4. A function $y(x)$ represented as an envelope of tangent lines.

tangent to each point on the curve $y(x)$ (see Figure 9.4). In this case every point (x, y) is uniquely determined by the pair (a, b) that comprises the slope and y intercept, respectively, of the line tangent to the curve in question at (x, y). Thus the curve is seen as the envelope of a family of tangent lines rather than the locus of a set of points. In Figure 9.5 we have shown the construction of a tangent line at a single point (x, y). It is evident there that, at the point of tangency, the line

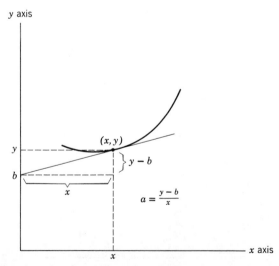

FIGURE 9.5. Construction of a tangent line at the point (x, y) on a curve.

may be expressed by

$$y = ax + b \qquad (9.39)$$

or

$$b = y - ax \qquad (9.40)$$

where a and b are the slope and y intercept of the line, as mentioned previously. Now, by the very definition of tangency, we know that

$$m(x) \equiv \frac{dy}{dx} = a \qquad (9.41)$$

at the point (x, y). Equation 9.41 is an implicit relation for x as a function of a provided that $y(x)$ is itself not a straight line.[7] (In that case the slope $y(x)$ does not depend on x.) Therefore we may express x as a function of the slope of the tangent line and introduce the result into (9.40). Moreover, we can do the same thing for $y(x)$ by combining it with (9.41). It follows that Equation 9.40 may be rewritten uniquely as

$$b = b(a) \qquad (9.42)$$

and that $y(x)$ may be replaced by the intercept of a tangent line expressed as a function of its slope. This construction and elimination of y and x can be carried through at every point on the curve in the cartesian plane to give an "envelope equation" having the general form of (9.42). Conversely, were we to be presented with (9.42) initially, we could recover $y(x)$ by noting that Equation 9.40 in differential form is

$$db = dy - x\, da - a\, dx = x\, da$$

since $dy = a\, dx$ at (x, y). Therefore

$$\frac{db}{da} = -x \qquad (9.43)$$

provides us with a as a function of x, in correspondence with the implication of Equation 9.41. This result we may put into Equation 9.40, using (9.42) to get b as a function of x, and thereby regain the function $y(x)$. The procedure and its converse are outlined in Table 9.1. The primary point we are discussing here is that a curve may be represented equally well either by considering a function y of a *point* variable x or a function b of a *slope* variable a. In the mathematical vernacular, this second function $b(a)$ is called the *Legendre transform* of y with respect to x.

As an example of these ideas, suppose we have the function

$$y(x) = x^2 \qquad (9.44)$$

---

[7] We exclude this possibility from consideration here with no real loss of generality. If $y(x)$ were a straight line our method would be redundant since Figures 9.3 and 9.4 would be identical. In this special instance the locus of points is equal to the envelope of the "family" of tangent lines.

TABLE 9.1   The Legendre Transform of a Function
y(x) and Its Inverse

Legendre Transformation	Inverse Transformation
Given	Given
$y = y(x)$	$b = b(a)$
Then employ	Then employ
$m(x) = \dfrac{dy}{dx} = a$	$\dfrac{db}{da} = -x$
$b = y - ax$	$y = ax + b$
Eliminate y and x to get	Eliminate a and b to get
$b = b(a)$	$y = y(x)$

which is graphed in Figure 9.6*a*. The slope of this function at any point x is

$$m(x) = 2x$$

which is equal to the slope of a line tangent to the parabolic curve $y(x)$ at (x, y):

$$2x = a$$

This equation tells us x as a function of the slope of a tangent line,

$$x = \tfrac{1}{2}a$$

It follows that Equation 9.40 for the intercept b here takes on the form

$$b = y - ax = \frac{a^2}{4} - \frac{a^2}{2} = -\frac{a^2}{4} \tag{9.45}$$

where we have used (9.44) to get y as a function of a. Equation 9.45 is plotted in Figure 9.6*b*. It tells us the intercept of every line that is tangent to the parabola $y = x^2$ as a function of its slope. Thus it permits us to construct the family of tangent lines whose envelope is an upward-curving parabola emanating from the origin of the cartesian plane. Conversely, were we to begin with Equation 9.45, we could write

$$a = 2x$$

according to Equation 9.43, and could introduce this relation into (9.40) to get (with a look at Equation 9.45)

$$-x^2 = y - 2x^2$$

or

$$y = x^2$$

   If the function under consideration depends on more than one variable, then its graph will be a surface rather than a curve. This surface, in keeping with the discussion we have just gone through, may be described either as the locus of

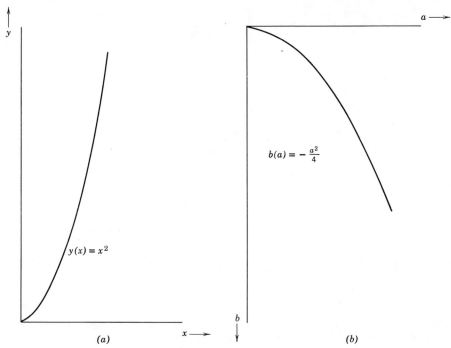

FIGURE 9.6. (*a*) The locus of points that represents the parabola $y(x) = x^2$. (*b*) The Legendre transform of $y(x) = x^2$.

points $(x_1, x_2, \ldots, x_n, y)$ in an $(n + 1)$-dimensional space or as the envelope of a set of tangent planes intersecting the y axis. These alternatives are illustrated for the case $y(x_1, x_2)$ in Figure 9.7. It may be evident from our comments here that a *third* possibility opens up in this more general situation. We need not necessarily transform y with respect to *all* of the variables $x_1, x_2 \ldots, x_n$. Instead we might create the function

$$y[a_1, a_2, \ldots, a_m] \equiv b(a_1, a_2, \ldots, a_m; x_{m+1}, \ldots, x_n)$$

$$= y - \sum_{k=1}^{m} a_k x_k \qquad (m \leqslant n) \tag{9.46}$$

where

$$a_k = \frac{\partial y}{\partial x_k} \qquad (k = 1, \ldots, m)$$

This is a *partial* Legendre transform of y with respect to $\{x_1, x_2, \ldots, x_m\}$. It allows for the possibility of representing a function of several variables partly in terms of the y intercept and slopes of the tangent planes and partly in terms of a locus

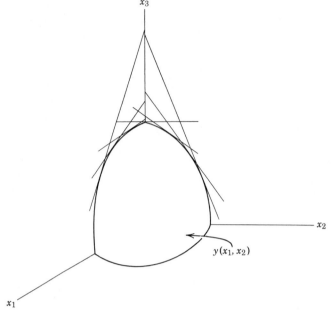

FIGURE 9.7. A function $y(x_1, x_2)$ may be represented either as a surface or as the envelope of a set of tangent planes. (Only the edges of the tangent planes are shown.)

of points. Although this possibility is difficult to envision geometrically,[8] it is mathematically legitimate and provides a quantity whose representation of the surface $y(x_1, x_2 \ldots, x_n)$ is as accurate as a locus of points alone. The primary value of the partial Legendre transform does not lie with its attendant geometric picture, at any rate, but instead with the fact that it permits us the alternative—which may be of great practical value—of representing a function *uniquely* in terms of a set of slope variables instead of a set of point variables. It is this property of the partial Legendre transform that we wish to exploit in the present case.[9]

Indeed, it is time we got back to dynamics! If we regard the Lagrangian of a single particle as a function of position, velocity, and the time, then we may portray it geometrically as a surface in a four-dimensional space. This surface, in parallel with our previous discussion, may be described by the locus of points (x, v, t, L) or as a set of L intercepts and slopes generated by an appropriate partial Legendre transform, along with the locus of points for the variables not transformed. In particular, consider the possibility of representing the projection of the

---

[8] For $y(x_1, x_2)$ it amounts to giving intercepts and slopes for the projection of the surface onto one (x, y) plane while maintaining a locus of points for the other.
[9] The application of the partial Legendre transform to thermodynamics is at least as important as the one to be made here. For an illuminating discussion, see H. B. Callen, *Thermodynamics*, Wiley, New York, 1960.

Lagrangian surface onto the L-v plane by an L intercept and slope of a tangent line. The slope in question here is

$$\frac{\partial L}{\partial v} = p_L \qquad (9.47)$$

according to Equation 9.16. If we create the partial Legendre transform with respect to the velocity, then, we shall have a quantity that is mathematically equivalent to $L(x, v, t)$ in every way, but that is a function of the variables x, p_L, and t instead. By Equation 9.46, this quantity is

$$L[v] = L - \frac{\partial L}{\partial v} v$$

or

$$H(x, p_L, t) \equiv p_L v - L \qquad (9.48)$$

where we have defined the *Hamiltonian function* H to be the negative of the partial Legendre transform of the Lagrangian with respect to velocity:

$$H \equiv -L[v]$$

The Hamiltonian function depends on the variable p_L, *not* v, despite the appearance of Equation 9.48, because Equation 9.47 is an implicit relation for the velocity in terms of the Lagrangian momentum p_L that can be introduced into $L(x, v, t)$ and thus into Equation 9.48. By virtue of all that has been said previously, we fully expect that $H(x, p_L, t)$ is as mathematically and physically complete as is the Lagrangian for describing the motion of a single particle.

We have already seen how the Hamiltonian figures importantly in the problem of energy conservation. In order to underscore its significance we shall, in fact, soon be restating that relation to the Energy Theorem. But first, we wish to pursue the newly realized property of physical equivalence to the Lagrangian that is possessed by the Hamiltonian. Just because $H(x, p_L, t)$ *is* a partial Legendre transform, we know that it is mathematically permissible to express it as the total differential

$$dH = \frac{\partial H}{\partial x} dx + \frac{\partial H}{\partial p_L} dp_L + \frac{\partial H}{\partial t} dt \qquad (9.49)$$

Now, this equation must be identical with the differential form of Equation 9.48,

$$dH = v \, dp_L + p_L \, dv - \frac{\partial L}{\partial x} dx - \frac{\partial L}{\partial v} dv - \frac{\partial L}{\partial t} dt$$

According to Equations 9.35 and 9.47, we may rewrite this latter expression in the form

$$dH = \left[Q - \frac{dp_L}{dt} \right] dx + v \, dp_L - \frac{\partial L}{\partial t} dt \qquad (9.50)$$

A comparison of (9.50) with (9.49) leads us immediately to state

$$\frac{\partial H}{\partial p_L} = v$$

$$-\frac{\partial H}{\partial x} = \frac{dp_L}{dt} - Q \tag{9.51}$$

$$\frac{\partial H}{\partial t} = -\frac{\partial L}{\partial t}$$

Equations 9.51 are known as *Hamilton's equations.*[10] The first of them is the Hamiltonian definition of velocity—a clear correspondent of Equation 9.47. The second equation is, in fact, a recasting of the Lagrangian equation of motion, since

$$\frac{\partial H}{\partial x} = -\frac{\partial L}{\partial x}$$

by virtue of Equation 9.48. The third equation is an obvious result of (9.48), since the Hamiltonian will be explicitly time dependent only if the Lagrangian is as well.

To get as sound as possible a perception of the physical meaning of Hamilton's equations, we shall consider them in three different special cases. In doing so, we shall be able to further illuminate the relation of the Hamiltonian to the concept of energy conservation.

1. *H does not depend explicitly on the time; there are no dissipative forces; U(x, v) is a potential energy.* In this case Equation 9.47 reduces to

$$p_L = mv = p$$

and the Hamiltonian is

$$H(x, p) = \frac{p^2}{m} - \frac{p^2}{2m} + V(x) = \frac{p^2}{2m} + V(x) \tag{9.52}$$

where $U(x, v) = V(x)$. According to the generalized Energy Theorem, stated in the previous section, the Hamiltonian is a constant of the motion and is equal to the total energy. (The latter conclusion, of course, is obvious from Equation 9.52.) Hamilton's equations are now just

$$\frac{\partial H}{\partial p} = \frac{dx}{dt}$$

$$-\frac{\partial H}{\partial x} = \frac{dp}{dt} \tag{9.53}$$

The symmetry of these equations with respect to x and p has inspired for them the name *canonical equations of motion.* Notice that the canonical equations of

---

[10] These equations were, in fact, first obtained by Lagrange, but they lay dormant until Hamilton showed them to be of fundamental importance in dynamics.

motion are first-order differential equations. They represent, in fact, a special case of a proposition in the theory of differential equations, that every second-order differential equation (such as the Lagrangian equation of motion) can be replaced by two first-order differential equations.

2. *H does not depend explicitly on the time; there are no dissipative forces.* In this case the velocity is a solution of

$$p_L = mv - \frac{\partial U}{\partial v}$$

and we have

$$H(x, p_L) = p_L v(p_L) - L(x, v(p_L)) \tag{9.54}$$

where $v(p_L)$ is determined finally by the precise form of $U(x, v)$. The generalized Energy Theorem now tells us that the Hamiltonian is a constant of the motion, but it is not necessarily equal to the total energy.[11] Hamilton's equations in this instance are

$$\frac{\partial H}{\partial p_L} = \frac{dx}{dt}$$

$$-\frac{\partial H}{\partial x} = \frac{dp_L}{dt} \tag{9.55}$$

which differ from Equations 9.53 because p_L is not the same as the Newtonian momentum.

3. *H does not depend explicitly upon the time; U(x, v) is a potential energy.* Here we have

$$H(x, p) = \frac{p^2}{2m} + V(x)$$

but the Hamiltonian is *not* a conserved quantity because dissipative forces are at work on the particle in addition to the conservative forces responsible for $V(x)$. Hamilton's equations are

$$\frac{\partial H}{\partial p} = \frac{dx}{dt}$$

$$-\frac{\partial H}{\partial x} = \frac{dp}{dt} - Q \tag{9.56}$$

where Q represents the dissipative forces. The second of Equations 9.56 is just a statement of the Second Law for nonconservative motion, which was discussed in Chapter 3. It is worthwhile to point out once more that, although H does not

---

[11] In the special case that $U(x, v) = V(x) + f(x, v)$, such that $f(x, v) = v(\partial f/\partial v)$, the Hamiltonian *would* be equal to the total energy. This special possibility qualifies our statement of the Generalized Energy Theorem. For a concrete example, see T. C. Bradbury, op. cit, p. 289.

depend explicitly upon the time, we must not conclude that dH/dt vanishes. Indeed,

$$\frac{dH}{dt} = \frac{\partial H}{\partial x}\frac{dx}{dt} + \frac{\partial H}{\partial p}\frac{dp}{dt} = \left(Q - \frac{dp}{dt}\right)\frac{dx}{dt} + \frac{dx}{dt}\frac{dp}{dt}$$

$$= Q\frac{dx}{dt} \tag{9.57}$$

all of the steps coming directly from Equations 9.56. Thus, for nonconservative motion, the Hamiltonian is a total energy whose time rate of change is not zero. Equation 9.57 has a natural interpretation as the *rate of energy dissipation* for the particle moving under the force Q. This very useful concept is awkward to introduce in the Newtonian scheme because we have there no obvious generalization of the definition of total energy. On the other hand, it emerges as a very reasonable inference from the Hamiltonian formulation, where the idea of total energy is a special case of the more comprehensive Hamiltonian function. This attribute alone should assure us that the present discussion is patently more than a pouring of old wine into new bottles.

9.6 THE LAGRANGIAN AND HAMILTONIAN PROGRAMS FOR DYNAMICS

Our investigation of Hamilton's Principle has developed to the point where it should prove worthwhile to pause and make an epitome of its dynamical content in relation to the Newtonian Program. This comparison is valuable, not only because it can increase our understanding of the fundamental notions in the Lagrangian and Hamiltonian formalisms, but also because it will illustrate the important fact that classical dynamics is to a surprising degree independent of any particular philosophical stance one might take toward its foundations. To elaborate on this latter point we begin our remarks by noting that Hamilton's Principle places emphasis on a *property* of a particle—its energy—whereas the Second Law presumes the most important ingredient of motion to be the external *agent*—a force—acting on the particle. Although the Lagrangian-Hamiltonian formalism is not entirely successful in its attempt to bring about this shift in emphasis (dissipative motion will not conform), it does, in large measure, realize the possibility of casting important dynamical questions into a language directed toward the intrinsic properties of a particle rather than toward its environment. Thus it appears that, for the most part, we may hold the concepts of kinetic energy and the potential function to be no less basic than that of force, notwithstanding the somewhat more abstract character of the former quantities.

A second important difference between Hamilton's Principle and the Second Law is that the former is *holistic* and *teleological* while the latter is *analytical* and *causal*. The holistic character of Hamilton's Principle is a natural outcome of its close relation with the concept of energy. Whenever the motion of a particle is conservative, we are afforded the opportunity of ignoring the details of its trajectory to state simply that, throughout the entire motion, whatever may be its complexities, the sum of kinetic and potential energies remains fixed. In this way we are supposing the motion to be a kind of irreducible whole with a single fundamental attribute. Similarly, Hamilton's Principle imparts to the entire motion of a particle between two points of time the property of making a certain integral stationary. The motion is not differentiated into parts and analyzed, but instead is considered to be an irreducible entity giving occasion for the general condition that S be invariant under infinitesimal virtual displacements. Conversely, the Second Law is an analytical principle since it predicates a motion upon a knowledge of the forces acting at each successive instant. We can perceive no irreducibility in the entire motion of a particle between two points of time, but rather we see it decomposed into the set of all responses to the forces exerted as time passes.

The teleological aspect of Hamilton's Principle was made apparent already in the first section of this chapter. Our argument there very strongly gave the impression that the motion of a particle derived from a *purpose*, namely, to make the average difference between kinetic and potential energies the least possible. In the most general statement of Hamilton's Principle, given in Section 9.4, this quality is still very much present and is brought forward clearly if we say that the purpose of any motion is to make the integral

$$\int_{t_1}^{t_2} \left[T(v) + W(x) \right] dt$$

have a stationary value with respect to virtual displacements. On the other hand, we are *not* saying that the motion has any particular cause. It is simply carried through to realize the end of satisfying Hamilton's Principle and we do not single out any dynamical agents to hold responsible for its existence. This conclusion stands in sharp contrast with the Second Law, which *defines* force to be the cause of all nonuniform motion. Indeed, at each instant a particle accelerating is constrained from moving as its inertia would drive it and instead must conform to the dictates of its perturbing environment. It is most remarkable that this causal attitude, despite its polar opposition to the teleological view contained in Hamilton's Principle, is not necessary to the dynamics of conservative motion. The same equation of motion results regardless of the philosophical premises.

It would be a mistaken impression, however, to think that the difference in motivation between Hamilton's Principle and the Second Law has any effect on the determinism inherent in classical dynamics. Because each of these tenets produces a differential equation for the position that is of second order in the time, each is subject to the purely mathematical requirement that two initial conditions be known in order that the solution be completely specified. Thus in either formalism we retain the precept that the motion of a particle is fully deter-mined by a knowledge of its initial position and velocity together with the appropriate equation of motion.[12]

To complete our summary of the Lagrangian and Hamiltonian formalisms we shall outline the programs for dynamics which emerge from them, just as we did for the Second Law in Chapter 1. A reference to Section 1.1 may be helpful as a guide in ascertaining the significance of what follows.

THE LAGRANGIAN PROGRAM

1. From the results of experiment, physical intuition, or just by educated guessing, we write down an expression for the Lagrangian as a function of the position, velocity, and the time. If any dissipative forces are present, they must also be expressed.

2. Next we introduce the Lagrangian into the equation of motion (9.35) and consider the result either as a first-order differential equation for $v(t)$ or a second-order differential equation for $x(t)$.

3. We solve the equation of motion for the velocity or the position as a function of the time. If the former is obtained we use

$$\frac{dx}{dt} = v(t)$$

to create a differential equation for $x(t)$. If the position is obtained first, we use the equation above to calculate $v(t)$ directly.

THE HAMILTONIAN PROGRAM

1. From the results of experiment, a knowledge of the Lagrangian, physical intuition, or just by educated guessing, write down an expression for the Hamiltonian as a function of the position, Lagrangian momentum, and the time. If any dissipative forces are present, they must also be expressed.

2. Next we introduce the Hamiltonian into Equations 9.51. The second of these is regarded as either a first-order differential equation for $p_L(t)$ or a second-order differential equation for $x(t)$.

3. We solve the second Hamilton equation for the Lagrangian momentum or the position as a function of the time. If the former is obtained we use

$$\frac{\partial H}{\partial p_L} = \frac{dx}{dt}$$

---

[12] It is, in fact for the very reason that, on the microscopic level, it proves to be impossible to specify the initial position and velocity of a particle with arbitrarily high precision, quantum dynamics is *not* a deterministic theory.

to create a differential equation for $x(t)$. If the position is obtained first, we use the equation to calculate $p_L(t)$.

The Newtonian and Lagrangian programs have in common that a prescription through a differential equation is made to calculate $x(t)$ and $v(t)$. In the Hamiltonian program we encounter two differential equations and a method for getting $x(t)$ and $p_L(t)$. The fact that position and either velocity or momentum remains central to all three formalisms has prompted their inclusion in a rather general definition of the concept of *state* in classical dynamics. *The state of a single particle at a given time t is defined to be its position and velocity or its position and momentum at that time.* Thus the equations of motion in classical dynamics prescribe the evolution in time of the state of a particle, given a knowledge of its initial state.

The question of whether to employ the Newtonian Program or the Lagrangian-Hamiltonian Program in a dynamical problem would appear to be largely a matter determined by what form in which the problem is posed, as well as by personal taste. The Second Law has the often decisive advantage of being expressed in a language of rather tactile concepts—force, acceleration, mass—and, therefore, it may be simpler to conceive of on a physical basis. Most of this advantage, of course, accrues from a predisposition to see phenomena in terms of cause and effect. When we observe the motion of a particle, we often wish to analyze that motion and reduce it to primary causes which are both tangible and demonstrable. The Lagrangian and Hamiltonian formalisms, on the other hand, readily offer us the opportunity to understand motion in terms of its general properties, such as conservation theorems, and in terms of its symmetry with respect to space and time transformations. Moreover, these formalisms are more readily extended to describe assemblies of particles, such as rigid bodies and fluids, and fields than is the Newtonian picture. If we admit that conserved quantities and the parallel structures in descriptions of motion are the ultimate aims of dynamics, then it follows that the Lagrangian-Hamiltonian Programs are the more significant. But this evolution and refining of our thinking should not in any sense subtract from the great value of the Newtonian Program in an initial encounter with a dynamical problem or in the setting of the stage for a more comprehensive analysis based on a formulation in terms of the concept of energy.

FOR FURTHER READING

R. P. Feynman, R. B. Leighton, and M. Sands   *The Feynman Lectures on Physics*, Addison-Wesley, Reading, Mass., 1964. Chapter 19 in Volume II of this textbook is a classic introduction to Hamilton's Principle.

W. Hauser   *Introduction to the Principles of Mechanics*, Addison-Wesley, Reading, Mass., 1965. Chapters 5 and 6 are a concise introduction to the Lagrangian and Hamiltonian in three-dimensional space.

J. B. Marion   *Classical Dynamics of Particles and Systems*, Academic Press, New York, 1970. Chapter 6 is an introduction to the calculus of variations with physical emphasis. Chapter 7 presents Hamilton's Principle for three-dimensional motion.

H. Goldstein   *Classical Mechanics*, Addison-Wesley, Reading, Mass., 1959. Chapters 1, 2, and 7 are an excellent advanced discussion of the Lagrangian and Hamiltonian Programs.

PROBLEMS

1. Consider the two examples of particle motion discussed in Section 9.1 under the condition that the initial speed is zero. Then, in both examples, the total energy of the particle is zero and is equally divided between kinetic and potential energy. Evidently, according to the teleological ideas developed in Section 9.1, the particle should be "satisfied" under this condition and should have no reason to move at all. Analyze what actually happens in the two cases (uniform field and harmonic force) with zero initial speed and evaluate the proposal that a particle always moves so as to make the difference between its kinetic and potential energies the least possible, on the average.

2. A particle adsorbed on the surface of a one-dimensional crystal may be supposed to possess the potential energy

$$V(x) = \tfrac{1}{2}V_0\left[1 - \cos\left(\frac{2\pi x}{a}\right)\right] \qquad (-\infty \leqslant x \leqslant +\infty)$$

where V_0 is the binding energy at an adsorption site and a is the distance between sites. Analyze the motion of a particle placed at $x = 0$ with the velocity $v = (V_0/m)^{1/2}$ (m is the mass of the particle) according to the ideas developed in Section 9.1.

3. In general, the position of a particle (in one spatial dimension) may be expanded in a Maclaurin series:

$$x(t) = x(0) + \left(\frac{dx}{dt}\right)_{t=0} t + \cdots + \frac{1}{n!}\left(\frac{d^n x}{dt^n}\right)_{t=0} t^n + \cdots.$$

Show that, for the case of motion in a uniform field under arbitrary initial conditions, the inclusion in $x(t)$ of any *more* than the first three terms of this series will not lead to a minimum value of the action integral, given by Equation 9.3. Show also that the retention of any *less* than the first and third terms will not produce a minimum value for S (unless $x(0) = 0$).

4. Three candidates for the position of an harmonic oscillator during the time interval $0 \leqslant t \leqslant \pi/2\omega_0$ are:

$$x_1(t) = A\left(1 - \frac{2\omega_0 t}{\pi}\right) \qquad x_2(t) = A\cos\omega_0 t \qquad x_3(t) = A\left(1 - \frac{4\omega_0^2 t^2}{\pi^2}\right)$$

Each of these trial functions satisfies the conditions $x(0) = A$ and $x(\pi/2\omega_0) = 0$. Show that $x_2(t)$ produces the least value of the action integral.

5. Consider the trial functions

$$x_1(t) = \frac{t(1 + at^2)}{1 + a} \qquad x_2(t) = \frac{t(1 + bt)}{1 + b}$$

which are to describe the motion of a particle in a uniform gravitational field in the time interval $0 \leqslant t \leqslant 1s$. Calculate S using each trial function, then minimize S by choosing values of the parameters a and b such that

$$\frac{dS}{da} = 0, \quad \frac{dS}{db} = 0$$

respectively. Which trial function produces the lesser value of S?

6. Write down the Lagrangian and Lagrangian equation of motion for the particle described in Problem 2. Solve the equation of motion for the case $(2\pi x/a) \ll 1$ ("localized adsorption").

7. A particle is subject to the force

$$F(x, v) = -kx - \alpha x^2 - \beta v^2$$

Calculate the Lagrangian and Lagragian equation of motion for the particle.

8. A particle is described by the Lagrangian

$$L(x, v) = \tfrac{1}{2}mv^2 - \alpha xv$$

where α is a positive constant.

(a) Calculate the Lagrangian momentum of the particle.

(b) Solve the Lagrangian equation of motion.

9. The Lagrangian of a single particle in special relativity may be expressed

$$L(x, u) = -m_0 c^2 \left(1 - \frac{u^2}{c^2}\right)^{1/2} - V(x)$$

where $V(x)$ is the potential energy and all other symbols are as defined in Chapter 8. Show that the Lagrangian equation of motion in this case is identical with Equation 8.38.

10. Show that the quantity $W(x(\alpha, t)) - W(x(t))$ represents the change in kinetic energy of a particle during the virtual displacement from the path $x(t)$ to the path $x(\alpha, t)$.

11. For an otherwise free particle subject to Stokes damping:

(a) Write down the Lagrangian and the Lagrangian equation of motion.

(b) Show that the "Lagrangian" $L(u) = \tfrac{1}{2}mu^2$, where $u = v + (1/\tau)x$, leads to the same equation of motion as found in (a).

(c) Explain the physical significance of the conservation of the "momentum" mu.

12. Find the Lagrangian for a particle subject to the force $F(x, t) = F(x) + F(t)$. Identify the generalized potential function.

13. With Table 3.2 and the definition of the Lagrangian, state a form of Hamilton's Principle for LC circuits, as described in Special Topic 3.

14. Give a teleological description of the behavior of the electric charge in an LC circuit that parallels the discussion in Section 9.1.

15. State a form of Hamilton's Principle for LRC circuits, as described in Special Topic 3. What is the relation between the equation of motion and Kirchoff's Law?

16. Write down the Lagrangian and the Lagrangian equation of motion for a driven LRC circuit.

17. Calculate the Hamiltonian and Hamilton's equations for the particle described in Problem 8. What are the conserved quantities? Is the Hamiltonian the total energy?

18. Calculate the Hamiltonian of a free, single particle in special relativity, using the Lagrangian of Problem 9. Is the Hamiltonian the total energy?

19. Write down Hamilton's equations for an LC circuit. (See Table 3.2 for guidance.)

20. Find Hamilton's equations for a driven, damped oscillator. Interpret in physical terms the third equation.

Mathematical Appendix

A.1 VECTORS IN THE FIRST THREE DIMENSIONS

The study of motion in three-dimensional space requires that the dynamical quantities—position, velocity, acceleration, force, and the like—be specified in terms of a set of three real numbers known as coordinates. As is shown in Chapter 4, the coordinates of a dynamical variable each may be taken to be an independent entity whose time variation is subject to prescription by a physical law. For example, each of the coordinates of position must be introduced separately into an appropriate version of the Second Law:

$$\frac{d^2 x_i}{dt^2} - \frac{F_i\,(x_1, x_2, x_3, v_1, v_2, v_3, t)}{m} = 0 \qquad (i = 1, 2, 3)$$

There are accordingly three differential equations to solve in order to know the motion. It is evident here that this requirement can lead to a cumbersome analysis unless some simplifying procedures are introduced. For that reason the concept of a *vector space* has found its way into classical dynamics.[1]

The motivation for the axioms that define a vector space comes from the fact that the three coordinates of position are used to locate a particle uniquely in *ordinary space* relative to three mutually perpendicular axes. If the coordinates

---

[1] The concept of a vector evolved with much controversy during the latter part of the nineteenth century. For a lively account of what happened, see M. J. Crowe, *A History of Vector Analysis*, University of Notre Dame Press, Notre Dame, Ind., 1967.

are prescribed by the respective points of intersection, with the axes, of three lines extending from the location of the particle to make right angles with those axes, the coordinates are said to be *cartesian*. We take these coordinates to be fundamental. Other kinds of coordinates, of course, are possible, *but they cannot be used to define a vector*. We shall consider certain of them and their relation to cartesian coordinates in Section A.3.

Formally a vector is defined to be an ordered triple of real numbers, $\mathbf{A} = \{A_1, A_2, A_3\}$, which satisfies the following axioms.

1. Any two vectors $\mathbf{A}$ and $\mathbf{B}$ can be added to produce a third vector $\mathbf{C}$. The addition, symbolized by $+$, is defined by

$$\mathbf{A} + \mathbf{B} \equiv \{A_1 + B_1, A_2 + B_2, A_3 + B_3\}$$
$$\equiv \{C_1, C_2, C_3\} = \mathbf{C} \tag{A.1}$$

where $+$ refers to the ordinary addition of numbers. The addition $+$ can be readily shown to have the properties of commutativity and associativity:

$$\mathbf{A} + \mathbf{B} = \mathbf{B} + \mathbf{A} \tag{A.2}$$

$$\mathbf{A} + (\mathbf{B} + \mathbf{C}) = (\mathbf{A} + \mathbf{B}) + \mathbf{C} \tag{A.3}$$

2. The vector $\mathbf{0} = \{0, 0, 0\}$ exists and results in

$$\mathbf{A} + \mathbf{0} = \mathbf{A}$$

for any vector $\mathbf{A}$. $\mathbf{0}$ is called the *zero vector*.

3. For each vector $\mathbf{A}$ there exists a vector $(-\mathbf{A})$ called the *inverse of* $\mathbf{A}$, which has the property

$$\mathbf{A} + (-\mathbf{A}) = \mathbf{0}$$

4. Every vector $\mathbf{A}$ can be multiplied by an ordinary number a. This process is called *multiplication by a scalar*. It is defined by

$$a \circ \mathbf{A} \equiv \{aA_1, aA_2, aA_3\} \tag{A.4}$$

where $\circ$ is a symbol for the multiplication. Usually we shall abbreviate $a \circ \mathbf{A}$ by the conventional $a\mathbf{A}$. Multiplication by a scalar has the following properties:

(i) $a \circ (b \circ \mathbf{A}) = (ab) \circ \mathbf{A}$ (association)   (A.5)

(ii) $1 \circ \mathbf{A} = \mathbf{A}$ (A.6)

(iii) $a \circ (\mathbf{A} + \mathbf{B}) = a \circ \mathbf{A} + a \circ \mathbf{B}$ (distribution of a)   (A.7)

(iv) $(a + b) \circ \mathbf{A} = a \circ \mathbf{A} + b \circ \mathbf{A}$ (distribution of $\mathbf{A}$)   (A.8)

The set of quantities that obey axioms 1 through 4 is called a *vector space*. With a little experience one can use the axioms to prove a number of interesting theorems about vectors. For example, a moment's thought shows that the definition of *subtraction* should be

$$\mathbf{A} - \mathbf{B} \equiv \mathbf{A} + (-\mathbf{B}) \tag{A.9}$$

that is, the addition of the inverse. Moreover, the inverse itself can be shown to be unique and identically the same as $(-1) \circ \mathbf{B}$. To see the truth of the latter contention, we write out the chain of equalities

$$
\begin{aligned}
(-1) \circ \mathbf{A} &= (-1) \circ \mathbf{A} + (\mathbf{A} + (-\mathbf{A})) && \text{(by axioms 3 and 4)} \\
&= (-1 + 1) \circ \mathbf{A} + (-\mathbf{A}) && \text{[by (A.3) and (A.6)]} \\
&= (0 + 1) \circ \mathbf{A} - \mathbf{A} + (-\mathbf{A}) && \text{[by axioms 3 and 4 and (A.9)]} \\
&= 0 + (-\mathbf{A}) && \text{(by axiom 4ii)} \\
&= (-\mathbf{A}) && \tag{A.10}
\end{aligned}
$$

We see from this result that Equation A.9 may be written

$$(\mathbf{A} - \mathbf{B}) \equiv \{A_1 - B_1, A_2 - B_2, A_3 - B_3\} \tag{A.11}$$

which certainly conforms to our intuitive expectations.

The fact that the coordinates of a vector are specified in *ordinary space* in terms of three axes suggests that a similar prescription may be possible in the *vector space*. We shall call a set of labeled vectors $\{\mathbf{A}_1, \mathbf{A}_2, \mathbf{A}_3\}$ a *sequence* and consider a *linear combination* of the sequence such that

$$a_1 \mathbf{A}_1 + a_2 \mathbf{A}_2 + a_3 \mathbf{A}_3 = 0 \tag{A.12}$$

If the only way to satisfy Equation A.12 is to set $a_1 = a_2 = a_3 \equiv 0$, we shall say that the sequence is *linearly independent*. (Note that, in arriving at Equation A.10, we proved that $0 \circ \mathbf{A} = \mathbf{0}$.) On the other hand, if, for example, $a_1 \neq 0$ and (A.12) still holds, we can write

$$\mathbf{A}_1 = \frac{-a_2}{a_1} \mathbf{A}_2 + \frac{-a_3}{a_1} \mathbf{A}_3$$

In this case we would say that $\mathbf{A}_1$ is *linearly dependent* upon $\mathbf{A}_2$ and $\mathbf{A}_3$.

Now consider the sequence comprising

$$\hat{\mathbf{x}}_1 = \{1, 0, 0\} \qquad \hat{\mathbf{x}}_2 = \{0, 1, 0\} \qquad \hat{\mathbf{x}}_3 = \{0, 0, 1\} \tag{A.13}$$

It follows from the definition of + that these three vectors must be linearly independent. Moreover, *any* vector can be written as the linear combination

$$\mathbf{A} = A_1 \hat{\mathbf{x}}_1 + A_2 \hat{\mathbf{x}}_2 + A_3 \hat{\mathbf{x}}_3 \tag{A.14}$$

if $\mathbf{A} = \{A_1, A_2, A_3\}$. We shall call the sequence $\{\hat{\mathbf{x}}_1, \hat{\mathbf{x}}_2, \hat{\mathbf{x}}_3\}$ the *fundamental basis* for the vector space. Note that the fundamental basis contains as many vectors as there are dimensions in ordinary space; this condition obviously is necessary in order to make a correspondence between a position vector and a position in ordinary space. To emphasize this point the *dimension of a vector space* is defined as the maximum number of linearly independent vectors in a sequence. We deal here with a vector space of dimension three. If we should consider four vectors, for example, then at most three of them can be linearly independent. This means that the linear combination

$$a_1\mathbf{A}_1 + a_2\mathbf{A}_2 + a_3\mathbf{A}_3 + a_4\mathbf{A}_4 = \mathbf{0}$$

for example, *can* hold without each of the a_i vanishing identically. Suppose $a_4 \neq 0$. Then we can write

$$\mathbf{A}_4 = \frac{-a_1}{a_4}\mathbf{A}_1 + \frac{-a_2}{a_4}\mathbf{A}_2 + \frac{-a_3}{a_4}\mathbf{A}_3 \qquad (A.15)$$

which is very reminiscent of Equation A.14. Because of this possibility, we call *any* linearly independent sequence of vectors a *basis* for the vector space. If $\{\mathbf{A}_1, \mathbf{A}_2, \mathbf{A}_3\}$ is such a sequence, then any other vector in the space must be related to them as is $\mathbf{A}_4$ in Equation A.15, according to what is meant by linear independence and dependence. Therefore, a basis may be used to write any vector in terms of its coordinates *with respect to that basis*. If the basis chosen is the fundamental basis, the coordinates are those of the vector relative to the cartesian axes with respect to which the $\hat{\mathbf{x}}_i$ are defined. Each of the vectors $\hat{\mathbf{x}}_i$, of course, corresponds to a position in ordinary space. A look at Equations A.13 should be enough to convince us that, in fact, $\hat{\mathbf{x}}_1$ is associated with a point at unit distance along the cartesian "x axis," $\hat{\mathbf{x}}_2$, the same along the "y axis," and $\hat{\mathbf{x}}_3$, the same along the "z axis." Therefore, the fundamental basis plays the same role for vectors as do the three cartesian axes for points in space. Because of this fact the vectors $A_i\hat{\mathbf{x}}_i$ are called the fundamental *components* of the vector $\mathbf{A}$. The components of a vector are completely analogous to the coordinates of a point.

It is worthwhile to mention in passing that Equation A.14 is unique. This characteristic would seem to be an essential one; it follows directly from the linear independence of the fundamental basis. To see that, let us write

$$\mathbf{A} = \sum_{i=1}^{3} A_i\hat{\mathbf{x}}_i = \sum_{i=1}^{3} A_i'\hat{\mathbf{x}}_i \qquad (A.16)$$

where Σ means a vector sum and we *assume* $A_i \neq A_i'$, that is, the expression for $\mathbf{A}$ is not unique. Now, by Equation A.16 and the definition of subtraction,

$$\sum_{i=1}^{3} (A_i - A_i')\hat{\mathbf{x}}_i = \mathbf{0}$$

But this equation cannot hold unless $A_i = A_i'$, since the $\hat{\mathbf{x}}_i$ are linearly independent. Therefore, we have a contradiction with our initial assumption, derived from that assumption. We must have $A_i = A_i'$ and the expression (A.14) is unique.

Examples

1. The sum and difference of the two vectors $\mathbf{A} = \{1, 3, 7\}$ and $\mathbf{B} = \{4, 0.5, 2\}$.
   $\mathbf{A} + \mathbf{B} = \{4 + 1, 3 + 0.5, 2 + 7\} = \{5, 3.5, 9\}$. $\mathbf{A} - \mathbf{B} = \{4 - 1, 3 - 0.5, 2 - 7\} = \{3, 2.5, -5\}$. Note that $\mathbf{B} + \mathbf{A} = \{1 + 4, 0.5 + 3, 7 + 2\} = \{5, 3.5, 9\} = \mathbf{A} + \mathbf{B}$.

2. The inverse of the vector $\mathbf{A} = \{1, 3, 7\}$. $(-\mathbf{A}) = \{-1, -3, -7\}$. Thus $\mathbf{A} + (-\mathbf{A}) = \{1 - 1, 3 - 3, 7 - 7\} = \{0, 0, 0\} = \mathbf{0}$.

3. Linear independence of the sequence $\mathbf{A}_1 = \{1, 3, 7\}$, $\mathbf{A}_2 = \{4, 0.5, 2\}$, $\mathbf{A}_3 = \{2, 5, 3\}$. We have to check the linear combination $a_1\mathbf{A}_1 + a_2\mathbf{A}_2 + a_3\mathbf{A}_3 = \{a_1 + 4a_2 + 2a_3, 3a_1 + 0.5a_2 + 5a_3, 7a_1 + 2a_2 + 3a_3\}$. If the sequence is linearly independent, we must have

$$a_1 + 4a_2 + 2a_3 = 0$$

$$3a_1 + 0.5a_2 + 5a_3 = 0$$

$$7a_1 + 2a_2 + 3a_3 = 0$$

satisfied *only* by $a_1 = a_2 = a_3 = 0$. Treating these equations as three simultaneous equations for the "unknowns" a_1, a_2, a_3, we may use the necessary and sufficient condition that nonvanishing solutions exist, that is that the determinant of the coefficients *vanish*, in order to test for linear independence. In the present case we have

$$\begin{vmatrix} 1 & 4 & 2 \\ 3 & 0.5 & 5 \\ 7 & 2 & 3 \end{vmatrix} = (1.5 - 10) - 3(12 - 4) + 7(20 - 1) = 100.5 \neq 0$$

Therefore, the three simultaneous equations have no solution other than $a_1 = a_2 = a_3 = 0$. The three vectors *are* linearly independent.

Although the defining axioms do not require it, the concept of vector as used in classical dynamics always involves an even more geometrical picture than we get by simply associating components with coordinates. What is done is to introduce geometry into the *vector space itself* by defining "length" and "angle" for that space. These quantities derive their meaning from the notion of a *scalar product*. The scalar product of two vectors is an ordinary number, denoted by the symbol $\mathbf{A} \cdot \mathbf{B}$ (the "dot product of $\mathbf{A}$ and $\mathbf{B}$"). It has the properties:

(i) $\mathbf{A} \cdot \mathbf{B} \equiv A_1B_1 + A_2B_2 + A_3B_3$
(ii) $\mathbf{A} \cdot \mathbf{B} = \mathbf{B} \cdot \mathbf{A}$
(iii) $\mathbf{A} \cdot \mathbf{A} \geqslant 0$ with equality if and only if $\mathbf{A} = \mathbf{0}$ (A.17)
(iv) $(a\mathbf{A} \cdot \mathbf{B}) = (\mathbf{A} \cdot a\mathbf{B}) = a(\mathbf{A} \cdot \mathbf{B})$
(v) $((\mathbf{A} + \mathbf{B}) \cdot \mathbf{C}) = \mathbf{A} \cdot \mathbf{C} + \mathbf{B} \cdot \mathbf{C}$

Property (iii) above states that the scalar product of a vector with itself is *positive-definite*. Because of this characteristic we may define the *distance* between two vectors **A** and **B** by[2]

$$\|\mathbf{A} - \mathbf{B}\| \equiv [(\mathbf{A} - \mathbf{B}) \cdot (\mathbf{A} - \mathbf{B})]^{1/2} \tag{A.18}$$

The distance between two vectors is the square root of the scalar product of their difference with itself. In the special case $\mathbf{B} = \mathbf{0}$ we have

$$\|\mathbf{A}\| = (\mathbf{A} \cdot \mathbf{A})^{1/2} \tag{A.19}$$

which formally defines the *length* of a vector **A**. If we use Equation A.16 to write (A.18) and (A.19) in terms of coordinates, we get

$$\|\mathbf{A} - \mathbf{B}\| = [(A_1 - B_1)^2 + (A_2 - B_2)^2 + (A_3 - B_3)^2]^{1/2}$$
$$\|\mathbf{A}\| = (A_1{}^2 + A_2{}^2 + A_3{}^2)^{1/2} \tag{A.20}$$

These are precisely the same as the expressions in geometry for the distance between two points and between the origin of coordinate axes and a single point, respectively. Thus we are led to associate a vector with a *directed line segment* emanating from the origin and ending in the point specified by its cartesian coordinates (see Figure A.1). To make the geometric correspondence complete, we shall define +

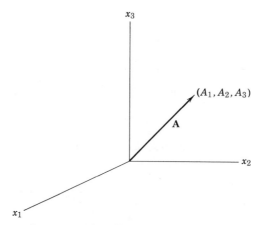

FIGURE A.1. The geometric representation of a vector.

for directed line segments as illustrated in Figure A.2. It is easy to see that the axioms 1 through 4 are met by this procedure and, in particular, that **0** is simply a point ("degenerate") line segment, that $a\mathbf{A}$ is a segment $|a|$ times as long as **A**, and that $(-\mathbf{A})$ points exactly opposite **A** but is of equal length. Moreover, a little

---

[2] Sometimes the symbol $|\mathbf{A} - \mathbf{B}|$ is used for the distance between **A** and **B**. We prefer $\|\;\|$ in order to avoid confusion with the symbol for an absolute value.

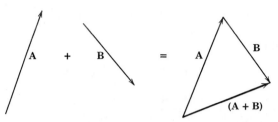

FIGURE A.2. The definition of + for directed line segments: The tail of vector **B** is placed at the head of vector **A**, preserving relative directions, and the sum (**A** + **B**) is drawn from the tail of **A** to the head of **B**.

thought shows that any three directed line segments are linearly independent if they do not all lie in the same plane. The geometric picture of Equation A.14 is then apparent (see Figure A.3).

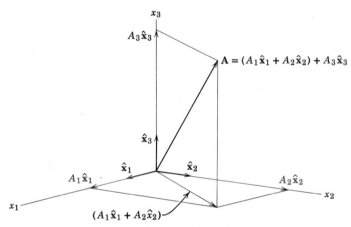

FIGURE A.3. The geometric representation of Equation A.14.

The concept of angle is introduced into the vector space by the definition

$$\cos \theta_{AB} \equiv \frac{\mathbf{A} \cdot \mathbf{B}}{\|\mathbf{A}\| \, \|\mathbf{B}\|} \tag{A.21}$$

The motivation for (A.21) is made clear when we note that, by the definition (A.17),

$$\hat{\mathbf{x}}_1 \cdot \hat{\mathbf{x}}_2 = \hat{\mathbf{x}}_1 \cdot \hat{\mathbf{x}}_3 = \hat{\mathbf{x}}_2 \cdot \hat{\mathbf{x}}_3 = 0$$

$$\hat{\mathbf{x}}_1 \cdot \hat{\mathbf{x}}_1 = \hat{\mathbf{x}}_2 \cdot \hat{\mathbf{x}}_2 = \hat{\mathbf{x}}_3 \cdot \hat{\mathbf{x}}_3 = 1$$

Since the $\hat{\mathbf{x}}_i$ lie along mutually perpendicular axes we know that the angle between any two of them is 90° and, therefore, that $\hat{\mathbf{x}}_i \cdot \hat{\mathbf{x}}_j (i \neq j)$ *should* vanish, as suggested by (A.21). In addition, the fact that the endpoint of the $\hat{\mathbf{x}}_i$ lie at unit distance from the origin requires that $\hat{\mathbf{x}}_i \cdot \hat{\mathbf{x}}_i = 1$ for all i; the same follows from Equation A.21. We can summarize these results by writing:

$$\hat{\mathbf{x}}_i \cdot \hat{\mathbf{x}}_j \equiv \delta_{ij} = \begin{cases} 1 & i \neq j \\ 0 & i \neq j \end{cases} \tag{A.22}$$

which is a convenient shorthand.

When $\mathbf{A}$ is regarded as a directed line segment, we can say that its components are its respective, geometric projections onto the fundamental basis. Thus we write, in the usual way,

$$A_1 = \|\mathbf{A}\| \cos \theta_{Ax_1} \qquad A_2 = \|\mathbf{A}\| \cos \theta_{Ax_2} \qquad A_3 = \|\mathbf{A}\| \cos \theta_{Ax_3}$$

or, by Equations A.21 and A.22,

$$A_i = \hat{\mathbf{x}}_i \cdot \mathbf{A} \qquad (i = 1, 2, 3) \tag{A.23}$$

The truth of Equations A.23 can also be seen by taking the scalar product of Equation A.14 with $\hat{\mathbf{x}}_k$ ($k = 1, 2,$ or 3) and referring to Equations A.22:

$$\mathbf{A} \cdot \hat{\mathbf{x}}_k = \left(\sum_{i=1}^{3} A_i \hat{\mathbf{x}}_i \right) \cdot \hat{\mathbf{x}}_k = \sum_{i=1}^{3} A_i (\hat{\mathbf{x}}_i \cdot \hat{\mathbf{x}}_k) = \sum_{i=1}^{3} A_i \delta_{ik} = A_k$$

Examples

1. The scalar product of the vectors $\mathbf{A} = \{1, 3, 7\}$ and $\mathbf{B} = \{4, 0.5, 2\}$. $\mathbf{A} \cdot \mathbf{B} = 1 \times 4 + 3 \times 0.5 + 7 \times 2 = 4 + 1.5 + 14 = 19.5$. Note that, although $\mathbf{A}, \mathbf{B},$ and the vector $\mathbf{C} = \{2, 5, 3\}$ form a basis, they are not mutually perpendicular, since $\mathbf{A} \cdot \mathbf{B} \neq 0, \mathbf{B} \cdot \mathbf{C} \neq 0,$ and $\mathbf{A} \cdot \mathbf{C} \neq 0$.

2. The distance between $\mathbf{A} = \{1, 3, 7\}$ and $\mathbf{B} = \{4, 0.5, 2\}$. $\|\mathbf{A} - \mathbf{B}\| = [(1 - 4)^2 + (3 - 0.5)^2 + (7 - 2)^2]^{1/2} = (9 + 6.25 + 25)^{1/2} = 6.34$.

3. The length of $\mathbf{A} = \{1, 3, 7\}$. $\|\mathbf{A}\| = (1^2 + 3^2 + 7^2)^{1/2} = 7.66$.

4. The components and coordinates of $\mathbf{A} = \{1, 3, 7\}$. The components are $\hat{\mathbf{x}}_1,$ $3\hat{\mathbf{x}}_2,$ and $7\hat{\mathbf{x}}_3$. The coordinates are $A_1 = 1, A_2 = 3, A_3 = 7$.

5. Perpendicularity of $\mathbf{A} = \{1, 3, 7\}$ and $\mathbf{B} = \{1, -5, 2\}$. We have $\mathbf{A} \cdot \mathbf{B} = 1 \times 1 - 3 \times 5 + 7 \times 2 = 1 - 15 + 14 = 0$. $\mathbf{A}$ is perpendicular to $\mathbf{B}$.

6. The angle between $\mathbf{A} = \{1, 3, 7\}$ and $\mathbf{B} = \{4, 0.5, 2\}$. $\mathbf{A} \cdot \mathbf{B} = 19.5, \|\mathbf{A}\| = 7.66,$ and $\|\mathbf{B}\| = (16 + 0.25 + 4)^{1/2} = 4.5$. Thus $\cos \theta_{AB} = 19.5/(7.66 \times 4.5) = 0.566$. $\theta_{AB} = 55.5°$.

Of great importance to applications in dynamics is the concept of an *axial vector*. The coordinates of this vector are defined by

$$C_i \equiv \sum_{j=1}^{3} \sum_{k=1}^{3} \varepsilon_{ijk} A_j B_k \qquad (i = 1, 2, 3) \tag{A.24}$$

where

$$\varepsilon_{ijk} = \begin{cases} 1 & \text{if } ijk = 123, 231, \text{ or } 312 \\ -1 & \text{if } ijk = 213, 132, \text{ or } 321 \\ 0 & \text{otherwise} \end{cases} \tag{A.25}$$

C is called "the axial vector created from **A** and **B**" and is usually written $\mathbf{C} = \mathbf{A} \times \mathbf{B}$ because of the appearance of products of coordinates in (A.24). This fact also has prompted the name "vector product" for axial vectors, although the term is somewhat misleading. From the basic definitions (A.24) and (A.25) we see that

$$C_1 = A_2 B_3 - A_3 B_2$$
$$C_2 = A_3 B_1 - A_1 B_3 \tag{A.26}$$
$$C_3 = A_1 B_2 - A_2 B_1$$

and

$$\varepsilon_{ijk} = \varepsilon_{jki} = \varepsilon_{kij} = -\varepsilon_{jik} = -\varepsilon_{ikj} = -\varepsilon_{kji}$$

Using Equations A.26 we can easily show that axial vectors have the following properties:

$$\mathbf{A} \times \mathbf{B} = -\mathbf{B} \times \mathbf{A} \qquad \mathbf{A} \times \mathbf{B} = (-\mathbf{A}) \times (-\mathbf{B}) \tag{A.27}$$

$$(a\mathbf{A}) \times \mathbf{B} = a(\mathbf{A} \times \mathbf{B}) \tag{A.28}$$

The invariance of $\mathbf{A} \times \mathbf{B}$ under space reversal implied by the second of Equations A.27 is the reason for the name "axial."

To get an ideal of the geometric properties of axial vectors we shall consider the quantities $\hat{\mathbf{x}}_i \times \hat{\mathbf{x}}_j$. According to Equations A.24:

$$(\hat{\mathbf{x}}_1 \times \hat{\mathbf{x}}_2)_1 = \sum_{j=1}^{3} \sum_{k=1}^{3} \varepsilon_{1jk}(\hat{\mathbf{x}}_1)_j(\hat{\mathbf{x}}_2)_k = \varepsilon_{112}(\hat{\mathbf{x}}_1)_1(\hat{\mathbf{x}}_2)_2 = 0$$

$$(\hat{\mathbf{x}}_1 \times \hat{\mathbf{x}}_2)_2 = \sum_{j=1}^{3} \sum_{k=1}^{3} \varepsilon_{2jk}(\hat{\mathbf{x}}_1)_j(\hat{\mathbf{x}}_2)_k = \varepsilon_{212}(\hat{\mathbf{x}}_1)_1(\hat{\mathbf{x}}_2)_2 = 0$$

$$(\hat{\mathbf{x}}_1 \times \hat{\mathbf{x}}_2)_3 = \sum_{j=1}^{3} \sum_{k=1}^{3} \varepsilon_{3jk}(\hat{\mathbf{x}}_1)_j(\hat{\mathbf{x}}_2)_k = \varepsilon_{312}(\hat{\mathbf{x}}_1)_1(\hat{\mathbf{x}}_2)_2 = 1$$

where in each case the initial reduction of the double sum comes from looking at the coordinates of the $\hat{\mathbf{x}}_i$ and the final result is determined by Equation A.25. We see here that $\hat{\mathbf{x}}_1 \times \hat{\mathbf{x}}_2 = \hat{\mathbf{x}}_3$. In the same way we can show that

$$\hat{\mathbf{x}}_i \times \hat{\mathbf{x}}_j = \varepsilon_{ijk}\hat{\mathbf{x}}_k \qquad (i, j, k = 1, 2, 3) \tag{A.29}$$

in general. Therefore we conclude that *an axial vector always points perpendicularly to the plane of the two vectors from which it is created*. The actual direction of the axial vector can always be determined with a "right-hand rule," as shown in Figure A.4. The rule, of course, comes directly from Equations A.29. We note in

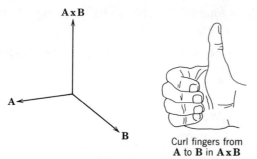

AxB

A◄

B

Curl fingers from
A to B in AxB

FIGURE A.4. The "right-hand rule" for determining the direction of an axial vector.

passing that two parallel vectors create **0** as an axial vector; this fact may be used as the algebraic definition of "parallel."

The quantities δ_{ij} and ε_{ijk} can be shown by direct calculation to be related through the identity

$$\sum_{k=1}^{3} \varepsilon_{ijk}\varepsilon_{lmk} \equiv \delta_{il}\,\delta_{jm} - \delta_{im}\,\delta_{jl} \tag{A.30}$$

Because of Equation A.30 it is possible to prove a number of very useful expressions involving axial vectors. The most frequently encountered of these are:

$$\|\mathbf{A} \times \mathbf{B}\| = \|\mathbf{A}\|\,\|\mathbf{B}\| \sin \theta_{AB} \tag{A.31}$$

$$\mathbf{A} \cdot (\mathbf{B} \times \mathbf{C}) \equiv \mathbf{C} \cdot (\mathbf{A} \times \mathbf{B}) \equiv \mathbf{B} \cdot (\mathbf{C} \times \mathbf{A}) \equiv (ABC) \tag{A.32}$$

$$\mathbf{A} \times (\mathbf{B} \times \mathbf{C}) \equiv (\mathbf{A} \cdot \mathbf{C})\mathbf{B} - (\mathbf{A} \cdot \mathbf{B})\mathbf{C} \tag{A.33}$$

$$(\mathbf{A} \times \mathbf{B}) \times (\mathbf{C} \times \mathbf{D}) \equiv (ABD)\mathbf{C} - (ABC)\mathbf{D} \tag{A.34}$$

To see how a typical proof might go, consider

$$(\mathbf{A} \times \mathbf{B}) \cdot (\mathbf{A} \times \mathbf{B}) = \sum_{k=1}^{3} (\mathbf{A} \times \mathbf{B})_k(\mathbf{A} \times \mathbf{B})_k$$

$$= \sum_{k=1}^{3} \sum_{i=1}^{3} \sum_{j=1}^{3} \sum_{l=1}^{3} \sum_{m=1}^{3} \varepsilon_{kij}\varepsilon_{klm}A_iB_jA_lB_m$$

$$= \sum_{i=1}^{3} \sum_{j=1}^{3} \sum_{l=1}^{3} \sum_{m=1}^{3} \left(\sum_{k=1}^{3} \varepsilon_{ijk}\varepsilon_{lmk}\right) A_iB_jA_lB_m$$

$$= \sum_{i,j,l,m} (\delta_{il}\,\delta_{jm} - \delta_{im}\,\delta_{jl})A_iB_jA_lB_m$$

$$= \sum_i \sum_l \sum_j \sum_m \delta_{il}\,\delta_{jm}\,A_iA_lB_jB_m - \sum_i \sum_m \sum_j \sum_l \delta_{im}\,\delta_{jl}\,A_iB_mB_jA_l$$

$$= \sum_i A_iA_i \sum_j B_jB_j - \sum_i A_iB_i \sum_j B_jA_j = (\mathbf{A} \cdot \mathbf{A})(\mathbf{B} \cdot \mathbf{B}) - (\mathbf{A} \cdot \mathbf{B})^2$$

Therefore

$$(\mathbf{A} \times \mathbf{B}) \cdot (\mathbf{A} \times \mathbf{B}) = \|\mathbf{A}\|^2\,\|\mathbf{B}\|^2(1 - \cos^2 \theta_{AB})$$
$$= \|\mathbf{A}\|^2\,\|\mathbf{B}\|^2 \sin^2 \theta_{AB}$$

and Equation A.31 is established. The reader may find it instructive to repeat this proof to show that

$$(\mathbf{A} \times \mathbf{B}) \cdot (\mathbf{C} \times \mathbf{D}) = (\mathbf{A} \cdot \mathbf{C})(\mathbf{B} \cdot \mathbf{D}) - (\mathbf{A} \cdot \mathbf{D})(\mathbf{B} \cdot \mathbf{C}) \tag{A.35}$$

Examples

1. $\mathbf{A} \times \mathbf{B} = -\mathbf{B} \times \mathbf{A}$. We have:

$$\begin{aligned}
(-\mathbf{B} \times \mathbf{A})_i &= \sum_j \sum_k \varepsilon_{ijk}(-B_j)A_k = \sum_j \sum_k (-\varepsilon_{ikj})A_k(-B_j) \\
&= \sum_k \sum_j \varepsilon_{ikj}A_kB_j = (\mathbf{A} \times \mathbf{B})_i \qquad (i = 1, 2, 3)
\end{aligned}$$

2. $\mathbf{A} \cdot (\mathbf{B} \times \mathbf{C}) = \mathbf{C} \cdot (\mathbf{A} \times \mathbf{B})$. We have:

$$\begin{aligned}
\mathbf{A} \cdot (\mathbf{B} \times \mathbf{C}) &= \sum_i A_i(\mathbf{B} \times \mathbf{C})_i = \sum_i A_i \sum_j \sum_k \varepsilon_{ijk}B_jC_k \\
&= \sum_{i,j,k} \varepsilon_{ijk}A_iB_jC_k = \sum_{k,i,j} \varepsilon_{kij}C_kA_iB_j \\
&= \sum_k C_k \sum_i \sum_j \varepsilon_{kij}A_iB_j = \sum_k C_k(\mathbf{A} \times \mathbf{B})_k = \mathbf{C} \cdot (\mathbf{A} \times \mathbf{B})
\end{aligned}$$

A.2 LINEAR TRANSFORMATIONS IN VECTOR SPACES

We are already familiar with the fact that the potential energy in one dimension, speaking mathematically, is a *function V* of a scalar *argument x* with the scalar *value V(x)*. This very useful concept can be extended to arguments or values that are members of a vector space (rather than being scalars) by defining an *operator*. An operator is a function that associates a vector argument with a vector value, a scalar argument with a vector value, or a vector argument with a scalar value. The operator is called a *linear transformation* or *linear operator* if, in addition, it satisfies

$$O(a\mathbf{A} + b\mathbf{B}) = aO(\mathbf{A}) \oplus bO(\mathbf{B}) \tag{A.36}$$

where O is the operator, a and b are arbitrary scalars, $\mathbf{A}$ and $\mathbf{B}$ are arbitrary vectors, and $\oplus$ is either $+$ or $+$, depending on whether the value of O is a vector or a scalar, respectively. If the argument of O is a scalar, Equation A.36 has the form

$$O(a\varphi + b\psi) = aO(\varphi) \oplus bO(\psi) \tag{A.37}$$

where φ and ψ are arbitrary scalars. In general, all of the operators we shall encounter in doing classical dynamics will be linear and will satisfy one or the other of Equation A.36 and A.37. To make this point evident, we cite several examples of linear operators.

1. *The time-derivative operator.* This operator takes vectors into vectors. It is defined by

$$D_t(\mathbf{A}) \equiv \frac{d\mathbf{A}}{dt} = \left\{ \frac{dA_1}{dt}, \frac{dA_2}{dt}, \frac{dA_3}{dt} \right\} \tag{A.38}$$

Because the time derivative of a scalar function is a linear operation[3], it is easy to show that Equation A.36 is satisfied here with $\oplus = +$. Examples of the time derivative operator at work are found with the velocity (value $\mathbf{v}$, argument $\mathbf{r}$) and the acceleration (value $\mathbf{a}$, argument $\mathbf{v}$) of a moving particle, as shown in Equations 4.4 and 4.6, respectively.

2. *The del operator.* This operator takes scalars into vectors. It is defined by

$$D_r(\varphi) \equiv \nabla\varphi = \left\{ \frac{\partial\varphi}{\partial x_1}, \frac{\partial\varphi}{\partial x_2}, \frac{\partial\varphi}{\partial x_3} \right\} \tag{A.39}$$

and is seen to be a generalization of the one-dimensional operator d/dx. Again, the linearity of spatial derivatives $\partial/\partial x_i$ guarantees the linearity of ∇, according to Equation A.37 with $\oplus = +$. The value of ∇ is called the *gradient* of the scalar function φ. An obvious example of this entity is the conservative force:

$$\mathbf{F} = -\nabla V$$

where $V(x_1, x_2, x_3)$ is the potential energy function. It should be noted that, in Equations A.38 and A.39, the operators are not "hungry," that is, they are displayed along with their arguments. Often it is convenient, however, to think of the del operator as a vector in itself, writing

$$\nabla = \hat{\mathbf{x}}_1 \frac{\partial}{\partial x_1} + \hat{\mathbf{x}}_2 \frac{\partial}{\partial x_2} + \hat{\mathbf{x}}_3 \frac{\partial}{\partial x_3} \tag{A.40}$$

This tack is a permissible one as long as every expression in which it appears is understood ultimately in terms of equations such as (A.39). Having written (A.40), we may immediately proceed to manipulate ∇ as we might any vector, with the important exception that quantities to the left of ∇ will be on a different footing from those on the right. (It is just a question of who is being eaten!) In particular we may define the axial vector

$$(\nabla \times \mathbf{A})_i = \sum_{j=1}^{3} \sum_{k=1}^{3} \varepsilon_{ijk} \frac{\partial A_k}{\partial x_j} \qquad (i = 1, 2, 3) \tag{A.41}$$

which is called the *curl* of A, by analogy with Equation A.24. The coordinates of the curl are then

$$(\nabla \times \mathbf{A})_1 = \frac{\partial A_3}{\partial x_2} - \frac{\partial A_2}{\partial x_3}$$

$$(\nabla \times \mathbf{A})_2 = \frac{\partial A_1}{\partial x_3} - \frac{\partial A_3}{\partial x_1}$$

$$(\nabla \times \mathbf{A})_3 = \frac{\partial A_2}{\partial x_1} - \frac{\partial A_1}{\partial x_2}$$

---

[3] For a discussion of the linear properties of derivatives, see J. M. H. Olmsted, *Calculus with Analytic Geometry*, Appleton-Century-Crofts, New York, 1966, Vol. 1, Chapter 11.

By the method outlined in Section A.1, it is straightforward to show that the del operator satisfies the following identities:

$$\mathbf{V} \times (\varphi \mathbf{A}) \equiv (\mathbf{V}\varphi) \times \mathbf{A} + \varphi \mathbf{V} \times \mathbf{A} \tag{A.42}$$

$$\mathbf{V} \times (\mathbf{V}\varphi) \equiv 0 \tag{A.43}$$

Equation A.43 figures importantly in establishing the necessary and sufficient conditions for a force $\mathbf{F}$ to be conservative. (See Equation 4.16.)

3. *The scalar product.* This operator takes vectors into scalars. Unlike the others we have considered, it is a *binary operator*, defined by

$$\mathbf{A} \cdot \mathbf{B} \equiv A_1 B_1 + A_2 B_2 + A_3 B_3 \tag{A.17}$$

as we very well know. The linearity of this operator is quickly verified after a look at its properties (iv) and (v) given in Section A.1. (Indeed, we have already used the linearity property in the demonstration of Equation A.23.) Two interesting special cases of (A.17) are the line integral, $\int \mathbf{A} \cdot d\mathbf{s}$, mentioned in Section 4.1, and the *divergence* of a vector,

$$\mathbf{V} \cdot \mathbf{A} = \frac{\partial A_1}{\partial x_1} + \frac{\partial A_2}{\partial x_2} + \frac{\partial A_3}{\partial x_3} \tag{A.44}$$

Here we have treated $\mathbf{V}$ as a vector in itself. With Equations A.41 and A.30 we can prove the following, commonly used identities involving $\mathbf{V}$:

$$\mathbf{V}(\mathbf{A} \cdot \mathbf{B}) \equiv (\mathbf{A} \cdot \mathbf{V})\mathbf{B} + (\mathbf{B} \cdot \mathbf{V})\mathbf{A} + \mathbf{A} \times (\mathbf{V} \times \mathbf{B}) + \mathbf{B} \times (\mathbf{V} \times \mathbf{A}) \tag{A.45}$$

$$\mathbf{V} \cdot (\mathbf{A} \times \mathbf{B}) \equiv \mathbf{B} \cdot (\mathbf{V} \times \mathbf{A}) - \mathbf{A} \cdot (\mathbf{V} \times \mathbf{B}) \tag{A.46}$$

$$\mathbf{V} \times (\mathbf{A} \times \mathbf{B}) \equiv \mathbf{A}(\mathbf{V} \cdot \mathbf{B}) - \mathbf{B}(\mathbf{V} \cdot \mathbf{A}) + (\mathbf{B} \cdot \mathbf{V})\mathbf{A} - (\mathbf{A} \cdot \mathbf{V})\mathbf{B} \tag{A.47}$$

$$\mathbf{V} \cdot (\mathbf{V} \times \mathbf{A}) \equiv 0 \tag{A.48}$$

A comparison of Equation A.47 with Equation A.33 illustrates the care that must be taken when $\mathbf{V}$ is manipulated as a vector in itself.

Examples

1. The gradient of $\varphi(x_1, x_2, x_3) = k(x_1{}^2 + x_2{}^2 + x_3{}^2)^{-1/2}$. $\partial\varphi/\partial x_1 = kx_1(x_1{}^2 + x_2{}^2 + x_3{}^2)^{-3/2}$, $\partial\varphi/\partial x_2 = kx_2(x_1{}^2 + x_2{}^2 + x_3{}^2)^{-3/2}$, $\partial\varphi/\partial x_3 = kx_3(x_1{}^2 + x_2{}^2) + x_3{}^2)^{-3/2}$. Then

$$\mathbf{V}\varphi = \frac{k}{(x_1{}^2 + x_2{}^2 + x_3{}^2)^{3/2}} (x_1\hat{\mathbf{x}}_1 + x_2\hat{\mathbf{x}}_2 + x_3\hat{\mathbf{x}}_3)$$

2. The curl of $\mathbf{A} = \{x_1{}^2 x_2, x_2 x_3{}^2, x_1{}^2 x_3\}$. We have:

$$(\mathbf{V} \times \mathbf{A})_1 = \frac{\partial A_3}{\partial x_2} - \frac{\partial A_2}{\partial x_3} = -2x_3 x_2$$

$$(\mathbf{V} \times \mathbf{A})_2 = \frac{\partial A_1}{\partial x_3} - \frac{\partial A_3}{\partial x_1} = -2x_1x_3$$

$$(\mathbf{V} \times \mathbf{A})_3 = \frac{\partial A_2}{\partial x_1} - \frac{\partial A_1}{\partial x_2} = -x_1{}^2$$

and

$$\mathbf{V} \times \mathbf{A} = -2x_3x_2\hat{\mathbf{x}}_1 - 2x_1x_3\hat{\mathbf{x}}_2 - x_1{}^2\hat{\mathbf{x}}_3$$

3. The divergence of $\mathbf{A} = \{x_1{}^2x_2, x_2x_3{}^2, x_1{}^2x_3\}$. $\partial A_1/\partial x_1 = 2x_1x_2, \partial A_2/\partial x_2 = x_3{}^2$, $\partial A_3/\partial x_3 = x_1{}^2$ and $\mathbf{V} \cdot \mathbf{A} = 2x_1x_2 + x_3{}^2 + x_1{}^2$.

4. $\mathbf{V} \cdot (\mathbf{V} \times \mathbf{A}) \equiv 0.$ $\quad \mathbf{V} \cdot (\mathbf{V} \times \mathbf{A}) = \sum_i \dfrac{\partial}{\partial x_i} (\mathbf{V} \times \mathbf{A})_i$

$$= \sum_i \frac{\partial}{\partial x_i} \sum_j \sum_k \varepsilon_{ijk} \frac{\partial A_j}{\partial x_k} = \sum_{i,\,j,\,k} \varepsilon_{ijk} \frac{\partial^2 A_j}{\partial x_i\,\partial x_k}$$

$$= \sum_{i,\,j,\,k} \tfrac{1}{2} \left[\varepsilon_{ijk} \frac{\partial^2 A_j}{\partial x_i\,\partial x_k} + \varepsilon_{kji} \frac{\partial^2 A_j}{\partial x_k\,\partial x_i} \right] = \sum_{i,\,j,\,k} \tfrac{1}{2}\varepsilon_{ijk} \left(\frac{\partial^2 A_j}{\partial x_i\,\partial x_k} - \frac{\partial^2 A_j}{\partial x_k\,\partial x_i} \right) \equiv 0.$$

Besides the operators we have just mentioned,[4] there is another very important kind of linear operator which transforms vectors into vectors in a special way. This operator is called the *rotation operator* $\mathscr{R}$ and is defined so as to preserve the length of a vector under transformation. Thus we have Equation A.36 and

$$\mathbf{A}' = R(\mathbf{A}) \equiv \mathscr{R}\mathbf{A} \qquad \text{with} \qquad \|\mathscr{R}\mathbf{A}\| \equiv \|\mathbf{A}\| \tag{A.49}$$

for any vector $\mathbf{A}$. We call $\mathbf{A}'$ "the rotated vector $\mathbf{A}$." The character of the transformation $\mathscr{R}$ can be understood clearly once we state that Equation A.49 will be applied only to fundamental basis vectors and not to arbitrary vectors. Thus we are taking a "passive" view of a rotation (the coordinate axes are rotated while the vector in question remains fixed) rather than an "active" view (the vector in question is rotated while the coordinate axes remain fixed).[5] If $\{\hat{\mathbf{x}}'_1, \hat{\mathbf{x}}'_2, \hat{\mathbf{x}}'_3\}$ is the result of rotating a fundamental basis, then, we have

$$\mathbf{A} = \sum_{i=1}^{3} A_i\hat{\mathbf{x}}_i = \sum_{k=1}^{3} A'_k\hat{\mathbf{x}}'_k \tag{A.50}$$

with

$$\hat{\mathbf{x}}'_k = \mathscr{R}\hat{\mathbf{x}}_k = \sum_{i=1}^{3} (\hat{\mathbf{x}}_i \cdot \mathscr{R}\hat{\mathbf{x}}_k)\hat{\mathbf{x}}_i \tag{A.51}$$

according to Equation A.23. Equation A.50 states that a vector $\mathbf{A}$ has coordinates A_i with respect to a given fundamental basis and coordinates A'_1 with respect to

---

[4] Strictly speaking, the energy functions $T(\mathbf{v})$, $V(\mathbf{r})$, and $U(\mathbf{r}, \mathbf{v}, t)$, and the force $\mathbf{F}(\mathbf{r})$ are operators. However, because they are generally not linear operators, they are usually treated simply as functions of several scalar arguments.

[5] For a discussion of the two views of a rotation and their interrelation, see F. W. Byron, Jr. and R. W. Fuller, *Mathematics of Classical and Quantum Physics*, Addison-Wesley, Reading, Mass., 1969, Chapters 1 and 4.

the rotated basis. We know already that the very definition of $\mathcal{R}$ requires $\|\hat{\mathbf{x}}'_k\| = \|\hat{\mathbf{x}}_k\|$. But, according to Equations A.51, this means

$$\mathcal{R}\hat{\mathbf{x}}_k \cdot \mathcal{R}\hat{\mathbf{x}}_k = \sum_{i=1}^{3}\sum_{l=1}^{3} (\hat{\mathbf{x}}_i \cdot \mathcal{R}\hat{\mathbf{x}}_k)(\hat{\mathbf{x}}_l \cdot \mathcal{R}\hat{\mathbf{x}}_k)(\hat{\mathbf{x}}_i \cdot \hat{\mathbf{x}}_l)$$

$$= \sum_{i=1}^{3}\sum_{l=1}^{3} (\hat{\mathbf{x}}_i \cdot \mathcal{R}\hat{\mathbf{x}}_k)(\hat{\mathbf{x}}_l \cdot \mathcal{R}\hat{\mathbf{x}}_k)\,\delta_{il}$$

$$= \sum_{i=1}^{3} (\hat{\mathbf{x}}_i \cdot \mathcal{R}\hat{\mathbf{x}}_k)^2 \equiv 1 \qquad (A.52)$$

If in addition we impose the requirement that $\{\hat{\mathbf{x}}'_1, \hat{\mathbf{x}}'_2, \hat{\mathbf{x}}'_3\}$ is still a fundamental basis, we have $\hat{\mathbf{x}}'_k \cdot \hat{\mathbf{x}}'_j = \mathcal{R}\hat{\mathbf{x}}_k \cdot \mathcal{R}\hat{\mathbf{x}}_j = \delta_{kj}$ and (A.52) may be generalized to

$$\sum_{i=1}^{3} (\hat{\mathbf{x}}_i \cdot \mathcal{R}\hat{\mathbf{x}}_k)(\hat{\mathbf{x}}_i \cdot \mathcal{R}\hat{\mathbf{x}}_j) = \delta_{kj} \qquad (A.53)$$

In consequence of Equation A.53, a rotation may be referred to simply as a *change of basis*. Now we are in a position to establish a relation between A'_i and A_i in Equation A.50. By substituting (A.51) into the second sum of (A.50) we get

$$A_i = \sum_{k=1}^{3} (\hat{\mathbf{x}}_i \cdot \mathcal{R}\hat{\mathbf{x}}_k)A'_k \qquad (i = 1, 2, 3)$$

The inverse of this expression—which is of greater value—is obtained by multiplying both sides of it with $(\hat{\mathbf{x}}_i \cdot \mathcal{R}\hat{\mathbf{x}}_j)$ and summing over i with a look at (A.53):

$$\sum_{i=1}^{3} (\hat{\mathbf{x}}_i \cdot \mathcal{R}\hat{\mathbf{x}}_j)A_i = A'_j$$

or

$$A'_j = \sum_{i=1}^{3} R_{ji}A_i \qquad (j = 1, 2, 3) \qquad (A.54)$$

where

$$R_{ji} \equiv (\mathcal{R}\hat{\mathbf{x}}_j \cdot \hat{\mathbf{x}}_i) \qquad (A.55)$$

Equations A.54 are a set of three linear, homogeneous equations for the coordinates of the rotated vector in terms of those of the original vector and the nine quantities R_{ji}. These latter quantities we must still interpret. This we do by adopting the geometric picture of a vector as a directed line segment. In that case a rotation is portrayed as in Figure A.5 and the R_{ji} for fixed j and $i = 1, 2, 3$ are readily seen to be the *direction cosines* of the rotated fundamental basis vector $\hat{\mathbf{x}}'_j$ relative to the original fundamental basis vectors $\hat{\mathbf{x}}_i$ ($i = 1, 2, 3$). Indeed, this is a direct geometric interpretation of Equation A.55 according to Equation A.21. Therefore, in order to know the coordinates of a rotated vector $\mathbf{A}'$, we must know its original coordinates as well as the angles of rotation prescribed for the fundamental basis.

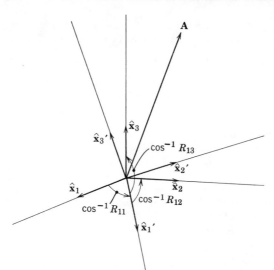

FIGURE A.5. The geometric picture of a rotation. The direction angles for $\hat{x}'_1$ are shown explicitly.

It is possible to make a strictly algebraic interpretation of Equations A.54 by writing them in the form

$$(\mathbf{A}') = (R) \times (\mathbf{A}) \tag{A.56}$$

where

$$(\mathbf{A}') = \begin{pmatrix} A'_1 \\ A'_2 \\ A'_3 \end{pmatrix} \qquad (\mathbf{A}) = \begin{pmatrix} A_1 \\ A_2 \\ A_3 \end{pmatrix}$$

$$(R) = \begin{pmatrix} R_{11} & R_{12} & R_{13} \\ R_{21} & R_{22} & R_{23} \\ R_{31} & R_{32} & R_{33} \end{pmatrix}$$

and the symbol $\times$ is *defined* by

$$A'_k = [(R) \times (\mathbf{A})]_k \equiv \sum_{i=1}^{3} R_{ki} A_i$$

Here the vectors $\mathbf{A}'$ and $\mathbf{A}$ and the operator $\mathscr{R}$ are represented as arrays of numbers called *matrices*. $(\mathbf{A})$ is called a *column matrix* and (R) is a (square) *rotation matrix*. The numbers R_{ki} given by Equations (A.55) are known as the *matrix elements* of (R) in the basis $\{\hat{x}_1, \hat{x}_2, \hat{x}_3\}$. We can extend these ideas further by extending the definition of matrix multiplication $\times$ to include

$$\mathbf{A} \cdot \mathbf{A} \equiv (\tilde{\mathbf{A}}) \times (\mathbf{A}) \equiv \sum_{i=1}^{3} \tilde{A}_i A_i = \sum_{i=1}^{3} A_i^{\,2}$$

where
$$(\tilde{A}) = (A_1 A_2 A_3)$$

is a row matrix, the *transpose* of (A). The transpose of a square matrix such as (R) is defined by

$$\tilde{R}_{ik} \equiv R_{ki} \tag{A.57}$$

that is, one exchanges columns for rows in the original matrix. The product of square matrices is defined by

$$[(R^2) \times (R^1)]_{ki} \equiv \sum_{j=1}^{3} R_{kj}{}^2 R_{ji}{}^1 \tag{A.58}$$

where (R^1) and (R^2) represent two different rotation operators, $\mathscr{R}^1$ and $\mathscr{R}^2$, in the same fundamental basis. Geometrically this product is interpreted as two rotations carried out consecutively on the same vector.

In the special case
$$R_{ki} = \delta_{ki} \equiv I_{ki}$$

we have a rotation matrix whose elements are zeroes except for those along the principal diagonal, which are ones. This corresponds to no rotation at all, that is, the *identity rotation*. Because every vector can be rotated back to its original orientation, we can write

$$(R) \times (R)^{-1} \equiv (I) \tag{A.59}$$

where $(R)^{-1}$ is the *inverse* of the rotation matrix (R). The matrix elements of (R^{-1}) are

$$R_{ki}{}^{-1} = \mathscr{R}^{-1}\hat{\mathbf{x}}_k \cdot \hat{\mathbf{x}}_i = \hat{\mathbf{x}}_k \cdot \mathscr{R}\hat{\mathbf{x}}_i = \mathscr{R}\hat{\mathbf{x}}_i \cdot \hat{\mathbf{x}}_k = R_{ik} \tag{A.60}$$

The first equality results from Equation A.55 (inverse rotations are still rotations); the second equality, from the fact that the projection of a fundamental basis vector rotated "back" to its original orientation upon another, untouched basis vector must be the same as the projection of the latter rotated "forward" upon the former left alone, and the third equality, from property (ii) of the scalar product given in Section A.1 (see Figure A.6). Equation A.60 may be rewritten

$$(\tilde{R}) = (R)^{-1} \tag{A.61}$$

with the help of Equation A.59. Equations A.61 and A.59 together imply

$$\sum_{i=1}^{3} R_{ki} R_{ji} = \delta_{kj} \tag{A.53}$$

which, as we have already seen, is a fundamental property of rotation matrix elements. Any matrix possessing an inverse whose relation to the transpose matrix is given by Equation A.61 is called an *orthogonal matrix*. Rotations, then, are one type of orthogonal transformation. Other types are considered in Chapter 8 (Lorentz transformations) and Chapter 6 (normal coordinate transformations).

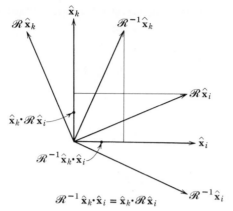

FIGURE A.6. A geometric demonstration of the relation $\mathscr{R}^{-1}\hat{\mathbf{x}}_k \cdot \hat{\mathbf{x}}_i = \hat{\mathbf{x}}_k \cdot \mathscr{R}\hat{\mathbf{x}}_i$.

All of these orthogonal transformations can be represented by matrices (R) that have the following properties:

(i) $(R) \times (R') = (R'')$

(ii) $(R) \times [(R') \times (R'')] = [(R) \times (R')] \times (R'')$

(iii) $(R)^{-1} \times (R) = (I)$

(iv) $(R)^{-1} = (\tilde{R})$

Property (i) guarantees that the product of two orthogonal matrices is an orthogonal matrix as well, while (ii) tells us that $\times$ has the associative property.[6]

Examples

1. Rotation by φ in the x_1-x_2 plane. By using trigonometry, one can show that the rotation is described by

$$A'_1 = \cos \varphi A_1 + \sin \varphi A_2$$

$$A'_2 = -\sin \varphi A_1 + \cos \varphi A_2$$

$$A'_3 = A_3$$

Upon comparing these results with Equation A.54, we can write $R_{11} = \cos \varphi$, $R_{12} = \sin \varphi$, $R_{13} = 0$; $R_{21} = -\sin \varphi$, $R_{22} = \cos \varphi$, $R_{23} = 0$; $R_{31} = 0$,

---

[6] Orthogonal matrices can further be distinguished by the values of their determinants. Those considered here all have determinant equal to $+1$. For a detailed discussion of orthogonal matrices, see J. B. Marion, *Principles of Vector Analysis*, Academic Press, New York, 1965, Chapter 1.

$R_{32} = 0, R_{33} = 1$. The new basis, expressed in terms of the old basis, accordingly has the form:

$$\hat{\mathbf{x}}_1' = \cos \varphi \hat{\mathbf{x}}_1 + \sin \varphi \hat{\mathbf{x}}_2$$

$$\hat{\mathbf{x}}_2' = \sin \varphi \hat{\mathbf{x}}_1 + \cos \varphi \hat{\mathbf{x}}_2$$

$$\hat{\mathbf{x}}_3' = \hat{\mathbf{x}}_3$$

as stipulated by Equation A.51.

2. Rotation of the vector $\mathbf{A} = \{1, 3, 7\}$ by $30°$ in the x_1-x_2 plane.

$$\begin{aligned} A_1' &= \textstyle\sum_i R_{1i}A_i = R_{11}A_1 + R_{12}A_2 + R_{13}A_3 \\ &= \cos 30° \times 1 + \sin 30° \times 3 + 0 \times 7 \\ &= 0.866 + 1.5 = 2.366 \end{aligned}$$

$$\begin{aligned} A_2' &= \textstyle\sum_i R_{2i}A_i = R_{21}A_1 + R_{22}A_2 + R_{23}A_3 \\ &= -\sin 30° \times 1 + \cos 30° \times 3 + 0 \times 7 = 2.098 \end{aligned}$$

$$A_3' = \textstyle\sum_i R_{3i}A_i = R_{31}A_1 + R_{32}A_2 + R_{33}A_3 = A_3 = 7$$

3. Orthogonality of the rotation matrix. Consider a rotation by φ in the x_1-x_2 plane. Then

$$\textstyle\sum_i R_{1i}R_{1i} = R_{11}{}^2 + R_{12}{}^2 + R_{13}{}^2 = \cos^2 \varphi + \sin^2 \varphi + 0 = 1$$

$$\begin{aligned} \textstyle\sum_i R_{1i}R_{2i} &= R_{11}R_{21} + R_{12}R_{22} + R_{13}R_{23} \\ &= -\cos \varphi \sin \varphi + \sin \varphi \cos \varphi + 0 = 0 \end{aligned}$$

$$\textstyle\sum_i R_{1i}R_{3i} = R_{11}R_{31} + R_{12}R_{32} + R_{13}R_{33} = 0 + 0 + 0 = 0$$

and so on, verifying in detail Equation A.53.

4. Inverse of the rotation matrix. The matrix

$$(R) = \begin{pmatrix} \cos \varphi & \sin \varphi & 0 \\ -\sin \varphi & \cos \varphi & 0 \\ 0 & 0 & 1 \end{pmatrix}$$

has the inverse

$$(R)^{-1} = (\tilde{R}) = \begin{pmatrix} \cos \varphi & -\sin \varphi & 0 \\ \sin \varphi & \cos \varphi & 0 \\ 0 & 0 & 1 \end{pmatrix}$$

For example, $[(R) \times (R)^{-1}]_{11} \equiv \sum_i R_{1i}R_{i1}{}^{-1} = \sum_i R_{1i}R_{1i} = 1$, as shown in (3) above. Similarly, $[(R) \times (R)^{-1}]_{12} \equiv \sum_i R_{1i}R_{i2}{}^{-1} = \sum_i R_{1i}R_{2i} = 0$. Therefore $(R) \times (R)^{-1} = (I)$.

As indicated in Figure A.5, a rotation generally is defined in terms of the angles of rotation of a vector counterclockwise about the three cartesian axes. In the special case that these angles of rotation are infinitesimal, it is possible to associate a vector $d\boldsymbol{\varphi}$ with the rotation, where

$$d\boldsymbol{\varphi} = \{d\varphi_1, d\varphi_2, d\varphi_3\} \tag{A.62}$$

The coordinates of $d\boldsymbol{\varphi}$ are the angles of rotation. The direction of $d\boldsymbol{\varphi}$ is prescribed in the usual way by writing

$$d\boldsymbol{\varphi} = d\varphi_1 \hat{\mathbf{x}}_1 + d\varphi_2 \hat{\mathbf{x}}_2 + d\varphi_3 \hat{\mathbf{x}}_3 \tag{A.63}$$

but it should be recognized that the components of $d\boldsymbol{\varphi}$ point along the *axes of rotation* for each of the angles $d\varphi_i$ not simply along the basis vectors $\hat{\mathbf{x}}_i$. Thus, the direction of $d\boldsymbol{\varphi}$ is that of the axis of rotation about which the rotation by $d\boldsymbol{\varphi}$ occurs, as shown in Figure A.7.

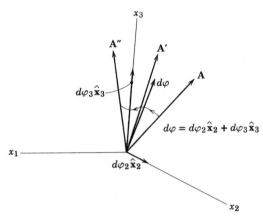

FIGURE A.7. A geometric picture of the vector $d\boldsymbol{\varphi}$.

Now, in order to say that $d\boldsymbol{\varphi}$ is a vector we must show that it follows the axioms (1) to (4) given in Section A.1. In particular, we must demonstrate that rotations, as represented by $d\boldsymbol{\varphi}$, can be added according to the definition of $+$. To do this we need only establish the truth of Equation A.63 and prove that the addition is commutative and associative. Since the coordinates of $d\boldsymbol{\varphi}$ are well defined in any rotation, the addition prescribed by (A.63) can always be carried out. The real question is whether or not

$$d\boldsymbol{\varphi} = \{d\varphi_1, 0, 0\} + \{0, d\varphi_2, 0\} + \{0, 0, d\varphi_3\} \tag{A.64}$$

regardless of the order of adding. To see the truth of Equation A.64, we consider first the rotation of a vector $\mathbf{A}$ counterclockwise about the x_1 axis by the angle $d\varphi_1$. In that case Equations A.54 reduce to

$$A_1' = A_1 = A_1$$

$$A_2' = \cos d\varphi_1 A_2 + \sin d\varphi_1 A_3 \simeq A_2 + d\varphi_1 A_3$$

$$A_3' = -\sin d\varphi_1 A_2 + \cos d\varphi_1 A_3 \simeq -d\varphi_1 A_2 + A_3$$

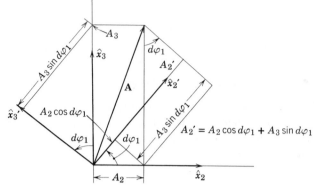

FIGURE A.8. Rotation by the angle $d\varphi_1$ about the x_1 axis.

to first order in the infinitesimal $d\varphi_1$ (see Figure A.8). The corresponding rotation matrix is therefore

$$(R^1) = \begin{pmatrix} 1 & 0 & 0 \\ 0 & 1 & d\varphi_1 \\ 0 & -d\varphi_1 & 1 \end{pmatrix}$$

In a similar way we deduce that

$$(R^2) = \begin{pmatrix} 1 & 0 & -d\varphi_2 \\ 0 & 1 & 0 \\ d\varphi_2 & 0 & 1 \end{pmatrix}$$

$$(R^3) = \begin{pmatrix} 1 & d\varphi_3 & 0 \\ -d\varphi_3 & 1 & 0 \\ 0 & 0 & 1 \end{pmatrix}$$

for rotations about the x_2 and x_3 axes, respectively. The matrix $(R) \equiv (R^1) \times (R^2) \times (R^3)$ represents the rotation by $d\varphi$ according to Equation A.58. Using this equation, it is straightforward to verify that

$$(R^1) \times (R^2) \times (R^3) = (R^2) \times (R^1) \times (R^3) = (R^3) \times (R^2) \times (R^1) \quad \text{(A.65)}$$

and so on. The matrix (R) can be generated by multiplying the matrices (R^i) *in whatever order we wish*. This result comes directly from our stipulation that the angles of rotation be infinitesimal; if the angles are finite, Equations A.54 cannot be reduced as shown above and the commutativity of $\times$ does not occur. The meaning of (A.65) is that the rotation by $d\varphi$ can be built up from the three basic rotations in any order. It follows that $+$ in (A.64) must be commutative as required. The associative character of $+$ follows from that of $\times$ in the same way. Therefore, we may conclude that $d\boldsymbol{\varphi}$ is indeed the *vector* sum of its components and that rotations can be represented by a *bona fide* vector.[7]

A.3 CURVILINEAR COORDINATE SYSTEMS

Often we find that a dynamical problem does not lend itself well to a description in terms of purely cartesian coordinates. Two obvious examples of this situation are the motion of a single planet relative to the sun and the motion of an isotropic oscillator in a uniform gravitational field. The reason for the difficulty invariably lies with the geometric symmetry of the problem in question. If the surfaces of constant potential energy for a particle are spheres, for example, there is no doubt that cartesian coordinates will be awkward to use and that some other coordinates that reflect this natural symmetry would be better. For this reason a rather large number of alternate, *curvilinear* coordinate systems have been introduced into dynamics. In general, these systems are related to the cartesian one by a *point transformation*:

$$x_i = x_i(q_1, q_2, q_3) \qquad (i = 1, 2, 3) \tag{A.66}$$

where x_i is the ith cartesian coordinate and $\{q_1, q_2, q_3\}$ is a set of curvilinear coordinates. The precise form of Equations A.66 depends on the geometric surfaces chosen to define the q_i. In the case of cartesian coordinates, a point in space is specified by the intersection of the three planes

$$x_1 = \text{constant} \qquad x_2 = \text{constant} \qquad x_3 = \text{constant}$$

These are shown in Figure A.9. We shall consider now two more possibilities that are of great utility in dynamics.[8]

A point is located in the *spherical polar coordinate system* by the intersection of the three surfaces

$$q_1 \equiv r = (x_1{}^2 + x_2{}^2 + x_3{}^2)^{1/2} = \text{constant} \tag{A.67}$$

$$q_2 \equiv \theta = \cos^{-1}\left[\frac{x_3}{(x_1{}^2 + x_2{}^2 + x_3{}^2)^{1/2}}\right] = \text{constant} \tag{A.68}$$

$$q_3 \equiv \varphi = \tan^{-1}\left(\frac{x_2}{x_1}\right) = \text{constant} \tag{A.69}$$

The first of these is a sphere centered at the origin, the second is a right circular cone centered on the cartesian x_3 axis, while the third is simply a half-plane

---

[7] Axioms 2 to 4 in Section A.1 are very easy to establish for $d\boldsymbol{\varphi}$ given the geometric picture of a rotation and the definition of the $d\varphi_i$. The *difficult* problem was the commutativity of $+$.

[8] For a general introductory discussion of the many curvilinear coordinate systems in use, see G. Arfken, *Mathematical Methods for Physicists*, Academic Press, New York, 1970, Chapter 2. For an exhaustive discussion, see P. M. Morse and H. Feshbach, *Methods of Theoretical Physics*, McGraw-Hill, New York, 1953, Chapter 5.

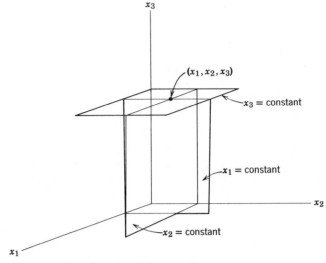

FIGURE A.9. The geometric surfaces that define cartesian coordinates.

through the x_3 axis (see Figure A.10). With Equations A.67 to A.69 at hand, we may immediately write down the appropriate form of Equations A.66:

$$x_1 = r \sin \theta \cos \varphi$$
$$x_2 = r \sin \theta \sin \varphi \tag{A.70}$$
$$x_3 = r \cos \theta$$

Here it should be noted that r is always positive, that θ is measured from the positive x_3 axis, and that φ is measured from the positive x_1 axis. Their respective ranges of variation are

$$0 \leqslant r < +\infty \qquad 0 \leqslant \theta \leqslant \pi \qquad 0 \leqslant \varphi \leqslant 2\pi$$

Equations A.70 are clearly of value when a dynamical problem exhibits spherical symmetry; that is, when the surfaces of constant potential are spheres. In the special case that the motion of a particle occurs in the x_1-x_2 plane, we may set $\theta = \pi/2$ to produce from (A.70) the *plane polar coordinates*:

$$x_1 = r \cos \varphi \qquad x_2 = r \sin \varphi \tag{A.71}$$

A point in the *cylindrical polar coordinate system* is specified by the intersection of the surfaces

$$q_1 \equiv \rho = (x_1{}^2 + x_2{}^2)^{1/2} = \text{constant} \tag{A.72}$$

$$q_2 \equiv \theta = \tan^{-1}\left(\frac{x_2}{x_1}\right) = \text{constant} \tag{A.73}$$

$$q_3 \equiv z = \text{constant} \tag{A.74}$$

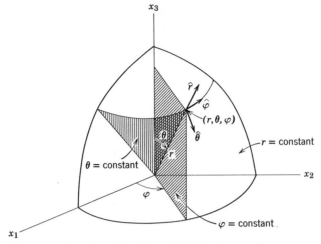

FIGURE A.10. The geometric surfaces that define spherical polar coordinates.

These surfaces are a right circular cylinder centered about the x_3 axis, a half-plane through the x_3 axis, and a plane parallel with the x_1-x_2 plane, respectively (see Figure A.11). The point-transformation equations are, accordingly,

$$x_1 = \rho \cos \theta$$
$$x_2 = \rho \sin \theta \qquad\qquad \text{(A.75)}$$
$$x_3 = z$$

The ranges of variation of the cylindrical polar coordinates are

$$0 \leqslant \rho < +\infty \qquad 0 \leqslant \theta \leqslant 2\pi \qquad -\infty < z < +\infty$$

In the special case $z = 0$ we can obtain from Equations A.75 the plane polar coordinates $\{\rho, \theta\}$.

Before ending this brief discussion of curvilinear coordinates, it is worthwhile to ask an important geometric question. In the cartesian system we know that the three axes along which the coordinates are marked are mutually perpendicular. If this were also the case for curvilinear systems, we would hold an obvious advantage in setting up a geometric representation of them in terms of directed line segments. For example, a vector whose coordinates were most conveniently expressed in the spherical polar system (through Equations A.70) could then be written

$$\mathbf{A} = A_r \hat{\mathbf{r}} + A_\theta \hat{\boldsymbol{\theta}} + A_\varphi \hat{\boldsymbol{\varphi}} \qquad\qquad \text{(A.76)}$$

where $\hat{\mathbf{r}}$ is a vector of unit length pointing along a radial line in the sphere defining the coordinate r, $\hat{\boldsymbol{\theta}}$ is a vector of unit length pointing along the tangent to the sphere (in the direction of increasing θ) where it is touched by the cone defining

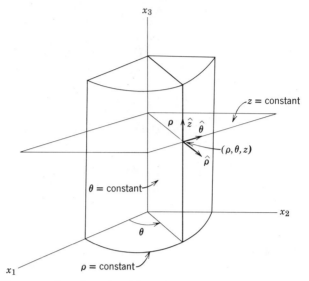

FIGURE A.11. The geometric surfaces that define cylindrical polar coordinates.

the coordinate θ, and $\hat{\varphi}$ is a vector of unit length pointing along a comparable tangent to the sphere where the sphere is touched by the half-plane defining the coordinate φ (see Figure A.10). The important point is that, because of the assumed mutual perpendicularity of $\{\hat{\mathbf{r}}, \hat{\boldsymbol{\theta}}, \hat{\boldsymbol{\varphi}}\}$, the definition of $+$ in Equation A.76 is *exactly* as given for vectors prescribed in terms of cartesian coordinates. In a similar way we can write

$$\mathbf{A} = A_\rho \hat{\boldsymbol{\rho}} + A_\theta \hat{\boldsymbol{\theta}} + A_z \hat{\mathbf{z}} \tag{A.77}$$

for a vector expressed in cylindrical polar coordinates under the assumption of mutual perpendicularity for $\{\hat{\boldsymbol{\rho}}, \hat{\boldsymbol{\theta}}, \hat{\mathbf{z}}\}$, shown in Figure A.11.

It is clear that the property of mutual perpendicularity cannot be a general one: it must depend importantly on the character of the point transformation defining a given set of curvilinear coordinates. To see this relationship in detail, let us consider an evident result of the special nature of cartesian coordinates, the expression for the square of the infinitesimal element of length:

$$(ds)^2 = (dx_1)^2 + (dx_2)^2 + (dx_3)^2 \tag{A.78}$$

Equation A.78 is a particular example of the relation

$$d\mathbf{s} \cdot d\mathbf{s} = \sum_{i=1}^{3} \sum_{j=1}^{3} \mathbf{q}_i \cdot \mathbf{q}_j \, dq_i \, dq_j \tag{A.79}$$

which specifies the square of the infinitesimal displacement,

$$d\mathbf{s} = \sum_{i=1}^{3} \mathbf{q}_i \, dq_i$$

in terms of an arbitrary basis and an arbitrary coordinate system. In the special case $\mathbf{q}_i = \hat{\mathbf{x}}_i$ ($i = 1, 2, 3$), Equation A.22 tells us immediately that (A.79) reduces to (A.78). For mutually perpendicular basis vectors the quantity $(ds)^2$ is simply a sum of squares. However, in general, the vectors $\mathbf{q}_i$ need not be of unit length nor will the scalar product $\mathbf{q}_i \cdot \mathbf{q}_j$ vanish when $i \neq j$. Because of the significance of this scalar product in establishing the geometry of a coordinate system, it is referred to as an element of the *metric tensor*,[9] g_{ij}:

$$g_{ij} \equiv \mathbf{q}_i \cdot \mathbf{q}_j \qquad (i, j = 1, 2, 3) \tag{A.80}$$

Thus Equation A.79 is written

$$(ds)^2 = \sum_{i=1}^{3} \sum_{j=1}^{3} g_{ij} \, dq_i \, dq_j \tag{A.81}$$

and ds is called the *metric form* for the coordinate system $\{q_1, q_2, q_3\}$. It is clear that a set of coordinates lies along mutually perpendicular axes if the matrix (g_{ij}) has nonzero elements only on its principal diagonal. To check this for the two curvilinear coordinate systems that we have been discussing, we incorporate the identity

$$dx_i \equiv \frac{\partial x_i}{\partial q_1} \, dq_1 + \frac{\partial x_i}{\partial q_2} \, dq_2 + \frac{\partial x_i}{\partial q_3} \, dq_3$$

into Equation A.78 and compare the result with Equation A.81. Since $(ds)^2$ must be invariant under a point transformation, we find

$$g_{ij} = \sum_{k=1}^{3} \frac{\partial^2 x_k}{\partial q_i \, \partial q_j} \qquad (i, j = 1, 2, 3) \tag{A.82}$$

Now, by direct calculation with Equations A.70, we have, for spherical polar coordinates,

$$g_{11} = (\sin \theta \cos \varphi)^2 + (\sin \theta \sin \varphi)^2 + \cos^2 \varphi = 1$$

$$g_{12} = -r \sin^2 \theta \sin \varphi \cos \varphi + r^2 \sin^2 \theta \sin \varphi \cos \varphi = 0$$

$$g_{13} = r \sin \theta \cos \theta \cos^2 \varphi + r \sin \theta \cos \theta \sin^2 \varphi - r \sin \theta \cos \theta = 0$$

$$g_{21} = r \sin \theta \cos \theta \cos^2 \varphi + r \sin \theta \cos \theta \sin^2 \varphi - r \sin \theta \cos \theta = 0$$

$$g_{22} = r^2 \cos^2 \theta \cos^2 \varphi + r^2 \cos^2 \theta \sin^2 \varphi + r^2 \sin^2 \theta = r^2$$

[9] For a complete discussion of g_{ij} see E. Butkov, *Mathematical Physics*, Addison-Wesley, Reading, Mass., 1968, §16.8.

$$g_{23} = -r^2 \sin \theta \cos \theta \sin \varphi \cos \varphi + r^2 \sin \theta \cos \theta \sin \varphi \cos \varphi = 0$$

$$g_{31} = -r^2 \sin^2 \theta \sin \varphi \cos \varphi + r \sin^2 \theta \sin \varphi \cos \varphi = 0$$

$$g_{32} = -r^2 \sin \theta \cos \theta \sin \varphi \cos \varphi + r^2 \sin \theta \cos \theta \sin \varphi \cos \varphi = 0$$

$$g_{33} = r^2 \sin^2 \theta \sin^2 \varphi + r^2 \sin^2 \theta \cos^2 \varphi = r^2 \sin^2 \theta$$

It follows that

$$(ds)^2 = (dr)^2 + (r \, d\theta)^2 + (r \sin \theta \, d\varphi)^2$$

and spherical polar coordinates have the property of mutual perpendicularity. In the same way we can show

$$(ds)^2 = (d\rho)^2 + (\rho \, d\theta)^2 + (dz)^2$$

Answers and Hints
for the Problems

CHAPTER 1

1. How is "cause" defined in the context of the Second Law?
2. Can the concept of inertia be derived from the Second Law?
3. The condition for equilibrium is: fictitious force + restraining force = 0. The fictitious acceleration is always opposite to that of the frame of reference.
4. Is the astronaut "weightless" relative to the gravity field of the Earth or that of the moon?
5. What is the velocity of the particle at the turning point?
6. $v(t) = v(0) - g_e t$, $x(t) = x(0) + v(0)t - \frac{1}{2}g_e t^2$, where $g_e = \left(1 - \frac{\rho_f V}{m}\right)g$. For uniform motion: $m = \rho_f V$, where m is the mass of the parcel.
7. $v(t) = v(0) - gt + g\tau(1 - e^{-t/\tau}) \simeq v(0) - \frac{1}{2}\left(\frac{g}{\tau}\right)t^2$.
   $x(t) = x(0) + v(0)t - \frac{1}{2}gt^2 + g\tau t - g\tau^2(1 - e^{-t/\tau})$.
8. $v(t) = v_0 + \frac{F_0 T}{2\pi m}\left[1 - \cos\left(\frac{2\pi t}{T}\right)\right]$.

$$x(t) = \left(v_0 + \frac{F_0 T}{2\pi m}\right)t - \frac{F_0 T^2}{4\pi^2 m}\sin\left(\frac{2\pi t}{T}\right).$$

To understand $\dfrac{F_0 T}{2\pi m}$, consult the First Law.

9. $v(t) = \dfrac{F_0 t}{m}$ $(0 < t < \frac{1}{2}T)$. $v(t) = \left(\dfrac{F_0 T}{m}\right)\left[1 - \left(\dfrac{t}{T}\right)\right]$ $(\frac{1}{2}T < t < T)$.

10. See Equations 1.24 and 1.6.

11. The two defining integrals sum to zero. Thus $T_2 + V(x_2) = T_1 + V(x_1)$. See Equation 1.26 to evaluate the role of the Second Law.

12. See Equations 1.18, 1.21, and 1.32.

13. Compute E/m to avoid having to know the mass of the sphere. Are the differences between values of E/m at different times greater than 1%?

14. (a) $V(x) = -\dfrac{e^2}{x}$, $\quad V(\infty) \equiv 0$

   (b) $V(x) = 4\varepsilon\left[\left(\dfrac{\sigma}{x}\right)^{12} - \left(\dfrac{\sigma}{x}\right)^6\right]$ $\quad V(\infty) \equiv 0$

   (c) $V(x) = \dfrac{e^{-\alpha x}}{x}$, $\quad V(\infty) \equiv 0$

   (d) $V(x) = \frac{1}{2}V_0\left[\cos\left(\dfrac{2\pi x}{a}\right) - 1\right]$ $\quad V(0) \equiv 0$

15. $E(t) = \frac{1}{2}m(g\tau)^2(1 - e^{-t/\tau})^2 - m(g\tau)^2(1 - e^{-t/\tau}) + mg^2\tau t e^{-t/\tau}$.
   $E(\infty) = -\frac{1}{2}m(g\tau)^2$ which is less than $E = \frac{1}{2}mv^2(0) + mgx(0) = 0$.
   $$\lim_{\tau\to\infty} E(t) = \lim_{\tau\to\infty}\left[\frac{1}{2}m(g\tau)^2\left(\frac{t}{\tau}\right)^2 - m(g\tau)^2\left[\frac{t}{\tau} - \frac{1}{2}\left(\frac{t}{\tau}\right)^2\right] + mg^2\tau t\left(1 - \frac{t}{\tau}\right)\right] = 0$$
   $= \frac{1}{2}mv^2(0)$ also.

CHAPTER 2

1. $E = \frac{1}{2}mv^2(0) - ma_0 x(0) = -ma_0 x_m$.

2. $a_0 = d^2x/dt^2$, according to the physical interpretation of this constant parameter. What happens to the higher-order derivatives of $x(t)$?

3. $d = 21.2$ m.

4. $a_0 = 0.70$ m/s^2.

5. $e/m = 1.76 \times 10^{11}$ coulomb/kg.

6. Equations 1.29 and 1.30 are the key.

7. (d) $v = \dfrac{c}{\pi x_e} \left(\dfrac{V_0}{2m} \right)^{1/2}$.

8. (c) $V(x) \simeq \dfrac{-V_0}{1+a} \left[1 - \dfrac{a}{1+a} \left(\dfrac{x}{x_0} \right) \right]$. $x(t) = v(0)t - \left(\dfrac{V_0 a/2mx_0}{(1+a)^2} \right) t^2$.

9. $V(x) \simeq \dfrac{-V_0}{1+a} \left[1 - \dfrac{a}{1+a} \left(\dfrac{x}{x_0} \right) + \dfrac{a^2}{(1+a)^2} \left(\dfrac{a-1}{2} \right) \left(\dfrac{x}{x_0} \right)^2 \right]$. The Second Law
is formally that for an harmonic oscillator in a uniform field.

10. $v_{\min} = \left(\dfrac{2V_0}{m} \right)^{1/2}$.

11. (a) $V_0 = 35$ MeV, $x_0 = 10^{-12}$ cm. (b) No. The explanation for this paradox comes from quantum mechanics (the tunnel effect).

12. (b) $V(x) \simeq V_B \left(1 - \dfrac{x}{2x_0} \right)$. $\qquad x(t) = \left(\dfrac{V_B}{4mx_0} \right) t^2$.

13. Adapt Equation 2.10 to the problem and apply Equation 2.13.

14. $\int (b - ax^2)^{1/2} \, dx = \dfrac{1}{\sqrt{a}} \sin^{-1} \left[x \left(\dfrac{a}{b} \right)^{1/2} \right]$.

15. (b) $v_{\text{crit}} = \dfrac{k^{3/2}}{\sqrt{3mb}}$.

16. $\dfrac{\omega_{AB_1}}{\omega_{AB_2}} = \left[m_{B_2} \left(1 + \dfrac{m_{B_1}}{m_A} \right) \Big/ m_{B_1} \left(1 + \dfrac{m_{B_2}}{m_A} \right) \right]^{1/2}$.

17. See 14 above.

18. What is the total energy at a turning point?

19. $\delta = \tan^{-1} \left[\dfrac{T(0)}{V(0)} \right]^{1/2}$. What is the physical implication of $\delta = 0$? of $\delta \neq 0$?

20. $\omega_0 t_m - \delta = n\pi$ determines the turning points, where $n = 0, 1, \ldots$

21. $\Delta z = -0.196$ cm.

22. (a) $T = 2\pi \left[\dfrac{l \rho_b}{g \rho} \right]^{1/2}$, where l is the length of the buoy, ρ_b is its density, and ρ is the density of water. One assumes that $\rho_b < \rho$ and that the buoy displacement is measured from its center.
(b) 0.38 g/cm^3

23. (a) $\dfrac{d^2 z}{dt^2} + g \left(1 - \dfrac{\rho_a}{\rho_p} \right) = 0$, where ρ_p is the parcel density.

(b) $\rho = \dfrac{Mp}{RT}$, where M is the molecular weight and p is the pressure.

(c) $1 - \left(\dfrac{T_p}{T_a}\right) = \dfrac{(\Gamma_p - \Gamma_a)z}{[T(0) - \Gamma_a z]}$.

24. Note that $V(t) = l\dfrac{d\theta}{dt}$ and that $F(0) = mg \sin \theta$.

25. Adapt Equation 2.10 to the problem.

26. How is the period defined in terms of $\theta(0)$?

27. Apply Equation 2.42.

28. (a) $\langle T \rangle = \frac{1}{4}mA^2\omega_0{}^2$ (b) $\langle V \rangle = \frac{1}{4}mA^2\omega_0{}^2$.

29. (b) What do Figures 2.5a and 2.5b imply about $\langle \xi \rangle$ and $\langle v \rangle$?

30. Does a conserved quantity deviate from its mean value during a period of motion?

31. (a) See Equation S.4.
    (b) Apply the relations following Equation S.5.

32. Equate the Fourier series for $m\dfrac{d^2\xi}{dt^2}$ to that for $F(\xi)$.

33. (a) $V(x) \simeq -V_0\left[1 - \left(\dfrac{x}{x_0}\right)^2 + \dfrac{1}{2}\left(\dfrac{x}{x_0}\right)^4\right]$.

    (b) $\tilde{F}(x) = \dfrac{2V_0}{x_0}\left(\dfrac{x}{x_0}\right)^3$. All the steps beginning with Equations S.18 and S.19 apply.

34. (a) $V(\xi) = V_0\left[\exp\left(\dfrac{-2c}{x_e}\xi\right) - 2\exp\left(\dfrac{-c\xi}{x_e}\right)\right] \simeq -V_0\left[1 - \dfrac{c^2}{x_e{}^2}\xi^2 + \dfrac{c^3}{x_e{}^3}\xi^3\right]$

    (b) $\tilde{F}(\xi) = \dfrac{3V_0 c^3}{x_e{}^3}\xi^2$. In the absence of a constant force, set $a_0 \equiv 0$. Then the coefficients a_1 and a_2 along with the frequency ω are determined by the equations:

$$\tfrac{1}{10}x_e = a_1 + a_2 \qquad \omega^2 = \omega_0{}^2\left(1 - \dfrac{3c}{2x_e}a_2\right)$$

$$a_2 = \dfrac{3c}{4x_e}\left(1 - \dfrac{4\omega^2}{\omega_0{}^2}\right)^{-1}a_1{}^2$$

35. $\tilde{F}(\theta) = \dfrac{mg}{6}\theta^3$. All the steps beginning with Equations S.18 and S.19 apply to this problem. The desired result is derived from the equation for a_1 under the assumption that $a_1 \simeq l\theta(0)$. Note that the Fourier series is an expansion of $l\theta(t)$ in this case.

CHAPTER 3

1. $v(t) = v(0)e^{-t/\tau}\left[1 + \dfrac{3\rho r}{8\eta}v(0)(1 - e^{-t/\tau})\right]^{-1}$

$t_{1/2} = \tau \ln\left[\dfrac{2 + (3\rho r v(0)/8\eta)}{1 + (3\rho r v(0)/8\eta)}\right]$

2. See Equation 3.16. $t_{1/2}(n) = (2^{n-1} - 1)\left[\dfrac{\alpha_n}{m}(n - 1)v^{n-1}(0)\right]^{-1}$

3. (a) $\dfrac{du}{dt} = \dfrac{-\beta F_0}{m}u^2(t)$ and see Equation 3.16.

   (b) $F \simeq -F_0 - F_0\beta v.$ $\dfrac{dv}{dt} + \dfrac{F_0\beta v}{m} = \dfrac{-F_0}{m}$ and see Equation 3.20.

4. $x(t) = x(0) + v(0)t_{1/2}\ln\left[1 + \left(\dfrac{t}{t_{1/2}}\right)\right].$

5. $v = v(0)\exp\left(\dfrac{-\Delta x}{v(0)t_{1/2}}\right).$ Consider a plot of Δx against $v/v(0)$ on semilog paper.

6. Consider the expression for x_{stop}.

7. $v(0) = 3.84$ m/s.

8. How does the half-life vary with $v(0)$?

9. $v(0) - v = \dfrac{\Delta x}{\tau} = 8.65 \times 10^{-2}$ cm/s.

10. (a) See Equation 3.19. (b) What is the relation between $t_{stop}{}^c$ and $F_C(v)$? Between τ and $F_S(v(0))$?

11. See Equation 3.22 and the definition of m_{eff}.

12. (b) 1.7×10^{-2} cm.

13. 10^{-2} cm.

14. 3×10^{-6} s.

15. The least time is 430 s.

16. (a) $\alpha_1 = 5.9 \times 10^{-3}$ g/cm; $\alpha_{100} = 1.3 \times 10^{-2}$ g/cm.
    (b) $\tau_1 = 4.5$ s; $\tau_{100} = 30$ s.
    (c) $x_{100} - x_1 = \dfrac{1}{48}gt^4\left(\dfrac{1}{\tau_{100}{}^2} - \dfrac{1}{\tau_1{}^2}\right) \simeq \dfrac{1}{12g}(\Delta x)^2\left(\dfrac{1}{\tau_{100}{}^2} - \dfrac{1}{\tau_1{}^2}\right),$ $\Delta x = 200$ ft.

17. $x(0) - x(t) = 2v_T\ln\cosh(t/2\tau) \simeq v_T t - \ln 2 + \cdots . v_T = 103$ m/s.

18. $\displaystyle\int \dfrac{dv}{v^2 - v_T{}^2} = \dfrac{1}{2v_T}\ln\dfrac{v - v_T}{v + v_T}.$

19. $v_T(\text{Stokes}) = 77$ cm/s. $v_T(\text{Newton}) = 234$ cm/s.

20. (a) $\tau = \dfrac{J}{B}, \omega_0 = \left(\dfrac{K}{J}\right)^{1/2}, Q = \left(\dfrac{KJ}{B^2}\right)^{1/2}.$

   (b) $\omega_d = \left(\dfrac{K}{J} - \dfrac{B^2}{4J^2}\right)^{1/2}, B^2 = 4KJ.$

21. See Equation 3.45.

22. (a) $\dfrac{d^2\theta}{dt^2} + \omega_0{}^2\theta(t) = -\left(\dfrac{\mu g}{R - r}\right) \qquad \dfrac{d\theta}{dt} > 0$

$$= \dfrac{\mu g}{R - r} \qquad \dfrac{d\theta}{dt} < 0$$

   (b) $\theta(t) = \mp \tfrac{7}{5}\mu + C \cos(\omega_0 t - \delta)$ In the interval $\left[0, \dfrac{\pi}{\omega_0}\right]$ set $C = \theta_0 - \tfrac{7}{5}\mu$

and $\delta = 0$. Continuity at $t = \pi/\omega_0$ requires $C = \theta_0 - \tfrac{21}{5}\mu$ and $\delta = 0$ in the

interval $\left[\dfrac{\pi}{\omega_0}, \dfrac{2\pi}{\omega_0}\right]$. In general, $\theta(t) = (-1)^{n-1}\tfrac{7}{5}\mu + [\theta_0 - (2n - 1)\tfrac{7}{5}\mu]$ in the

interval $\left[(n - 1)\dfrac{\pi}{\omega_0}, n\dfrac{\pi}{\omega_0}\right], n = 1, 2, \ldots.$

   (c) The graph is a cosine curve, with a linearly decreasing amplitude, which takes on a constant value after n satisfies the inequality $5\theta_0/14\mu < (n - 1)$.

23. See Equations 3.56.

24. Substitute $\theta_s(t) = b_1 e^{-t/2\tau_E} \cos \omega t + b_2 e^{-t/2\tau_E} \sin \omega t$ into Equation 3.52.

25. $\phi(t) = b_1 \left[\left(\dfrac{t}{\tau_D} - \dfrac{t}{\tau} - 1\right)e^{-t/\tau} + e^{-t/\tau_D}\right]$ for $\tau \neq \tau_D$, with b_1 determined by substitution of $\phi(t)$ into Equation 3.52.

26. See Equation 3.66.

27. (a) See Equation 3.55. (b) See Equations 3.56.

28. (a) See Figure 3.8.

29. $k = 10^5$ N/m. $\beta = 2.9 \times 10^3$ kg/s

30. $\beta = 1.1 \times 10^4$ kg/s

31. $m_{\text{crit}} = \dfrac{3\pi l \eta^2}{\rho_f g R}.$

32. The shift method!

33. (a) $\dfrac{d^2\xi}{dt^2} + \dfrac{[B + F'(\omega)]}{ml^2}\dfrac{d\xi}{dt} + \dfrac{g}{l}\cos\theta_E\xi(t) = 0.$

34. See Equations 3.55, 3.56, and those following (S.5) in Special Topic 2.

35. See the equations following (S.5) in Special Topic 2.

36. $\langle P \rangle_{max} = \frac{1}{2}F_0 a_0 \tau$. $1 = \frac{2(\omega_+/\tau)^2}{[(\omega_0^2 - \omega_+^2)^2 + (\omega_+/\tau)^2]}$.

37. $\tau = \frac{\langle E_s \rangle}{\langle P \rangle}$. See also Equations 3.55, 3.56, and those following (S.5) in Special Topic 2. τ represents a "time of evolution" for the oscillator to deviate from its "mean energy" $\langle E_s \rangle$ by $\langle P \rangle \tau$.

38. See Problem 29. $C = 500 \ \mu F$, $R = \sqrt{10} \ \Omega$.

39. $i(t) = i(0) \exp \left(\frac{-Rt}{L} \right)$.

40. $\tau = 880$ yr. $t_{1/2}(\text{calc.}) = 610$ yr. $t_{1/2}(\text{obs.}) = 10^3$ yr.

CHAPTER 4

1. (b) $V(x_1, x_2) = F_0(x_1^2 + x_2^2)^{1/2}$.

2. $F(x_1, x_2, x_3) = \frac{-\mu g}{(x_1^2 + x_2^2 + x_3^2)^{3/2}} [x_1 x_3 \hat{\mathbf{x}}_1 + x_2 x_3 \hat{\mathbf{x}}_2 - (x_1^2 + x_2^2)\hat{\mathbf{x}}_3]$.

3. $F(x_1, x_3) = -mg \left[\frac{2kx_1 \hat{\mathbf{x}}_1}{1 + (2kx_1)^2} + \frac{2kx_3 \hat{\mathbf{x}}_3}{1 + 2kx_3} \right]$. At any point on its path, the particle moves on an infinitesimal "inclined plane."

4. Write down the mth coordinate of $\mathbf{V} \times \nabla V$.

5. Trifles make perfection.

6. How is "parallel" defined for vectors?

7. $\nabla \times \mathbf{r} = \mathbf{0}$ is the key relationship.

8. $\varphi(t) = \varphi(0) + \tan^{-1} \left[\frac{\phi(0)}{\omega_0} \tan \omega_0 t \right]$.

$\int \frac{dx}{a + b \cos x} = \frac{2}{(a^2 - b^2)^{1/2}} \tan^{-1} \left[\left(\frac{a - b}{a + b} \right)^{1/2} \tan \frac{x}{2} \right]$.

9. (a) Consider the relation between r_{min}, a, and e along with Equation 4.47.
   (b) Energy is conserved!

10. (b) Consider the theory developed in Section 3.4 as applied independently to the motions in the x_1 and x_2 directions. The phase relation between $x_1(t)$ and $x_2(t)$ is important.

11. $E_T = \frac{1}{2}MV^2 + mgX_3 + \frac{1}{2}\mu v^2 + V(r)$.

12. $V(r) = E - \frac{(J^2 r_0/\mu)}{r^3}$.

13. $V(r) = E - \dfrac{J^2}{2\mu}\left(\dfrac{1 + B^2}{A^2} - \dfrac{2}{Ar}\right)$.

14. The possible circular orbit has the radius $r = r_0\left[\dfrac{1}{2} + \dfrac{1}{2}\left(1 - \dfrac{12K/r_0}{\mu c^2}\right)^{1/2}\right]$

   where r_0 is the radius of a circular Keplerian orbit.

15. $E = -\dfrac{1}{2}\left(\dfrac{g}{r_0}\right)(1 - 2\alpha r_0)$. $J = (\mu r_0 g)^{1/2}$.

16. The discussion of Section 4.5 applies. Can $U(r)$ be greater than E?

17. The method of Section 4.3 applies. $r(t) = \left[r^2(0) + \dfrac{2E}{\mu}t^2\right]^{1/2}$ where $E =$

   $\dfrac{(J^2/2\mu) - A}{r^2(0)}$. $t_{\text{fall}} = \mu r^2(0)(2\mu A - J^2)^{-1/2}$.

18. $\displaystyle\int \dfrac{x\,dx}{(c + bx - ax^2)^{1/2}} = -\dfrac{1}{a}(c + bx - ax^2)^{1/2} + \dfrac{b}{2a^{3/2}}\sin^{-1}\dfrac{2ax - b}{(b^2 + 4ac)^{1/2}}.$

   $T = 2\pi\left(\dfrac{\mu}{g}\right)^{1/2}\left(-\dfrac{g}{2E'}\right)^{3/2}$, where $E' = E - \alpha g$.

19. $r(\varphi) = \left(\dfrac{J^2}{\mu g}\right)(1 - e\cos\varphi)$, where $e = \left(1 + \dfrac{2J^2 E'}{g^2\mu}\right)^{1/2}$ and $E = E' - \alpha g$.

20. Remember that $\mathbf{J} = J\hat{\mathbf{x}}_3$ in Equation 4.78.

21. Calculate the time derivative of T_{ij} taking into account the Second Law.

22. Express $V(r)$ in cartesian coordinates and calculate the coordinates of F.

23. (a) Consider the relation between e and $J^2/K\mu$. (b) Energy is conserved!

24. $J_{\text{Tiros}}/J_{\text{OSO}} = 0.595$.

25. $V(\text{apogee}) = 1650$ km/hr (OGO 1), 1055 km/hr (Explorer 33).

26. $M_E \simeq 6 \times 10^{24}$ kg.

27. The volume of a sphere of radius R is $4\pi R^3/3$. $\rho \simeq 2.8 \times 10^{11}$ g/cm^3.

28. (a) Stable. (b) Stable if $\alpha r \ll 1$. (c) Stable if $K > 0$.

29. $a = 7617$ km. $e = 0.1154$. $T = 62.2$ min.

30. $\Delta a/a = 5.27 \times 10^{-8}$. The time to fall is about 3.15×10^3 yr.

CHAPTER 5

1. The law of cosines is relevant.

2. For a given θ_R, what value of m_A/m_B makes $\Delta E_R/T_{\text{LAB}}$ a maximum?

3. Equation 5.7 is the key.

4. See Equation 5.8.

5. $m_A \simeq 1.2$ amu. $v \simeq 3.1 \times 10^7$ m/s.

6. $m_B = 10$ amu.

7. θ_D is a maximum when $2\theta_R = \dfrac{\pi}{2} - \theta_D$. $\theta_D = 0.031°$.

8. Compare Equations 5.7 and 5.11.

9. $m_B = 92$ amu.

10. 16 collisions.

11. $\theta_D = 20°$. $T_{LAB} = 200$ MeV.

12. (a) $p(\beta) = \dfrac{2\pi}{\sigma_t} \displaystyle\int_\beta^\pi \sigma(\theta_D) \sin \theta_D \, d\theta_D$.

    (b) $p(\theta_D) = \dfrac{\sigma(\theta_D)}{R^2}$.

13. $\sigma_{LAB}/\sigma_{CM} = 3.76$.

14. $p(60°) = 0.25$.

15. $\left\langle \dfrac{\Delta E_D}{T_{LAB}} \right\rangle = 0.50$. $(m_A = m_B)$.

16. $\sigma(0 \leqslant 0 < 45°) = 221$ Mbarn.

17. $\sigma(\theta) = \dfrac{d^2}{4}\left(1 + \dfrac{V_0}{T_\infty}\right)$.

18. See Equation 5.45.

19. See Problem 12b and Equations 5.32, 5.33, and 5.45. Rate of detection $\simeq 10^{-2}$ particle/s-cm². ($\theta = 31.8°$, $\sigma(30°) = 8.9$ barn/ster.)

20. $r_{min} = 8.4 \times 10^{-13}$ cm. The head-on collision is in the CM frame.

21. $\sigma(\theta_D) = \left(\dfrac{e^2}{T_{LAB}}\right)^2 \cot \theta_D \csc^3 \theta_D$.

22. Rate of detection $= 6.7 \times 10^4$ particle/s.

23. $\sigma(\theta) = \dfrac{k}{\pi T_\infty} \dfrac{1 - (\theta/\pi)}{(\theta/\pi)^2[2 - (\theta/\pi)]^2} \csc \theta$.

24. 2.5×10^{-12} cm.

25. Equation 5.51 is the key.

CHAPTER 6

1. $\omega_1{}^2 = \frac{1}{2}(\omega_A{}^2 + \omega_B{}^2 + 2\omega_{AB}{}^2) + [\frac{1}{4}(\omega_A{}^2 - \omega_B{}^2)^2 + \omega_{AB}{}^4]^{1/2}$

$\omega_2{}^2 = \frac{1}{2}(\omega_A{}^2 + \omega_B{}^2 + 2\omega_{AB}{}^2) - [\frac{1}{4}(\omega_A{}^2 - \omega_B{}^2)^2 + \omega_{AB}{}^3]^{1/2}$

$a_{11} = \left[\dfrac{\omega_A{}^2 + \omega_{AB}{}^2 - \omega_2{}^2}{\omega_1{}^2 - \omega_2{}^2} \right]^{1/2}$ $\qquad$ $a_{12} = \left[\dfrac{\omega_1{}^2 - \omega_A{}^2 - \omega_{AB}{}^2}{\omega_1{}^2 - \omega_2{}^2} \right]^{1/2}$

$a_{21} = -\left[\dfrac{\omega_B{}^2 + \omega_{AB}{}^2 - \omega_2{}^2}{\omega_1{}^2 - \omega_2{}^2} \right]^{1/2}$ $\qquad$ $a_{22} = \left[\dfrac{\omega_1{}^2 - \omega_B{}^2 - \omega_{AB}{}^2}{\omega_1{}^2 - \omega_2{}^2} \right]^{1/2}$

2. Express the coefficients a_{ij} in terms of an "angle of rotation" about an axis perpendicular to the "$\xi_A - \xi_B$ plane" and use Equations 6.14.

3. Set $(\omega_0')^2 = \omega_0{}^2 + \omega_{AB}{}^2$ in Equation 6.3. What term in this equation is absent in the new "unperturbed state"?

4. $V(\theta_A, \theta_B) = \frac{1}{2}mgl^2(\theta_A{}^2 + \theta_B{}^2) + \frac{1}{2}kh^2(\theta_A - \theta_B)^2$, where k is the spring constant.
$\omega_1{}^2 = \left(\dfrac{g}{l}\right) + \left(\dfrac{2kh^2}{l^3}\right)$. $\omega_2{}^2 = \dfrac{g}{l}$.

5. $\omega^2 = \dfrac{1}{2}\left[\omega_0{}^2 + \left(\dfrac{M}{m} - 1\right)\omega_p{}^2 \right] \pm \dfrac{1}{2}\left\{ \left[\omega_0{}^2 + \left(\dfrac{M}{m} - 1\right)\omega_p{}^2 \right]^2 + 4\omega_0{}^2\omega_p{}^2\dfrac{M}{m} \right\}^{1/2}$.
$\omega_0{}^2 \equiv \dfrac{k}{m}$. $\omega_p{}^2 \equiv \dfrac{g}{l}$.

6. Identify the analogs in Kirchoff's voltage law and in Equations 6.4 and apply the method of Section 6.1.

7. $m_{11} = m_A$, $m_{22} = m_B$, $m_{33} = m_A$. $\phi_{11} = k$, $\phi_{12} = -k$, $\phi_{13} = 0$, $\phi_{21} = -k$, $\phi_{22} = 2k$, $\phi_{23} = -k$, $\phi_{31} = 0$, $\phi_{32} = -k$, $\phi_{33} = k$. Equation 6.50, applied to vibrations in one dimension, then yields $\omega_1{}^2 = 0$, $\omega_2{}^2 = k/m$, and $\omega_3{}^2 = \omega_2{}^2\left(1 + 2\dfrac{m_A}{m_B}\right)$. The ω_1 mode corresponds to uniform translation of the molecule; the ω_2 mode is an in-phase vibration of A and C relative to B, and the ω_3 mode is an out-of-phase vibration of A and C.

8. $\dfrac{\omega_3}{\omega_1} = \left(1 + 2\dfrac{m_A}{m_B}\right)^{1/2}$.

9. $k = 1.4 \times 10^6$ dyne/cm. $\kappa/l^2 = 0.6 \times 10^5$ dynes/cm. Consistency requires
$\omega_1{}^2 + \omega_2{}^2 = \dfrac{3}{11}\omega_3{}^2 + \dfrac{\omega_1{}^2\omega_2{}^2}{3\omega_3{}^2/11}$.

10. $k = 7.94 \times 10^5$ dyne/cm. $\dfrac{\kappa}{l^2} = 0.70 \times 10^5$ dynes/cm.

11. (a) See Figures 6.5 and 6.7.
    (b) $V_{11} = 2k \sin^2 \alpha + 4k_1$ $\qquad$ $V_{12} = 2ak \sin \alpha \cos \alpha = V_{21}$
    $V_{22} = 2a^2k \cos^2 \alpha$ $\qquad$ $V_{33} = 2b^2k$

(c) $\omega_3{}^2 = \left(1 + \dfrac{2m_H}{m_O}\sin^2\alpha\right)\dfrac{k}{m_H}$

$\omega_1{}^2 + \omega_2{}^2 = \dfrac{2k_1}{m_H} + \dfrac{k}{m_H}\left(1 + \dfrac{2m_H}{m_O}\cos^2\alpha\right)$

$\omega_1{}^2\omega_2{}^2 = 2\left(1 + \dfrac{2m_H}{m_O}\right)\cos^2\alpha\,\dfrac{kk_1}{m_H{}^2}$

12. $k = 7.76 \times 10^5$ dyne/cm. $k_1 = 1.85 \times 10^5$ dyne/cm. $m_H \times$ LHS $= 9.4, m_H \times$ RHS $= 11.8$.

13. $\alpha = 90°$ leads to $\omega_2 = \infty$.

14. Apply Equation 6.80 and the fact that $\mathbf{F}_{\alpha\beta}$ is parallel to $\mathbf{r}^\alpha$.

15. Each interparticle "bond" subtracts a degree of freedom.

16. The principal moments are the solutions of det $\left|I_{ij} - I\,\delta_{ij}\right| \equiv 0$: $I^3 - 750I^2 + 183,466I - 14,585,225 = 0$. $I = 309.2; 182.9; 257.9$ kg-m^2.

17. $I^3 - 350I^2 + 36,875I - 1,093,750 = 0$. I_{22} remains unchanged.
    $I = 175; 125; 50$ kg-m^2.
    $\bar{J} = 728.9$ joule-s.

18. The proof comes directly from Equations 6.88 and 6.89.

19. $I_{11} = 4Mb^2$. $I_{22} = 4Ma^2$. $I_{33} = 4M(a^2 + b^2)$.

20. $I_{11} = \rho \displaystyle\int_0^{2\pi}\int_0^{\pi}\int_0^{R} r^2(1 - \sin^2\theta\cos^2\varphi)r^2\,dr\,\sin\theta\,d\theta\,d\varphi = \dfrac{2MR^2}{5}$.

$I_{22} = \dfrac{2MR^2}{5}$. $I_{33} = \dfrac{2MR^2}{5}$. $M =$ mass of the sphere.

21. $I_{11} = \rho \displaystyle\int_{-L/2}^{L/2}\int_{-L/2}^{L/2}\int_{-L/2}^{L/2}(x_2{}^2 + x_3{}^2)\,dx_1 dx_2 dx_3 = \tfrac{1}{6}ML^2 = I_{22} = I_{33}$.

22. Period $= 303$ days. What is the magnitude of θ?

23. $I_C = 171$ kg-m^3.

24. In the absence of forces, the time derivatives of $\dot\phi$ and $\dot\psi$ are equal to 0.

25. $\omega_0 = 4$ rad/s. $I_C/I_A = 1.5$.

CHAPTER 7

1. $\mathbf{F}_a = \mathbf{F}_0 - m\mathbf{a}_0 + m\boldsymbol{\omega} \times (\mathbf{r} \times \boldsymbol{\omega}) + 2m(\mathbf{v} \times \boldsymbol{\omega})$.

2. $\mathbf{F}_a = \mathbf{F}_0 + m\boldsymbol{\omega} \times (\mathbf{r} \times \boldsymbol{\omega}) \,\substack{\circ\\+}\, 2m(\mathbf{v} \times \boldsymbol{\omega}) + m\mathbf{r} \times \dfrac{d\boldsymbol{\omega}}{dt}$.

3. Replace $\mathbf{r}$ by $\mathbf{A}$ in Figure 7.2.

4. (a) Period $\simeq 69$ s. (b) $g(z) = 0.098 \, (r_{in} + z)$, where r_{in} is the inner radius in meters and $0 \leqslant z \leqslant (100 - r_{in})$, also in meters.

5. $F_{Cor} = 2mv\omega_E \sin \phi = 25N$.

6. Take $\hat{\mathbf{x}}_1$ to point south so $dx_2 = 0$. Then slope $= dx_3/dx_1 = -a_1/a_3$. Elevation difference $= \left| \dfrac{dx_3}{dx_1} \right| \Delta x_1 = (2v\omega_E \sin \phi \times \text{width})/g = 0.0158$ m. The surface builds up on the north bank.

7. $F_{Cor} = 3.76$ N. $\dfrac{F_{Cor}}{mg} = 0.4$ at $r = 97$ m.

8. (a) $\dfrac{\partial V_{cen}}{\partial x_i} = m[(\mathbf{r} \cdot \boldsymbol{\omega})\omega_i - \omega^2 x_i]$ $(i = 1, 2, 3)$.

   (b) $\dfrac{\partial V_{Cor}}{\partial x_i} = 2m(\boldsymbol{\omega} \times \mathbf{v})_i$ $(i = 1, 2, 3)$.

9. The analogies $V_0(\mathbf{r}) \longleftrightarrow q\,\mathbf{E}_0 \cdot \mathbf{r}$ and $\dfrac{q}{c}\,\mathbf{H}_0 \longleftrightarrow 2m\boldsymbol{\omega}$ are the keys.

10. $g_0 - g = 220$ cm/s^2 (J), 174 cm/s^2 (S), 63 cm/s^2 (U).

11. 2.7×10^{-4}.

12. $\sin x \simeq x - \dfrac{x^3}{3!} + \dfrac{x^5}{5!}$. $\cos x \simeq 1 - \dfrac{x^2}{2!} + \dfrac{x^4}{4!}$. In the optimum case, $\phi = 0°$, one would need to know the apparent g to within one part in 10^8.

13. Equation 7.30b applies with $\phi = 90°$, $\omega = 0.313$ rad/s.
   $\dfrac{x_2}{x_3} = \tfrac{2}{3}\omega t \simeq 0.107$ and $x_2 \simeq 0.32$ m.

14. $x_2 = 0.0154$ m.

15. $T_C = 24$ hr. $C = 69$ km.

16. $T_C = 25.6$ hr and 26.2 hr.

17. $C(t) = C(0)e^{-t/\tau}$.

18. $\tau = 1.74$ day.

19. 8.5 m/s.

20. Apply the "shift method" to remove the constant terms in Equations 7.38, then solve the resulting homogeneous differential equations.

CHAPTER 8

1. Equations 8.6 to 8.8 are the key.

2. Length$' \equiv x_2'(t') - x_1'(t') = \alpha(v)[x_2(t) - x_1(t)]$ for any relativity principle.

3. $\mathbf{u}' = \dfrac{d\mathbf{r}'}{dt'}$, with $d\mathbf{r}'$ taken from Equation 8.23.

4. Use Equation 8.24 to express dt', then integrate to get $\Delta t'$. How is $d\mathbf{r}$ related to dt in this case?

5. If event A occurs at time t_A and event B occurs at time t_B, what determines if the two events are simultaneous?

6. Apply Equation 8.24 as done in Problem 4. $\beta = 10^{-4}$ in this case.

7. Apply the result in Problem 3. u' is the speed of light relative to the moving water and $u = u_{LAB}$. Note that $u' = c/n$, where n is the index of refraction, and that v/c is very small.

8. Consider the result in Problem 3 in the limit $c \to \infty$. Again $u' = c/n$ and $u = u_{LAB}$. Can n be infinitely large?

9. 242 m.

10. Calculate $\tan \theta'(v)$ using Equations 8.19 and the appropriate x_2 transform.

11. $e/m = 1.752, 1.761, 1.759, 1.763 \times 10^{11}$ C/kg.

12. A "neutral" system with moving internal constituents is needed.

13. Calculate the RHS of Equation 8.38.

14. $T(u) \simeq \frac{1}{2}m_0 u^2 \left[1 + \dfrac{3}{4}\left(\dfrac{u}{c}\right)^2 + \dfrac{5}{8}\left(\dfrac{u}{c}\right)^4\right]$. $u/c \simeq 0.346$.

15. $x_1(t) = \dfrac{p_0 c}{qE_{02}} \ln\left\{\dfrac{qE_{02}ct}{\varepsilon} + \left[\left(\dfrac{qE_{02}ct}{\varepsilon}\right)^2 + 1\right]^{1/2}\right\} \simeq \dfrac{p_0 c^2}{\varepsilon} t = v(0)t.$

   $x_2(t) = \{[\varepsilon^2 + (qE_{02}ct)^2]^{1/2} - \varepsilon\}/qE_{02} \simeq \dfrac{1}{2}\dfrac{qE_{02}c^2}{\varepsilon}t^2 = \dfrac{1}{2}\dfrac{qE_{02}}{m_0}t^2.$

16. $\tau(t) = \dfrac{m_0 c}{qE_{02}} \ln\left\{\dfrac{qE_{02}ct}{\varepsilon} + \left[\left(\dfrac{qE_{02}ct}{\varepsilon}\right)^2 + 1\right]^{1/2}\right\}.$

17. $T_{LAB} = m_A c^2 - m_{0A}c^2$. What is ΔE_D?

18. Equations 8.71 and 8.72 are the key.

19. 87 percent.

20. $\theta_D = 11.6°$.

CHAPTER 9

1. The essential point to recognize is that the minimum principle does not specify the *cause* of any kind of particle motion, but only stipulates the general condition met by a motion once it commences. The minimum principle says nothing about what "satisfies" a particle.

2. Does the particle leave the potential well?

3. Calculate $S(\alpha = 1)$ for $\varepsilon(t) = \sum\limits_{n \geqslant 3} \dfrac{1}{n!}\left(\dfrac{d^n x}{dt^n}\right)_{t=0} t^n$ and apply Equation 9.19. Consider also trial paths of the form $x(t) = v(0)t$, etc.

4. $S_1 = m\omega_0 A^2\left(\dfrac{1}{\pi} - \dfrac{\pi}{12}\right)$. $S_2 = 0$. $S_3 = 4m\omega_0 A^2\left(\dfrac{4}{\pi} - \dfrac{\pi}{15}\right)$.

5. $x_2(t)$.

6. $L(x, v) = \frac{1}{2}mv^2 - \frac{1}{2}V_0\left[1 - \cos\left(\dfrac{2\pi x}{a}\right)\right]$. $m\dfrac{dv}{dt} = \dfrac{-V_0\pi}{a}\sin\left(\dfrac{2\pi x}{a}\right) \simeq \dfrac{-2V_0\pi^2}{a^2}x$.

7. $L(x, v) = \frac{1}{2}mv^2 - \frac{1}{2}kx^2 - \frac{1}{3}\alpha x^3$. $m\dfrac{dv}{dt} = -kx(t) - \alpha x^2(t) - \beta v^2$.

8. (a) $p_L = mv - \alpha x$. (b) $v(t) = v(0)$. $x(t) = x(0) + v(0)t$.

9. $\dfrac{d}{dt}\left(\dfrac{\partial L}{\partial u}\right) - \dfrac{\partial L}{\partial x} = \dfrac{d}{dt}\left\{\dfrac{m_0 u}{[1 - (u^2/c^2)]^{1/2}}\right\} + \dfrac{dV}{dx} = 0$.

10. See Equations 9.27 and 1.25.

11. (a) $L(x, v) = \frac{1}{2}mv^2$. $m\dfrac{dv}{dt} = -\beta v^2$. (b) $\dfrac{d}{dt}\left(\dfrac{\partial L}{\partial u}\right) = 0 = m\dfrac{dv}{dt} + \dfrac{m}{\tau}v$ where $\tau = m/\beta$.

12. $U(x, v) = -\int_0^x F(x')\,dx' - F(t)x$.

13. $S = \int_{t_1}^{t_2}\left[\frac{1}{2}L(i(t))^2 - \dfrac{1}{2}\dfrac{(q(t))^2}{C}\right]dt$.

14. Table 3.2 is the key.

15. $S = \int_{t_1}^{t_2}\left[\frac{1}{2}L(i(t))^2 + \left(i(t)R + \dfrac{q(t)}{C} + V(t)\right)q(t)\right]dt$.

16. $L(q, i, t) = \frac{1}{2}L(i(t))^2 - \dfrac{1}{2}\dfrac{(q(t))^2}{C} + V(t)q(t)$. $Q(i) = iR$.

17. $H(x, p_L) = \dfrac{1}{2m}(p_L + \alpha x)^2$. $\dfrac{(p_L + \alpha x)}{m} = \dfrac{dx}{dt}$. $\dfrac{-\alpha}{m}(p_L + \alpha x) = \dfrac{dp_L}{dt}$.

18. $H(x, p) = mc^2$.

19. $H(q, iL) = \dfrac{(iL)^2}{2L} + \dfrac{q^2}{2C}$. "Momentum" $= i(t)L$.

20. $H(q, iL, t) = \dfrac{(iL)^2}{2L} + \dfrac{q^2}{2C} - V(t)q$. $Q(i) = iR$.

Index